Perspectives in Neural Computing

Springer
London
Berlin
Heidelberg
New York
Barcelona
Hong Kong
Milan
Paris
Santa Clara
Singapore
Tokyo

Also in this series:

Adrian Shepherd
Second -Order Methods for Neural Networks
3-540-76100-4

Dimitris C. Dracopoulos
Evolutionary Learning Algorithms for Neural Adaptive Control
3-540-76161-6

John A. Bullinaria, David W. Glasspool and George Houghton (Eds)
4th Neural Computation and Psychology Workshop, London,
9-11 April 1997: Connectionist Representations
3-540-76208-6

Maria Marinaro and Roberto Tagliaferri (Eds)
Neural Nets - WIRN VIETRI-97
3-540-76157-8

Gustavo Deco and Dragan Obradovic
An Information-Theoretic Approach to Neural Computing
0-387-94666-7

Thomas Lindblad and Jason M. Kinser
Image Processing using Pulse-Coupled Neural Networks
3-540-76264-7

L. Niklasson, M. Bodén and T. Ziemke (Eds)
ICANN 98
3-540-76263-9

Maria Marinaro and Roberto Tagliaferri (Eds)
Neural Nets - WIRN VIETRI-98
1-85233-051-1

Amanda J.C. Sharkey (Ed.)
Combining Artificial Neural Nets
1-85233-004-X

Dirk Husmeier
Neural Networks for Conditional Probability Estimation
1-85233-095-3

Achilleas Zapranis and Apostolos-Paul Refenes
Principles of Neural Model Identification, Selection and Adequacy
1-85233-139-9

Mark Girolami
Self-Organising Neural Networks
1-85233-066-X

Maria Marinaro and Roberto Tagliaferri (Eds)

Neural Nets
WIRN Vietri-99

Proceedings of the 11th Italian Workshop on Neural Nets, Vietri sul Mare, Salerno, Italy, 20-22 May 1999

Società Italiana Reti Neuroniche (SIREN)
•
International Institute for Advanced Scientific Studies
"E.R. Caianiello" (IIASS)
•
IEEE NNC
•
INNS - SIG Italia
•
Department of Mathematics and Computer Science,
University of Salerno
•
Department of Information Science, University of Milan
•
Department of Physics "E.R. Caianiello", University of Salerno
•
Elsag S.p.A. (Genoa)
•
Istituto Italiano per gli Studi Filosofici (Naples)
•
Provincia di Salerno

Springer

Professor Maria Marinaro
Dipartimento di Scienze Fisiche, "E.R. Caianiello"
Universita di Salerno, 84081 Baronissi (SA), Italy

Dr Roberto Tagliaferri
DMI, Universita di Salerno, 84081 Baronissi (SA), Italy

Series Editor
J.G. Taylor, BA, BSc, MA, PhD, FlnstP
Centre for Neural Networks, Department of Mathematics, King's College,
Strand, London WC2R 2LS, UK

ISBN-13: 978-1-4471-1226-6 e-ISBN-13: 978-1-4471-0877-1
DOI: 10.1007/978-1-4471-0877-1

Springer-Verlag London Berlin Heidelberg

British Library Cataloguing in Publication Data
Neural nets-Wirn Vietri-99 : proceedings of the 11th
Italian workshop on neural nets, Vietri sul Mare, Salerno,
Italy, 20-22 May 1999. - (Perspectives in neural computing)
1. Neural networks (Computer science) - Congresses
I. Marinaro, M. II. Tagliaferri, Roberto III. WIRN Vietri 99
(11th : 1999 : Vietri sul Mare, Italy)
006.3'2

Library of Congress Cataloging-in-Publication Data
A catalog record for this book is available from the Library of Congress

Softcover reprint of the hardcover 1st edition 1999

Typesetting: Camera ready by contributors

34/3830-543210 Printed on acid-free paper SPIN 10689791

Preface

This volume contains the Proceedings of the 11th Italian Workshop on Neural Nets WIRN VIETRI-99, jointly organized by the International Institute for Advanced Scientific Studies "Eduardo R. Caianiello" (IIASS), the Società Italiana Reti Neuroniche (SIREN), the IEEE NNC Italian RIG and the Italian SIG of the INNS.

Following the tradition of previous years, we invited two foreign scientists to the workshop, Professor M. Niranjan and Professor J.G. Taylor, who respectively presented the lectures "On Sequential Bayesian Logistic Regression" and "Simple Reverberations and the Mind".

In addition, two review talks were presented, dealing with very up-to-date topics: "Computational Intelligence in Hydroinformatics: a Review" and "Theory, Implementation, and Applications of Support Vector Machines".

A large part of the book contains original contributions approved by referees as oral or poster presentations, which have been assembled for reading convenience into four sections: Mathematical Models, Pattern Recognition and Signal Processing, Architectures and Algorithms, Applications.

The last part of the books contains the papers dealing with Neural Net Applications in Economics: this section has been edited, organized and reviewed by Professors S. Giove and M. Salzano whom we should like to thank. It contains, in particular, four review talks about Statistical Neural Networks and their Applications in Economics, Neural Networks for Economic Forecasting, Fuzzy Local Algorithms for Time Series Analysis and Forecasting, Regional Economic Policy and Computational Economics.

The editors would like to thank the invited speakers, the review lecturers and all the contributors whose highly qualified papers helped with the success of the workshop.

Finally, special thanks go to the referees for their accurate work.

Maria Marinaro
Roberto Tagliaferri

Organizing - Scientific Committee:

B. Apolloni *(Univ. Milano)*, A. Bertoni *(Univ. Milano)*, N.A. Borghese *(CNR Milano)*, P. Campadelli *(Univ. Milano)*, D.D. Caviglia (*Univ. Genova*), A. Colla *(ELSAG S.p.A. - Genova)*, A. Esposito (*IIASS*), M. Frixione *(Univ. Salerno)*, C. Furlanello *(IRST - Trento)*, M. Gori *(Univ. Siena)*, G.M. Guazzo *(IIASS)*, F. Lauria *(Univ. Napoli)*, M. Marinaro *(Univ. Salerno - IIASS)*, F. Masulli *(Univ. Genova)*, F.C. Morabito (*Univ. Reggio Calabria*), P. Morasso *(Univ. Genova)*, G. Orlandi *(Univ. Roma "La Sapienza")*, T. Parisini *(Politecnico di Milano)*, E. Pasero *(Politecnico Torino)*, A. Petrosino *(INFM)*, V. Piuri (*Politecnico di Milano*), S. Rampone *(Univ. Sannio)*, R. Serra *(CRA Montecatini - Ravenna)*, F. Sorbello *(Univ. Palermo)*, R. Tagliaferri *(Univ. Salerno)*.

Referees:

Andretta M., Apolloni B., Bertoni A., Borghese N.A., Bottigli U., Burattini E., Burrascano P., Campadelli P., Caprile B., Castellani G., Chella A., Colla A.M., Di Claudio E.D., Esposito A., Frattale Mascioli F.M., Furlanello C., Giove S., Gori M., Lauria F., Marangi C., Marinaro M., Masulli F., Morabito F.C., Morasso P., Orlandi G., Palmieri F., Palmisano D., Parisi R., Parisini T., Pasero E., Perna C., Petrosino A., Piuri V., Salzano M., Sanguineti V., Serra R., Sona D., Sperduti A., Tagliaferri R., Uncini A.

The sponsorship and support of:

- Società Italiana Reti Neuroniche (SIREN)
- International Institute for Advanced Scientific Studies "E.R. Caianiello" (IIASS)
- IEEE NNC
- INNS - SIG Italia
- Dept. of Mathematics and Computer Science, University of Salerno
- Dept. of Information Science, University of Milan
- Dept. of Physics "E.R. Caianiello", University of Salerno
- Elsag S.p.A. (Genova)
- Istituto Italiano per gli Studi Filosofici (Naples)
- Provincia di Salerno

are gratefully acknowledged

Contents

Section 1
Invited Papers

On Sequential Bayesian Logistic Regression

Mahesan Niranjan
Department of Computer Science
The University of Sheffield
Regent Court, 211 Portobello Street,
Sheffield, UK
M.Niranjan@dcs.shef.ac.uk

Abstract

The Extended Kalman Filter (EKF) algorithm for identification of a state space model is shown to be a sensible tool in estimating a Logistic Regression Model sequentially. A Gaussian probability density over the parameters of the Logistic model is propagated on a sample by sample basis. Two other approaches, the Laplace Approximation and the Variational Approximation are compared with the state space formulation. Features of the latter approach, such as the possibility of inferring noise levels by maximising the 'innovation probability' are discussed. Experimental illustrations of these ideas on a synthetic and a real world problems are shown.

1 Introduction

The logistic regression model is widely used in Bayesian Inference problems. There are several examples of experimental illustrations of new, and supposedly more powerful, ideas that perform worse than the simple logistic model. In Bayesian graphical modelling, the logistic is often used as the parametric model of conditional probabilities.

Many problems in inference are characterised by data that arrives sequentially. This is particularly true in applications in time series analysis and control of dynamical systems. Sequential estimation might be of use in classification problems too if one is in a nonstationary environment and is interested in online reestimation, application and refinement of the models. Computational simplicity in the form of not having to store all the data might also be an additional motivating factor for sequential learning.

In this paper, I consider the logistic regression model, with Bayesian training performed sequentially.

2 Sequential Logistic Regression

The logistic regressor is given by

$$p(s\ \boldsymbol{\theta}) = g\left((2s-1)\,\boldsymbol{\theta}^{\mathrm{t}}\boldsymbol{x}\right),$$

where $g(.)$ is the sigmoid $g(\alpha) = 1/(1+\exp(-\alpha)$. The uncertainty in parameters $\boldsymbol{\theta}$ is represented as a Gaussian probability density function $\mathcal{N}(\boldsymbol{\mu}, \boldsymbol{\Sigma})$. In sequential estimation, we consider examples $\{\boldsymbol{x}_n, s_n\}$ arriving one at a time, The task at the arrival of each example is to compute a Gaussian *posterior* probability distribution. We denote the parameters in this *posterior* distribution by $\mathcal{N}(\boldsymbol{\mu}_p, \boldsymbol{\Sigma}_p)$. Computing the true *posterior* is analytically not feasible. We assume the Gaussian approximation to the *posterior* is sufficient for our purposes.

2.1 Laplace Approximation

Spiegelhalter and Lauritzen derive an update algorithm for $\mathcal{N}(\boldsymbol{\mu}_p, \boldsymbol{\Sigma}_p)$, by making the Laplace approximation of fitting a quadratic function to the Log *posterior.*

The corresponding update equations are

$$\begin{aligned}\boldsymbol{\Sigma}_p^{-1} &= \boldsymbol{\Sigma}^{-1} + g(1-g)\,\boldsymbol{x}_n\,\boldsymbol{x}_n^{\mathrm{t}}\\ \boldsymbol{\mu}_p &= \boldsymbol{\mu} + (s_n - g)\boldsymbol{\Sigma}_p\,\boldsymbol{x}_n\end{aligned}$$

2.2 Variational Approximation

This leads to the following update equations

$$\begin{aligned}\boldsymbol{\Sigma}_p^{-1} &= \boldsymbol{\Sigma}^{-1} + 2\lambda(\xi)\,\boldsymbol{x}_n\,\boldsymbol{x}_n^{t}\\ \boldsymbol{\mu}_p &= \boldsymbol{\Sigma}_p\left[\boldsymbol{\Sigma}^{-1}\,\boldsymbol{\mu} + (s_n - 1/2)\,\boldsymbol{x}_n\right]\end{aligned}$$

The tuning papрrameter in this variational approximation is defined in a particular functional form $\lambda(\xi) = [0.5 - g(\xi)]/2\xi$. It can be optimised at the arrival of every example by running the EM algorithm. The update for ξ is given by

$$\xi^2 = \boldsymbol{x}^{\mathrm{t}}\boldsymbol{\Sigma}_p\boldsymbol{x} + \left(\boldsymbol{x}^{\mathrm{t}}\boldsymbol{\mu}_p\right)^2$$

2.3 Extended Kalman Filter

Consider the simplified dynamical system given by the following equations:

$$\begin{aligned}\boldsymbol{\theta}(n+1) &= \boldsymbol{\theta}(n) + \boldsymbol{w}(n)\\ y(n) &= f\left(\boldsymbol{\theta}(n), \boldsymbol{x}(n)\right) + v(n)\end{aligned}$$

For simplicity, the evolution of the parameter vector $\boldsymbol{\theta}$, is taken as a random walk model, and we consider a single output case. Mainstream literature on

Kalman filtering is more general than this, allowing for some known dynamics of the state vector, deterministic exogenous inputs etc. We stay with the simple model for brevity, extending any ideas reported here to the more general case is straightforward. The process noise $\boldsymbol{w}(n)$ has covariance matrix $\boldsymbol{Q}$, and the measurement noise has variance R, both assumed Gaussian. In general these parameters hold the key the success or otherwise of applying the Kalman filtering framework to any practical problem. In control applications, for example, one might have knowledge about these noise processes via additional measurements. In parameter estimation problems, such as the one considered here, there are systematic ways of tuning these noise parameters (Jazwinski, 1970; De Freitas, Niranjan & Gee, 1997). We return to this poit later.

The update equations for linear output case, *i.e.* $y(n) = \boldsymbol{\theta}^{\mathrm{t}}\,\boldsymbol{x} + v(n)$, consist of the elegant structure of a sequence of prediction and correction phases.

Prediction
$$\begin{aligned}\boldsymbol{\theta}(n|n-1) &= \boldsymbol{\theta}(n-1|n-1)\\ \boldsymbol{P}(n|n-1) &= \boldsymbol{P}(n-1|n-1) + \boldsymbol{Q}\end{aligned}$$

Data at Time n
$$\{\boldsymbol{x}(n),\ y(n)\}$$

Error / Innovation
$$e(n) = y(n) - \boldsymbol{\theta}^{\mathrm{t}}(n|n-1)\,\boldsymbol{x}(n)$$

Kalman Gain
$$\boldsymbol{k}(n) = \frac{\boldsymbol{P}(n|n-1)\,\boldsymbol{x}(n)}{\{R + \boldsymbol{x}^{\mathrm{t}}(n)\boldsymbol{P}(n|n-1)\boldsymbol{x}(n)\}}$$

Correction
$$\begin{aligned}\boldsymbol{\theta}(n|n) &= \boldsymbol{\theta}(n|n-1) + \boldsymbol{k}(n)e(n)\\ \boldsymbol{P}(n|n) &= \boldsymbol{P}(n|n-1) - \boldsymbol{k}(n)\boldsymbol{x}^{\mathrm{t}}(n)\boldsymbol{P}(n|n-1)\end{aligned}$$

In the case of a nonlinear model, such as a neural network, we expand the function $f\,(.)$ by Taylor series in the space of the parameters $\boldsymbol{\theta}$, and truncate it. It is common to use only terms upto the first order, In the options price tracking application (Niranjan, 1997), I found extending to the second order is well worth the effort. Many authors have looked at training a Neural Network in this setting. In doing this, the dynamical system remains unchanged, and the function $f\,(.)$ becomes the neural network. The update equations will have $\boldsymbol{x}$ replaced by the gradient vector of the output $f_{\theta}y(n)$ with respect to the unknown parameters. This, of course, can be calculated by error propagation. In work reported in Kadirkamanathan & Niranjan (1993), we considered a Radial Basis Functions setting. Puskorius & Feldkamp (1994) reports a highly successful application that uses a large Recurrent Neural Network.

While the material presented so far is all straightforward and a couple of decades old, what makes this framework interesting today is the flexibility it offers to a number of problems addressed by the Neural Network community. These have to do with regularisation in a sequential (or on-line) setting. When we restrict ourselves to looking at one example at a time, many of the well known tricks in training neural networks, such as cross validation to select learning parameters and model size, or bootstrap to deal with model uncertainty are no longer available to us.

That a Kalman filter can become handy in this environment can be seen from a Bayesian derivation of the filter. Recall that we are interested in propagating a Gaussian density in the parameter space. At some time n, we can write Bayes rule as follows:

$$p\left(\boldsymbol{\theta}(n)|Y(n)\right) = \frac{p\left(y(n)|\boldsymbol{\theta}(n)\right)\, p\left(\boldsymbol{\theta}(n)|Y(n-1)\right)}{p\left(y(n)|Y(n-1)\right)}$$

Here, $p\left(\boldsymbol{\theta}(n)|Y(n)\right)$ is the *posterior* probability distribution, having seen all the data for $n = 1, ..., N$.

The denominator term in the above expression is known as the innovation probability. A Bayesian way of describing this quantity is that this represents the only additional piece of information available in the new data item, having absorbed all knowledge upto that point in time via the dynamical system as well as integrating out all uncertainties of the parameters. From a practical perspective, this term gives us a handle to systematically tune the unknown noise processes. Jazwinski (1970) has derived a number of algorithms to estimate these noise parameters by maximising the innovation probability. These are also applicable to the regularisation of neural networks in a sequential training framework (De Freitas, Niranjan & Gee, 1999).

In this study of the logistic regression model, I reestimate the measurement noise variance R from data, to show that this can make the Extended Kalman filter achieve better convergence results than the Laplace approximation. For details of the algorithm and its derivations see De Freitas *et al* (1999).

2.4 Logistic through the EKF

Let us take the logistic regression through the extended Kalman filter. For the gradient of the logistic output $y(n)$ with respect to the parameter vector $\boldsymbol{\theta}$, we have

$$\boldsymbol{f}_\theta = g(1-g)\boldsymbol{x}_n$$

Substituting in the the covariance update equation, we have

$$\begin{aligned} P(n|n) &= P(n|n-1) - \frac{P(n|n-1)\boldsymbol{f}_\theta \boldsymbol{f}_\theta^{\mathrm{t}} P(n|n-1)}{R + \boldsymbol{f}_\theta^{\mathrm{t}} P(n|n-1)\boldsymbol{f}_\theta} \\ &= P(n|n-1) - \frac{P(n|n-1)\boldsymbol{x}\boldsymbol{x}^{\mathrm{t}} P(n|n-1)}{R/\left(g(1-g)\right)^2 + \boldsymbol{x}^{\mathrm{t}} P(n|n-1)\boldsymbol{x}} \end{aligned}$$

Setting $R = g(1-g)$ and applying matrix inversion lemma, we have

$$P^{-1}(n|n) = P^{-1}(n|n-1) + g(1-g)\boldsymbol{x}\boldsymbol{x}^{\mathrm{t}}$$

Taking this one step furter, we can look at what is known as the iterated Extended Kalman filter, which allows one to make repeated local linearisations at the operating point (Bar Shalom & Fortman, 1988). For this we need the following iterations:

$$\begin{aligned}
\boldsymbol{\theta}^{i+1}(n|n) &= \boldsymbol{\theta}^{i}(n|n) + \frac{1}{R} P^{i} f_{\theta\theta}^{i\,t}(n) \left\{ y(n) - f\left(\boldsymbol{\theta}^{i}(n|n)\right) \right\} \\
&\quad - P^{i}(n|n) P^{-1}(n|n-1) \left\{ \boldsymbol{\theta}^{i}(n|n) - \boldsymbol{\theta}^{i}(n|n-1) \right\} \\
P^{i}(n|n) &= P(n|n-1) \qquad\qquad (1) \\
&\quad - \left[R + f_{\theta}^{i}(n) P(n|n-1) f_{\theta}^{i\,t}(n) \right]^{-1} f_{\theta}^{i}(n) P(n|n-1)
\end{aligned}$$

where $\boldsymbol{f}_{\theta}$ is the gradient and $f_{\theta\theta}$ is the Hessian of the output with respect to the state vector. The superscript i refers to the iteration.

Taking the logistic model through these equations, and looking for the fixed points of the state update equation above gives the following results:

$$\begin{aligned}
[P(n|n)]^{-1} &= [P(n|n-1)]^{-1} + g^{i}(1 - g^{i}) \boldsymbol{x}\boldsymbol{x}^{t} \\
\theta(n|n) &= \theta(n|n-1) + (y(n) - g^{i}) P(n|n-1) \boldsymbol{x}.
\end{aligned}$$

These equations are identical to the update equations derived by the Laplace approximation. We have arrived at the same update equations in a roundabout way. What is interesting about this result is the set of simplifications we had to do along the way. Recall that we used

- local linearisation of the output equation at the operating point, and
- set the measurement noise variance to $g(1-g)$ at every example.

Obviousely, then, there can be situations where we could do better with a Kalman filtering framework by not making these restrictions. For instance we could do a quadratic approximation about the operating point, or set the noise variance to some other value to give us greater flexibility.

3 Illustrations

3.1 Synthetic Problem

The first test problem considered here is a simple classification task in two dimensions, where the two classes were Gaussian distributed with distinct means and equal covariance matrices. For this case, we know that the Bayes optimal class boundary is a straight line and the corresponding posterior probability is a logistic function. When estimating a logistic model to classify data from such a problem, we have a perfect match between the optimum solution and the functional form of the model. As one might expect in such a case, all three

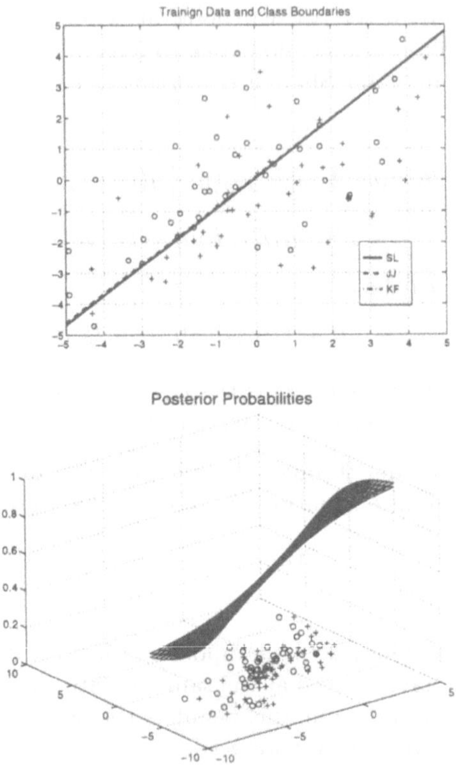

Figure 1: Class boundaries and posterior probabilities for a simple two dimensional example; in this easy problem, all three update algorithms converge to the same solution after seeing about 50 examples.

algorithms converge to the correct answer after presentation of a small number of examples. Fig. 1shows the typical behaviour; after the presentation of about 40 examples, the solutions are almost identical.

3.2 Australian Credit Data

This dataset originates from Quinlan (1987) and can be obtained from the benchmark STATLOG project archives. The problem comes from a credit card applications domain and the features are a mixture of continuous and discrete valued variables.

Fig. 2 shows the effect of automatically setting the measurement noise variance, R, to maximize the innovation probability $p\left(y(n)|Y(n-1)\right)$. The test set classification error is plotted as a function of iteration. We see that, though asymptotically both the Laplace approximation and the extended Kalman algorithm prodice results similar to the batch solution, the Kalman algorithm achieves a smoother convergence. Examples in the training sequence that are large outliers from what has been seen upto some point in time, tend to cause sudden jumps in the solution produced by the Laplace approximation.

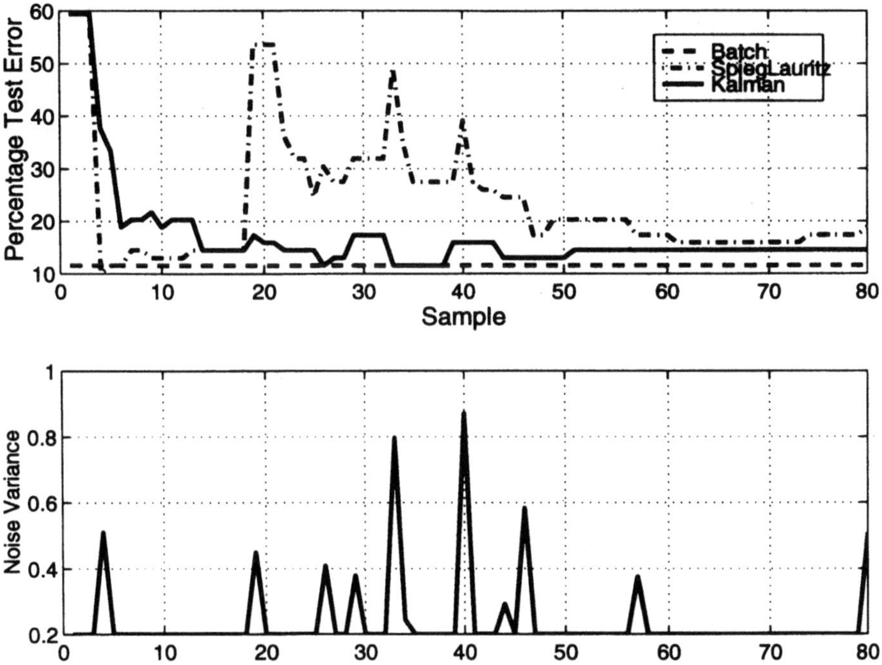

Figure 2: Comparing the behaviour of the Laplace approximation and the extended Kalman filter with measurement noise variance R inferred from data. On the Australian credit card problem, the upper plot shows the classification performance on the test data at the presentation of each example. The lower plot shows the estimated value of R. Note that in a number of places where R is increased by the Kalman algorithm are cases that cause the Laplace approximation to make sudden jumps. Increased R corresponds to not 'trusting' the the new data much, hence smaller update.

3.3 Medical Risk Prediction

Lovell *et. al* (1997) describe the *QAMC* project, which is an attempt to predict the risk of adverse outcome in pregnancy using a large number of features forming the patient profile. The data consists of 50 features relating to the mother and information about adverse outcome in pregnancy. The data set is very large, consisting of 700000 patients. It was found in this work that the use of multilayer perceptron type neural network architectures did not produce substantial improvement over the logistic model and the focus of Lovell *et. al* (1997) was in feature selection algorithms that used the area under the Receiver Operating Characteristic (ROC) curve as an objective measure.

I ran sequentially estimating the logistic model in predicting the probability of adverse outcome in this data with the Variational Approximation method and the EKF. The resulting learning curves computed over the entire training set for one of the trials are shown in Fig. 3. This is a typical picture in a large number of runs (starting from different initialisations), and the convergence for the EKF is consistently faster.

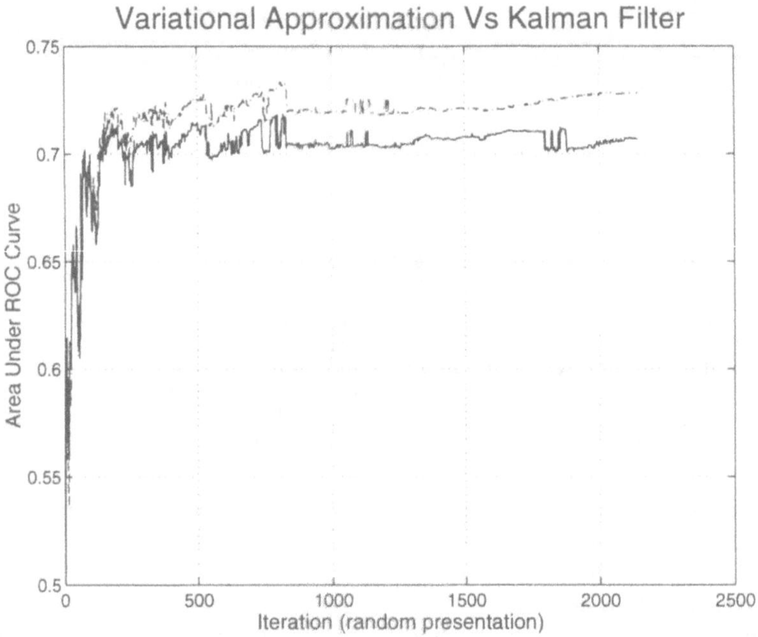

Figure 3: A comparison of the variational approximation and the extended Kalman filter on the large medical problem of predicting adverse outcome in pregnancy. The upper curve (faster convergence) is the EKF.

4 Discussion

This paper shows the use of Kalman filtering in the sequential Bayesian computation of the logistic regression model. The state space formulation and the subsequent Bayesian derivation of the Kalman filter enables us to automatically estimate the noise levels. This gives us a more powerful set of tools in sequentially training nonlinear models.

References

De Freitas, G.F.J., Niranjan, M. & Gee, A.H. (1999), Hierarchical Kalman-Bayes models for Regularisation and Sequential Learning, Neural Computation (In Press); Preprint Available in `http://www-svr.eng.cam.ac.uk/~jfgf`

Jaakkola, T.S. & Jordan, M.I. (1996), 'A variational approach to Bayesian logistic regression models and their extensions', Available from `psyche.mit.edu`

Jazwinski, A.H. (1970), *Stochastic Processes and Filtering Theory*, Vol. 64 in Mathematics in Science and Engineering, Academic Press.

Kadirkamanathan, V. & Niranjan, M. (1993), 'A Function Estimation Approach to Sequential Learning with Neural Networks', *Neural Computation* **5**, pp. 954-975.

Lovell, D.R., Scott, M.J.J., Niranjan, M, Prager, R.W., Dalton, K.J. B& Derom, R. (1997), 'On the use of expected attainable discrimination for feature selection in large scale medical risk prediction problems', Report `CUED/F-INFENG/TR.299`, Available from `http://www-svr.eng.cam.ac.uk/projects/qamc`.

Niranjan, M. (1997), 'Sequential Tracking in Pricing Financial Options using Model Based and Neural Network Approaches', In Mozer, M.C., Jordan, M.I. & Petsche, T., Ed., *Advances in Neural Information Processing Systems* **9**, MIT Press, pp: 960-966.

Niranjan, M., Cox, I.J. & Hingorani, S. (1994), 'Recursive estimation of formants in speech', Proceedings of the International Conference on Acoustics Speech and Signal Processing, ICASSP 94, Adelaide.

Puskorius, G.V. & Feldkamp, L.A. (1994), 'Neurocontrol of Nonlinear Dynamical Systems with Kalman Filter Trained Recurrent Networks', *IEEE Transactions on Neural Networks*, 5(2), pp. 279-297.

Quinlan, R. (1987), 'Simplifying decision trees', Int J Man-Machine Studies 27, pp. 221-234.

Spiegelhalter, D. & Lauritzen, S. (1990), 'Sequential updating of conditional probabilities on directed graphical structures', Networks **20**: 579-605.

SIMPLE REVERBERATIONS AND THE MIND

JG Taylor & N Taylor

Department of Mathematics, King's College, Strand, London WC2R2LS, UK
email: john.g.taylor@kcl.ac.uk; ntaylor@mth.kcl.ac.uk

Abstract

The manner in which simple reverberations may be used to explain higher-order cognitive processes is developed from the ideas of Caianiello on reverberations [1]. In Part I, a preliminary description is given as to how such reverberations can be created, held and annihilated in a controlled manner. A discussion is then given, in Part II, of the nature of, and problems associated with, modelling the frontal lobes. In particular the manner they may be used to learn and generate temporal sequences - involving the manipulation of reverberations - is studied as part of analysis of working memory. An earlier model, that of 'crumbling histories', is reviewed, and problems it contains considered. A simplified model of the architecture of the frontal lobes, the ACTION network, is then considered. This in its turn is applied to study the task of temporal sequence storage and generation. A hard-wired version of the ACTION network and the representative dynamics of some of its neurons is shown. We then analyse the manner in which the model is affected by damage corresponding to loss of dopamine. A further section analyses the essential neural activity of creation and annihilation of multi-neuron reverberations in these simulations by bifurcation methods. Part III then considers how simple reverberations are relevant in the posterior cortex to support the emergence of consciousness and reaches the conclusion that this takes place in the inferior parietal lobes. Part IV gives conclusions.

Part I. Controlling Reverberations.

I.1.Introduction

In 1960/61 Eduardo Caianiello developed a mathematical formulation of neural networks which is still the basis of present simple neural network systems [1]. One of the questions he addressed himself there, and also later, was as to the nature of the possible 'reverberations' - the periodic motions - that are sustainable by a given network architecture and set of connection weights. That neural networks can sustain periodic motions is now part of the folklore of the subject, and has been explored explicitly as part of the problem of binding through 40-Hz oscillations in the brain [2]. However such oscillations are very likely involved in binding activated representations at various levels in the brain, and not those solely at a given cognitive or conscious level. This leaves to one side how such higher level processes can be achieved by a set of coupled modules. For example how can consciousness emerge from neural activity? Our purpose here is to explore how reverberations can support a given level of higher cognition in the brain.

In this paper we delve into various aspects of the simplest of all of the reverberations of a neural system, that of its fixed points, in their relation to mental processes. These fixed points are those involving temporally extended neural activity, a feature naturally to be regarded as apposite to higher cognitive processes. Thinking and consciousness are known to involve extended neural activity, seen by both single cell [3] and brain imaging techniques [4]. We consider here the possibility that such continued activity is due to the existence of fixed points in the coupled neural networks involved.

The main ideas concerning temporally extended activity are associated with the phenomenon of working memory [5]. This is the ability to transiently store and manipulate information on-line to be used for cognition and for behavioural guidance. According to the original ideas [5] there are two components of working memory:

a) Buffer (or slave) stores for passively holding neural activity representing some suitably coded input. These stores are now known to be sited in the posterior cortex (see Part III for a fuller discussion).

b) Executive systems able to refresh and manipulate the passively held posterior material. These executive systems are sited in the frontal lobes.

These two types of working memory system are involved in two processes:

1) transient storage of information;

2) manipulation on-line of this information.

We will mainly consider the processes of type (1) above and leave to elsewhere a development of neural systems able to perform processes of type (2). However we do consider here how it is possible to combine various fixed points together so as to lead to temporal sequence storage and regeneration. This requires the creation of a new fixed point, representing one element of the sequence, by another one, which is itself annihilated in the process. The learning of temporal sequences in this manner is crucial to understand the notion of 'chunking', one which is expected to be at the basis of language learning and of thinking. Thus whilst we do not consider arbitrary manipulations of fixed points into other ones we do develop here systems, based on frontal lobe architecture, able to perform a limited amount of such manipulation.

To enable the use of the fixed points of a neural net in working memory, it is clear, as we have noted above, that these fixed points must be able to be created and destroyed by external input. Furthermore they must also be able, in the case of learning how to store and regenerate temporal sequences, to create and annihilate other fixed points of the system by suitable lateral connections. In order to understand how such processing can occur, we need first to develop a simple model. We will do this for the simplest recurrent system possible: the single neurone with feedback. On varying the input to this neurone we can show, using dynamical systems methods, how a fixed point can appear or disappear by a saddle-node bifurcation [6] [7]. This simple model is analysed in Part II so as to underpin the storage and regeneration of temporal sequences in a controlled and understandable manner. In this way we are able to give a principled explanation of executive working memory in the frontal lobes.

The problem of the source of the continued activity of posterior working memory and its relevance to consciousness is considered in Part III. The temporal extension is considered here also as arising from a reverberation of activity in a more cortically-based system, as compared to the frontal cortex with its associated basal ganglia and thalamus. The nature of the information required for consciousness, and a possible site for such information, is developed in Part III.

The paper finishes, in part IV, with conclusions about the principles in which fixed point reverberations of a neural network, and the controlled transitions between them, may be at the basis of higher order cognitive processes including consciousness; this is a natural progression from the ideas of Caianiello [1].

I.2. Controlling Reverberations.

Before we turn to the details of the ACTION models and of consciousness, which make up the next section, we will consider a more general discussion of the dynamics to be expected of single neurones involved in a recurrent fashion in temporal sequence generation. This is in terms of the simplest model of all, that of a single neurone with recurrence, as shown in figure 1(a).

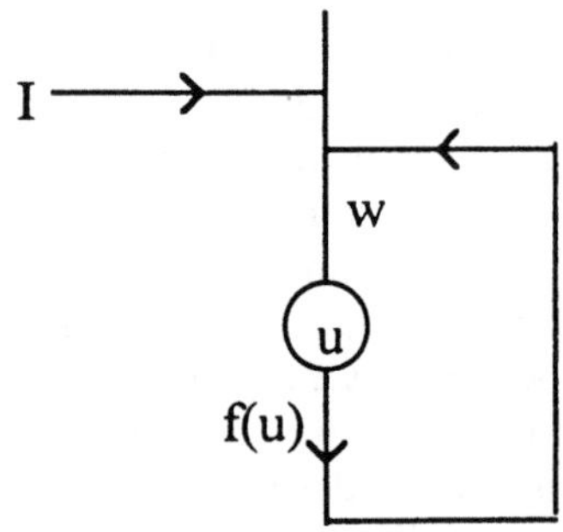

Figure 1(a). Recurrent neurone at the basis of the model. I is the external input, f(u) the graded output for a membrane potential u, and w the weight value of the self-recurrent connections.

The neurone of figure 1(a) has self-recurrent connection strength w and response in terms of its membrane potential u given by its mean firing rate f(u), for some sigmoidal response function f. There is also assumed to be an input of strength I driving the neurone, so that the membrane activity level is I+wf(u). Thus the membrane potential, assumed to reach equilibrium faster than the temporal dynamics of the input, satisfies

$$u = wf(u) + I = F(u).$$

The possible solutions to this equation, for the values of I=-1, 0, 1 are shown as the curves A, B and C in figure 1(b).

The temporal dynamics of a neurone responding to a variation of I can be seen, from figure 1(b), as follows. Initially there is no input I, so u takes its fixed point equal to the lower of the stable fixed points u_{s1}, where the subscript s denotes stable, and 1 the lower of the two stable fixed points. As I increases, say from some other neurone's activity or from external intervention, then u_{s1} increases smoothly, along curve (1) of figure 1b until I reaches the critical value I_{CL} when the fixed points u_{s1} and u_{ui} (the unstable fixed point) fuse. For higher values of I following a reverse saddle-node bifurcation, only the upper fixed point u_{s2} can persist. The transition to that value is rapid, as observed by the jump along the curve (2) needed to be made from the fused values of u_{s1} and u_{ui} at the reverse bifurcation point up to the upper fixed point u_{s2}. As I increases to the value corresponding to curve C, u_{s2} moves gradually upwards along curve (3). When the input returns to its original zero value then the fixed point is smoothly reduced by a small amount, along curve (4) even though a later saddle-node bifurcation occurs to recreate the unstable and the lower stable fixed points. The persistence of the neurone activity in the upper fixed point corresponds to the onset of the working memory, holding a memory of the input during the delay period in the task. Finally a reverse saddle- node bifurcation can occur, when suitably strong inhibitory input is present, to remove this higher fixed point level by a sharp transition at the critical input value I_{CL}, when the upper stable fixed point u_{s2} fuses with the unstable fixed point and disappears, as part of the reverse bifurcation. This proceeds with an initial smooth reduction of u along curve (5) of figure 1(b) until I reaches I_{CL}, then a sharp reduction arises along curve (6). Finally a further smooth reduction of u occurs along curve (7) until curve A is reached. Removal of the inhibitory input allows u to rise smoothly along curve (8) to the fixed point u_{s1}.

We emphasise both the onset of the working memory neurone and the different components of the time courses of its onset and offset. The onset has a slower, gradual rise and then a sudden sharp increase (along curve (2)) till maximum activity, with a consequent slower reduction of activity on offset of the input. The onset of inhibitory input causes a further gradual decrease of neurone response until the critical reverse bifurcation value is reached, when there is a rapid reduction of activity to the lower stable fixed point (along curve (6)), and then a gradual rise back to the original lower fixed point level of activity as occurred in the absence of input. In all, then, the driving of a working memory node on or off involves both gradual and more rapid responses. The gradual response component is that expected for a graded response; the more rapid response is a signal of bifurcation due to recurrence causing a reverse bifurcation, as figure 1(b) indicates. Experimental observation of such rapid changes, in companionship with the more gradual ones, are a clear signal of the presence of the mechanism we are considering.

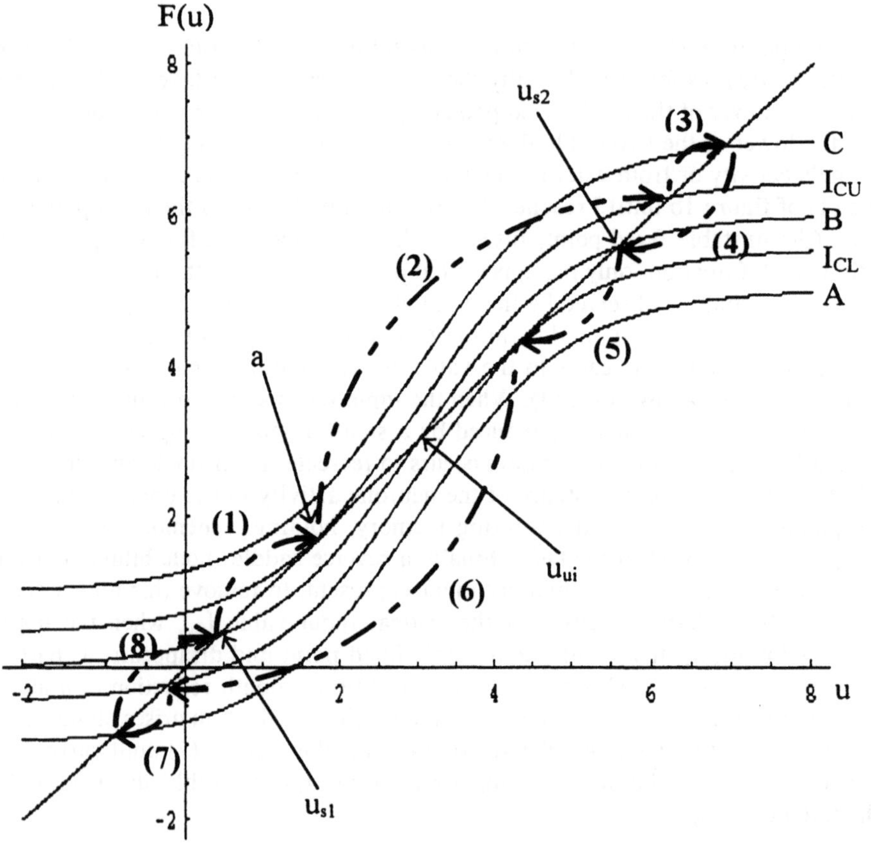

Figure 1(b). Bifurcations of the neurone with respect to varying external inputs. Curve A has input I=-1, B has I=0, C has I=1; the curve I_{CL} has I=-0.45 where the upper stable fixed point (u_{s2}) and the unstable fixed point (u_{ui}) have merged to form a bifurcation, whereas curve I_{CU} has I=0.45 and the bifurcation is formed from the lower stable fixed point (u_{s1}) and the unstable fixed point (u_{ui}). The movement of the lower fixed point u_{s1} (when I=0 for curve B representing F(u)) is along the dotted curve (1) to point a, where the curve now has a point of tangency. For higher I there is a rapid shift of the neurone activity along curve (2) to the upper fixed point, and a more gradual movement along curve (3) till I=1 (curve C). The reverse transition along curve (4) gradually, (5) gradually, (6) rapidly and (7) gradually, occurs as I is reduced back to -1 (curve A for F) from 1; removal of I (moving to curve B) brings u gradually to u_{s1} along curve (8).

We note that the effective time constants of neurons involved in buffer working memory sites have the values of about 1.5 seconds observed psychophysically [5]. Such a hundred-fold increase over the time constants of single living neurones is obtained by observing that in the simple model of the previous section such time constants are governed by the strength of the connection weight w. This can be seen very simply in the linear approximation to the single neurone equation, leading to the modified time constant $[1-wf'(0)]^{-1}$. Thus increase of w causes an increase in the time constant. Since the weights can be trained, this allows for adaptivity of the effective time constants.

The interaction between different possible sequences, can be achieved in such a model by means of lateral inhibitory interactions between the recurrent circuits encoding different working memories. We will see that in terms of the frontal lobes there is an existent set inhibitory neurones, the striatum and globus pallidus, which can immediately support this by lateral connections..

The above simple recurrent model of working memory activity has been used recently in a simulation [7]. However this was not done by relating to specific frontal lobe architecture, and could have as well been achieved by any recurrent system in the brain. Thus to answer the question as to how the frontal lobes are so crucially involved in working memory, as is well supported by experimental data [3], [4], we must turn to look at the frontal architecture, at least at a cartoon level; that we will do in terms of the ACTION network in the next section.

Part II. Modelling the Frontal Lobes.

II.1. Introduction

There is great importance in understanding the detailed structure and function of the frontal lobes so as to comprehend both the normal production of desired temporal sequences of actions and that under deficits such as Parkinson's disease (PD). To achieve such understanding increasingly more sophisticated models are being constructed [8]. We have developed a model of the frontal lobes - the ACTION network [9], [6] - which includes a simplified form of the architecture of the frontal system, but which can be extended if desired to incorporate ever more sophisticated versions of the biological structures known to be present there [10], [11], [12], [13]. It can also be applied to model various delayed response tasks [14]. We apply this model to the important questions of (1) how temporal sequences are stored and regenerated by frontal lobe systems, and (2) how the effects of dopamine depletion may be partially modelled so as to understand the interactions between the different modules present in the real system. With respect to these two questions, our model supports (1) the ACTION network framework for temporal sequence storage and generation (TSSG), with its several recurrent loops for holding activity, and (2) that for understanding the effects of dopamine the importance of the sub-thalamic nucleus [15].

The section begins with a brief review of the problem of TSSG and a specific system able to achieve it: the 'crumbling history' system [16]. The ACTION network is then described, and in particular how it enables temporal sequence storage and generation (TSSG) to be achieved. In the following section we present a hard-wired model which achieves TSSG of a sequence of 3 'actions', and show the firing patterns of some of the neurones. These are found to be very similar to those already observed in behaving monkeys [17], [18]. We then explore the behaviour of the ACTION network when there are modifications of its outputs, as considered as brought about by alterations of dopamine (DA) concentration. These are explored in the following section mathematically in terms of the way the fixed points of the basic non-linear mappings describing the frontal lobes' working memory are modified by the change. The section concludes with a discussion.

II. 2. The 'Crumbling History' Model of TSSG

The problem of storage and generation of sequences of actions is one which has interested numbers of researchers in neural networks. Storage of sequences for their classification can be achieved by an extension [19] of the topographic map [20], so generating a topographic map which allows more effective use of the map than would a static one. This is achieved by the use of leaky integrator neurones, holding a history of past activity on their surfaces, rather than those with no past history at all. This has been extended more recently [21] so as to be effective for longer sequences. However the problem of a temporal topographic map is that it is good for storage but not for regeneration of the sequence. That is what the crumbling history model solves.

The model [16] uses the architecture of figure 2. This is recurrent system in which a history of the past elements of the sequence are stored on the CH (crumbling history) module through the range of time constants possessed by its neurones.

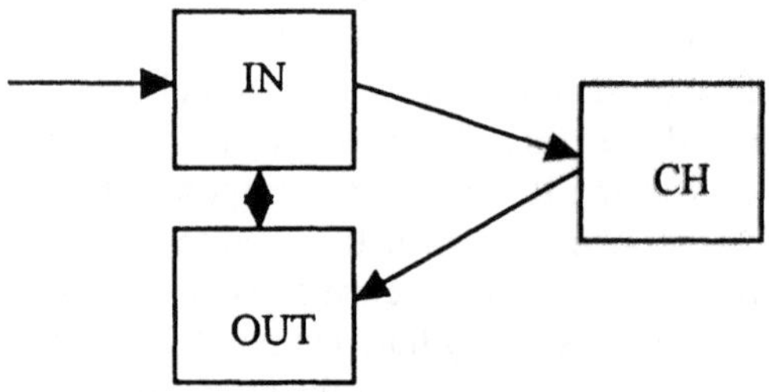

Figure 2. Architecture of the Crumbling History Model of [16]. The module CH is composed of a set of neurones with a range of time constants. The synapses from CH to OUT are adaptive, modified so as to learn to produce the next element of the sequence from the module OUT. The recurrent arrows from the OUT to the IN modules are used in the direction IN to OUT in the learning stage, and from OUT to IN in the regeneration stage.

A learning law was developed which was able to learn successive elements of the sequence, even in the presence of ambiguity. The learning law of the synapse w_{ij} from the i'th neurone in the CH module to the j'th neurone in the OUT module is:

$$\delta w_{ij}(n) = \varepsilon\,[P_j(n+1) - out_j(n)]out_i(n)$$

where ε is the learning rate, $P_j(n+1)$ the pattern at time-step n+1, $out_j(n)$ and $out_i(n)$ the outputs of the neurones j and i . The system is developed in two stages:
Stage1 (learning) in which learning proceeds according to the above equation. It ceases when the output of the OUT module at time n is the input to the IN module at time n+1. To achieve the knowledge at the OUT module of the pattern at time n+1, the two-way arrow in figure 1 is from IN to OUT.
Stage 2 (regeneration) in which generation of the sequence occurs by seeding the input module with the first (or first two) patterns and this is allowed to circulate.
The model was shown to be effective in solving the problem of TSSG, being able to store sequences with ambiguities (such as with the patterns ABAC, where A is not always followed by B); it was shown that it has capacity of order N, there being N neurones in the OUT module, which could also be shown by statistical mechanical methods [22]. Furthermore the system was shown to be able to be improved considerably, as far as error of recall is concerned, by introducing dynamical channel variables into the neurone synapses so as to provide further temporality.
Such a two-stage process of learning and regeneration is not, however, biological as it stands; it cannot be used as a model of the frontal lobes without further development. There are additional defects it possesses which are more serious:
1) it is not possible to modify the time constants of the neurones in the CH module so as to be able to handle variable delays;
2) it is not possible to easily allow for interaction between different stored temporal sequences, such as might occur in contention scheduling [23];
3) it gives little insight into how the frontal lobes achieve their powers of TSSG as part of working memory processing. In particular flexibility of processing is needed.

Working memory is presently of great present interest in brain imaging and single cell recording, so needs to be addressed. We will proceed to do so in the next section, starting with analysis of the simplest system - a single neurone with recurrence - so as to understand the basic mechanism which we will then put to work in constructing and training more powerful and biologically realistic ACTION network models of TSSG.

II.3. The ACTION Network Model of the Frontal Lobes.

We consider first the nature of the frontal lobes themselves. They have both cortical and sub-cortical components; the former we will not discuss, since their parcellation is similar to that of 6-layered posterior sites, with a similar hierarchy [24]. However there are no comparable subcortical structures accompanying the posterior cortex to those for frontal regions. The basic feature of the sub-cortical components of these frontal areas are the basal ganglia. They are two-layer systems activated unidirectionally from the cortex, the upper layer being the striatum (from its appearance), being decomposable into caudate and putamen. Beneath these, and close to the relevant part of thalamus (the mediodorsal nucleus) is the globus pallidus, comprising two components, termed internal and external. The unidirectional flow of information is :

cortex → striatum (STR)→ globus pallidus (internal) (GPi) → thalamus (TH) → cortex

[25] so there is a feedback of activity round the above 'long' loop. Such a directional flow is rather special to the frontal system, in that most interactions between brain centres are two-way. However the cortex only feeds to the striatum, with no recurrence, and nor are there any other known recurrences in the above flows, other than possibly from cortex back to thalamus. Thus there may also be similar feedback more rapidly round the 'short' cortex ↔ TH loop, and although there is known topography in these connections, the presence of any actual reverberation of such feedback activity is unknown. There is also the presence of the indirect loop, involving the globus pallidus external (GPe) as well as the sub-thalamic nucleus (STN), as shown in figure 3, a schematised version of the frontal lobe architecture we have termed elsewhere the ACTION network, since it is so important for the generation of actions [6], [9].

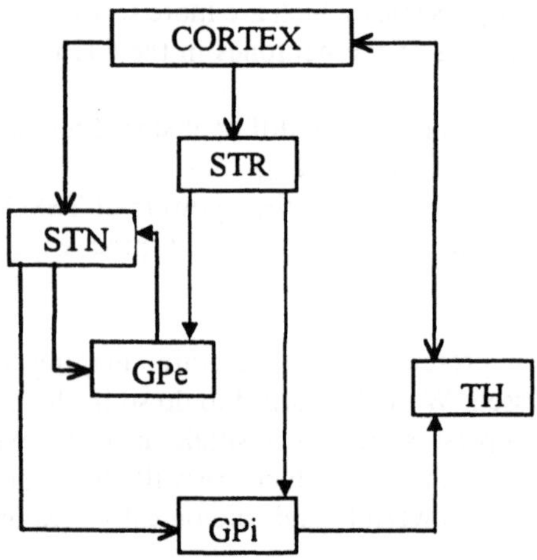

Figure 3. The ACTION network. Excitatory connections indicated by open arrowheads and inhibitories by closed.

Our purpose in this part of the paper is to explore the relevance of the ACTION network in understanding the underlying source of deficits in PD and similar diseases in which loss of DA causes increasing difficulty in making movements. In particular we explore here, as modelled in the ACTION network, of normal TSSG and of the effects of DA deficiency and excess. As part of this we will consider how the analysis in section I.2 of recurrent activity in the simplest possible single neurone model (figure 1) can be extended to the much more complex system of figure 3.

The ACTION network has been used to model TSSG elsewhere [10], [11], [12], [13]. In these papers it has been shown how, either in suitably hard-wired architectures or in trained ones, there appear a set of neurones which achieve chunking of the temporal sequences, and which agree with the activity patterns of neurones observed by researchers in single cell recordings in monkeys [17], [18]. These neurones have continued activity during delay periods. We classified them as various sorts of initiator, transition and complex memory neurones. These types of neurones are sequence-specific; in addition pre-movement and movement neurones are recorded [17], [18]. We turn in the next section to a detailed simulation of a specific hard-wired model of these processes.

II.4. System Dynamics

Neurones are modelled as leaky integrators, with a mean firing-rate response given by the positive part of a sigmoidal function. The cortical region used is the supplementary motor area (SMA) which is preferentially activated for the generation of movements from memory [17], [18], [25], [26]. In addition to the regions of figure 3 an auditory input layer is included. For this model two auditory inputs are used: the first is a mode signal which indicates that the sequence is to be performed from memory; the other is a go-signal which is used to control the timings of the movements of the sequence. The task is to generate a sequence of three different movements when the go-signal is presented successively at times 100, 200 and 300.

The weights of the neural system of figure 3 are defined as:

w^1 the connections from the auditory layer to SMA;
w^2 the lateral connections in SMA;
w^3 the connections from SMA to STR;
w^4 the lateral connections in STR.

The potentials are: s for SMA; a for auditory; m for TH; n for STN; u for GPe; r for STR; and d for GPi. The time constant of the neurones is indicated by λ and is chosen to be 0.5. The neurone equations are:

$s(t+1) = \lambda s(t) + w^1 * f(a(t),0) + w^2 * f(s(t),H) + 1.5 f(m(t),-0.2)$ (1)
$n(t+1) = \lambda n(t) + 0.5 f(s(t),0) - f(u(t),0) + T$ (2)
where T = 0.5 and is a tonic input;
$r(t+1) = \lambda r(t) + 0.5 f(s(t),0) + w^4 * f(r(t),\theta_3)$ (3)

$d(t+1) = \lambda d(t) + 0.5f(n(t),0) - 0.6f(r(t),\theta_1)$ (4)
$u(t+1) = \lambda u(t) + 0.5f(n(t),0) - 0.6f(r(t),\theta_2)$ (5)
$m(t+1) = \lambda m(t) + 0.5f(s(t),0) - 0.75f(d(t),0)$ (6)

where * indicates convolution for lateral connections in a given module and auditory input to SMA, with appropriate kernels for the weights. The numerical constants in equations (1) - (6) are introduced to achieve sustained recurrent activity.

The output function f is defined as:

$$f(x,\alpha) = \begin{cases} \tanh(x-\alpha), & \text{if } x-\alpha>0 \\ 0, & \text{otherwise} \end{cases} \quad (7)$$

where α is the threshold. The threshold H governs the use of the cortical lateral outputs. The different θs in equations (3), (4) and (5) effect the threshold of striatal output laterally (θ_3), to GPi (θ_1) and to GPe (θ_2). The negative threshold for thalamic neurones allows them to be spontaneously active without total disinhibition.

The weights in this model have been chosen to support suitable sustained activity, which are the extension of the reverberations already considered in the simplest recurrent neurone model of section I.2. The SMA to STR weights used range from 0.75 (used for the complex memory neurones) to 1.0 (for pre-movement and movement neurones). This weight is 0.9 for transition neurones and initiator neurone. The lateral connections in SMA range from -0.7 to 0.7, and the lateral connections in STR range from -0.5 to -1.0. The SMA to STR weights can vary from 0.35 to 1.0 for a given simulation and still produce sustained activity. For most neurone types a value of 0.35 for this weight would still allow the system to work, with appropriate changes to lateral SMA and STR weight values. The SMA to STR weights cannot be varied by this amount for complex memory cells they can take values from 0.5 upwards and work properly.

For the normal case, where striatal output is not effected, all θs are set to zero; some neuronal responses of SMA neurones are shown in figure 4, for the generation of a sequence of three different movements form memory (thereby indicating the effectiveness of the system and for later comparison with the PD and Huntington's chorea cases). Figure 4(a) indicates a complex memory neurone that is responsible for carrying the memory of the second and third movements throughout the sequence generation. The transition neurone (figure 4(b)) links together movements 2 and 3. The pre-movement neurone (figure 4(c)) is activated to some extent by each presentation of the go-signal, but is maximally activated for the third presentation when the movement required is generated by it exciting the related movement neurone (figure 4(c)). Other pre-movement and movement neurones exist that generate movement 1 or movement 2.

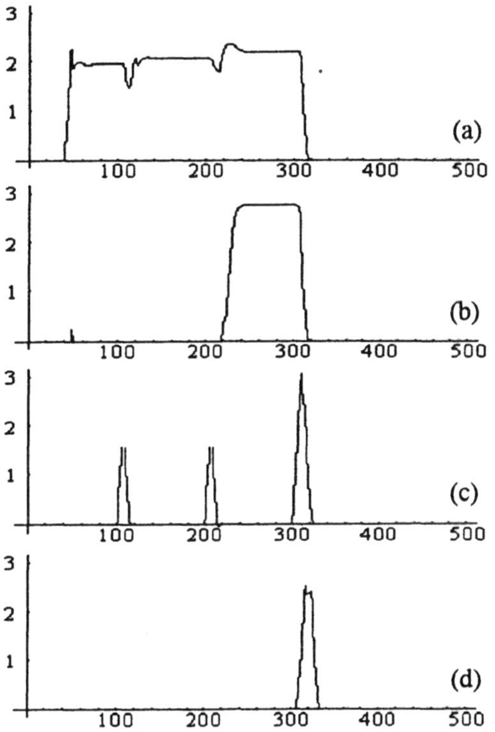

Figure 4. Normal activities of the modelled cortical neurones leading to the completion of the final movement of the sequence. The neurones are: (a) complex memory neurone; (b) transition neurone; (c) pre-movement neurone and (d) movement neurone.

We sum the activity of all GPe, GPi and STN neurones used for the model over the time course of the task; the values are: GPe 1051, GPi 1051 and STN 1763. This gives an average level of activity which sill be useful for comparison later with brain imaging data.

II.5. Effects of DA Modification

From studies of monkeys with induced PD symptoms [27] we know that changes occur to the total activity in GPe, Gpi and STN regions. GPe activity is noted as decreasing, whilst Gpi and STN increases. We can model these changes by altering the θ values, the threshold levels of striatal output. In the first case $\theta_1 = 0.1$, $\theta_2 = -0.1$ and $\theta_3 = 0.5$. Activities of the neurones of equations (1) to (7) for these values are plotted in figure 5.

For these values the final movement of the sequence is performed repetitively with the pre-movement neurone and movement neurone being reactivated (figures 5(c) and 5(d), respectively). There is little disruption to the complex memory neurone and transition neurone (figures 5(a) and 5(b), respectively) in comparison to the normal case figures 4(a) and 4(b). The first movement of the sequence is performed correctly, the second movement is made twice.

The total activities recorded from the model over the same time-course as previously are: GPi 1211, GPe 863 and STN 1968. These results show a 16% increase in GPi activity, an 18% decrease in GPe and an 11% increase in STN activity from the normal case.

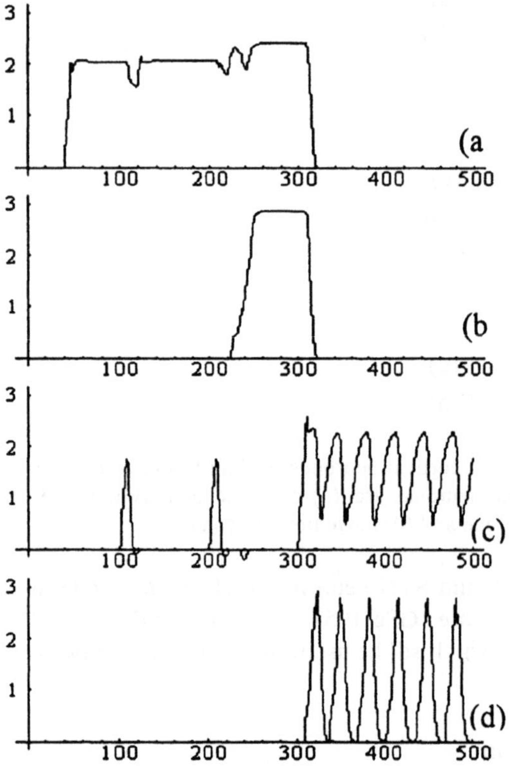

Figure 5. Neurone activities of: (a) complex memory neurone; (b) transition neurone; (c) pre-movement neurone and (d) movement neurone, for case $\theta_1 = 0.1$, $\theta_2 = -0.1$ and $\theta_3 = 0.5$.

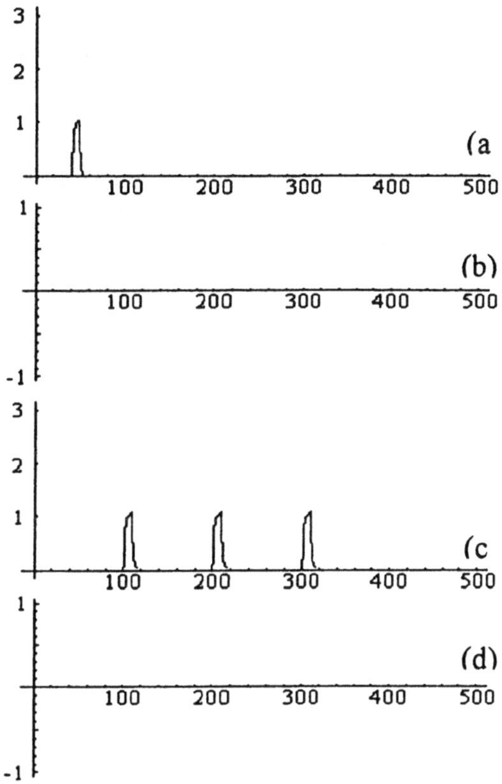

Figure 6. Neurone activities of: (a) complex memory neurone; (b) transition neurone; (c) pre-movement neurone and (d) movement neurone, for case $\theta_1 = 0.4$, $\theta_2 = -0.4$ and $\theta_3 = 0.5$.

The next choice of θ values indicates a more severe form of PD: by choosing $\theta_1 = 0.4$, $\theta_2 = -0.4$ and $\theta_3 = 0.5$ no movements are made by the system (figure 6). Although certain neurones are active (figures 6(a) and 6(c)) the movement neurone (figure 6(d)) is not activated and hence a movement is not made.

Here the total activities are: GPi 1939, GPe 525 and STN 2206. These represent changes (to the normal system) of: an increase of 84% in GPi activity, a decrease of 50% in GPe and an increase of 28% in STN activity. These are consistent with the changes in spontaneous activity observed in STN due to MPTP treatments leading to PD symptoms (from 19 to 26 Hz) and in GPi (from 53 to 76 Hz) [27].

We should note that our results arise during movement or attempted movement, so are not strictly comparable to those of [27]. However the changes in TSSG ability shown in figures 5 and 6 correspond to Parkinsonian symptoms.

We can also model Huntington's chorea by using new values for the θs. In Huntington's, spontaneous movements are made; we model this by reversing the changes in θs used in the PD cases. Figure 7 shows the neurone activities with $\theta_1 = -0.2$, $\theta_2 = 0.2$ and $\theta_3 = -0.2$. The complex memory neurone (figure 7a) and pre-movement neurone (figure 7c) are not active whilst the transition neurone (figure 7b) and movement neurone (figure 7d) are. Movements are now made in this model spontaneously without need for an active pre-movement neurone, nor does the go-signal control the timings of movements in any way.

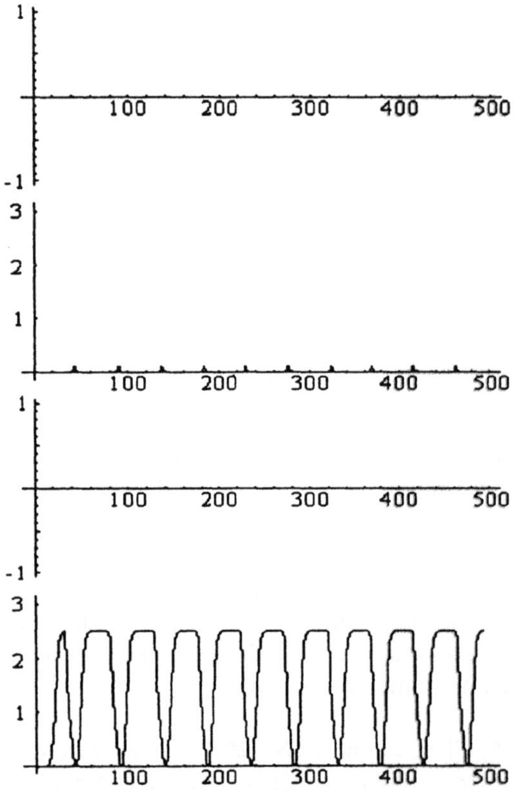

Figure 7. Huntington's chorea results. Neurone activities of: (a) complex memory neurone; (b) transition neurone; (c) pre-movement neurone and (d) movement neurone, for case $\theta_1 = -0.2$, $\theta_2 = 0.2$ and $\theta_3 = -0.2$.

Now the total activities are: GPi 806, GPe 1040 and STN 1841. This represents a decrease of 23% in GPi activity, a negligible change in GPe and a 4% increase in STN activity, a prediction that could be tested.

II.6. Fixed Point Analysis

The mathematical source of these dynamical behaviours, using ideas of section I.2, will now be considered. We analyse here the asymptotic state of equations (1) to (6). As part of this process we will only consider a single neurone in each module, so drop all lateral connections as well as auditory input and set all remaining weights to 1. The equations thus reduce to the fixed point equation:

$$s = F_1(s, n) \qquad (11)$$
$$n = F_2(s, n) \qquad (12)$$

where:

$$F_1(s, n) = 3[f[f(s,0) - 1.5\, f(f(n,0) - 1.2\, f(f\,(s,0),\theta_1)),0.2] \qquad (13)$$

and

$$F_2(s, n) = f(s,0) - 2f[f(n,0) - 1.2f(f(s,0),\, \theta_2),0] + 1 \quad (14)$$

By finding the fixed points of F_2 for a number of discrete values of s (so the function n(s) satisfying (14)), F_1 can be plotted for the same s values knowing the corresponding F_2 values; the shape of the resulting $F_1(s, n(s))$ for the normal system and two PD cases are shown in figure 8.

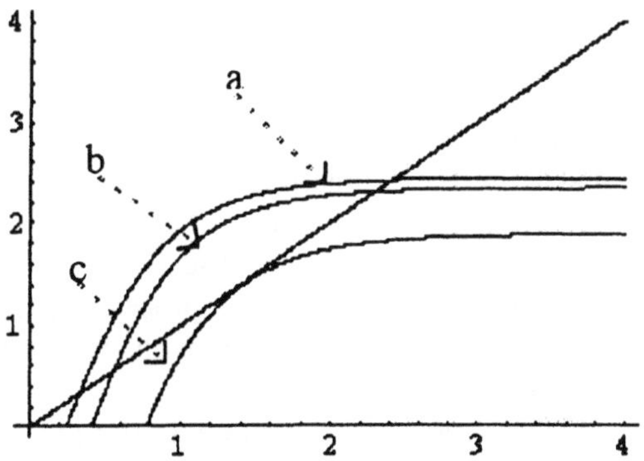

Figure 8. Parkinson's disease case. Fixed points of (a) normal system; (b) system that produces repeated movements; (c) system that produces no movements.

The normal case has two stable fixed points: 0 and 2.412; and a n unstable fixed point at 0.33. The first PD case also has two stable fixed points 0 and 2.298 with an unstable one at 0.56; the second PD case only has a fixed point at 0, which is stable. The results for the PD cases show that as the changes in θs increase the model cannot produce the sustained activity necessary to generate a movement.

The fixed points of the normal and Huntington's systems are shown in figure 9.

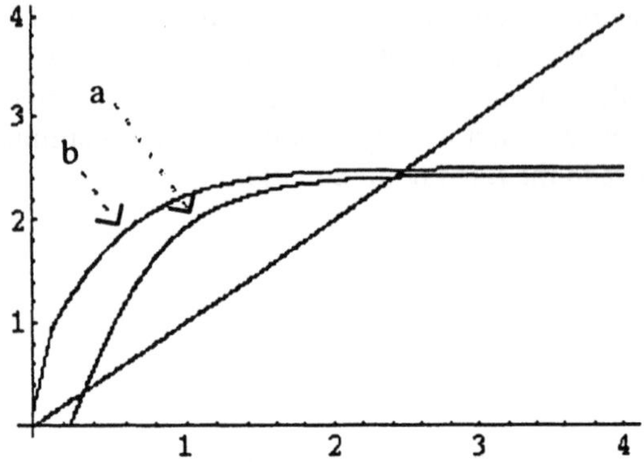

Figure 9. Huntington's chorea. Fixed points of (a) the normal system, and (b) the Huntingdon model.

The Huntington's model only has one fixed point at 2.488 which is stable. This leads to spontaneous activity in the model without an external input, and it is only the inhibitory connections between loops in the STR which prevent every SMA neurone being maximally active (hence the results of figure 7).

II.7. Summary and Discussion

The work presented in this part has achieved our overall aim of showing that the ACTION network can form a basis for modelling temporal sequence learning and generation (TSSG) with parameters which can be manipulated to mimic some of the effects of dopamine depletion or increase, as occur in Parkinson's disease or in Huntington's chorea. This takes the process of temporal sequence storage and generation beyond the capabilities possessed by the crumbling history model of section II.2. It allows relationship to be made to particular frontal components, as well as determining the effects on the TSSG capacities of modifications of the modules or their parameters. It is also possible to develop further processing the ACTION network, especially in modelling processes involved with working memory. More specifically, we have shown that:

a) a suitably hard-wired version of the ACTION network can simulate the phasic and tonic cells recorded in monkey SMA during TSSG tasks with a manipulandum [17], [18]. The detailed temporal character of the cells observed is reproduced as part of the dynamics of the ACTION network, with clear differentiation of the tonic and phasic neurones as is their use in chunking the temporal sequence of movements involved;

b) suitable modelling of DA reduction can be achieved by modification of the striatal output thresholds to the GPe and GPi as well as laterally. These are known to correspond to D2 and D1 receptors in the first and to the second and third cases

respectively. Thus DA reduction can be modelled as a reduction in the first and an increase in the second and third of these threshold values. This causes a reduction of activity in GPe and increases in STN and GPi. Such modifications are observed in monkeys treated with MPTP, which causes PD-like symptoms and destroys cells in the SNc. Modifications of TSSG for learnt sequences were also consistent with PD-like symptoms. Our results were consistent with the PD model of [15], in which excess STN activity causes strong inhibition of TH, and thereby rigidity and reduction of movement;
c) in a converse manner we modelled DA increase, as occurs in Huntington's chorea, by opposite changes to the thresholds for striatal to GPe, GPi and laterally on STR compared to those made in b) above. As to be expected, opposite changes occurred to the GPi, STN and GPe, with the first having decreased activity and little changes in the other two. It is the decreased inhibition from GPi, brought about by these parameter changes, which allows for increased spontaneous activity of TH and hence corresponding movements. It is these which, we suggest, are related to the symptoms of a patient suffering from Huntington's chorea.
d) Finally we analysed mathematically the source of these changes, in terms of the underlying working memory ability of the ACTION network. This had been analysed earlier [11] in terms of a reverse saddle-node bifurcation of the fixed point equations for the asymptotic cortical activity arising from the equations (1) - (10). The fixed points, shown in figure 8 and 9 in the normal and PD cases respectively, have two stable and one unstable fixed points without external input. On input above a certain level there is a loss of one of the stable and the unstable fixed points, and the activity increases up to the remaining stable fixed point. On removal of the input the system then remains in the upper stable fixed point. For the PD case (figure 9c) there is only the lower fixed point without any input. Thus input will cause a bifurcation but loss of it will lead to exactly the reverse bifurcation with activity settling back to the original low value. Thus no movement is supported, nor can TSSG be attained by the system any more after this change of the threshold parameters. In the Huntington's chorea case the fixed point curves move in the opposite direction to the PD case, and has only a single high value for the fixed point. This leads to excessive and spontaneous movement, as corresponds to one of the symptoms of the disease.

These results are only obtained by enormous simplifications of the living system. However they seem to be interesting enough to attempt to build on further. That could be done in a number of ways:
1) inclusion of more realistic neurones (shunting, spiking, noisy, compartmental, adaptive, realistic ionic channels, and so on);
2) extension to larger numbers of neurones, including populations of excitatory and inhibitory cells (as required for shunting);
3) explicit addition of dopamine, as arising from SNc, and including the main D1 and D2 types and their detailed neurochemical effects on cell responsiveness;
4) addition of learning with any or all of the above extensions included. Learning ACTION, as primitive in their main functions as the one used here, have already

been simulated for normal operating conditions [10], [13]. These were able to be analysed along the lines of section II.5, with similar insight being obtained;
5) development of the detailed predictions to be expected in brain imaging studies for such patients (or normals). Thus the time courses of neural activity predicted in figures 2, 3, 4 and 5 should be observable in real time by MEG, while the changes of average activity in all the regions involved, not only STN, GPe and GPi, can be observed by PET or fMRI. There is still some controversy about how to relate the latter BOLD signal, as well as the rCBF seen in PET, to the underlying neural activity producing it. However the predictions can be made by the rule of thumb that total synaptic activity (inhibitory or excitatory) is what is measured by either of these blood flow effects. Predictions concerning the temporal flow of activity both round the loops of the ACTION network and across its spatial extent can be developed on the basis of the present results.

Part III. Reverberations and Consciousness.

III.1. Introduction

The quest for the neural correlates of consciousness in the cerebral cortex (NCC) is now proceeding apace [28]; many attempts have been made to track down the possible sites which are necessary for consciousness to be present, and whose neural activity is well correlated with conscious experience. However there is still uncertainty as to sufficient conditions needing to be observed for neural activity for the network of sites involved in the brain to be claimed as *the* sites of consciousness. Here we consider mainly the question of where the NCC are most likely sited in cortex, as well as consideration of how reverberations are involved in that search.

The problem we presently face in the search for the NCC is that the nature of consciousness is subtle, and its characteristics seem to change from moment to moment. In order to prevent the NCC from wandering all over the brain, be it now in the primary sensory or unimodal associative cortices [29], or now in the prefrontal cortex [30], we will initially explore evidence in support of various possible siting of the NCC.

III. 2. Experimental Data

The data we consider are of a variety of sorts: from single cells, lesion effects on behaviour, and from brain imaging (using especially PET and fMRI). The single cell data involves the presence or absence of significant activity observed in an animal under anaesthesia, which has little effect on early sensory cortical responses (in V1 and MT, for example [31]) nor in inferotemporal cortex TE [32]. However there is considerable effect of anaesthesia on parietal lobe single cell responses, leading to great difficulty in making measurements of such cells before the advent of the ability to measure from awake behaving monkeys [33].

As a start to considering lesion effects, we note the singular lack of loss of consciousness due to frontal deficits, brought about either by disease or injury. The cases supporting this claim are numerous. For example there is the famous case of Phineas Gage [7], who had a tamping iron blown through his frontal lobes with considerable loss of frontal cortex but without successive loss of consciousness as he was carried to the local doctor. There is also the case of the young man who was born bereft of most of his frontal lobes [34], but yet, apart from great social problems, lived a normal conscious existence.

One area of brain lesions of particular relevance to the NCC is that of neglect. Patients usually suffer a loss of the right parietal lobe, and subsequently lose awareness of input from their left hemifield. This loss, for example, is observed in the inability to cross out lines on the left of their field of view. However their lot can be ameliorated by vibration of the neck muscles, and in general shows a dependence on body signals. Thus neglect is a loss of awareness in a body-centred frame of reference (although there may also be object-frame dependence of their loss [35]). It is now agreed that neglect arises specifically from damage to the inferior parietal lobe [36].

Much is also being discovered about the siting in the cortex of buffer working memories. Those for spatial vision are in the right [37], those for language and temporal estimation in the left, inferior parietal lobes [38]. On the other hand extinction, involving loss of awareness of the right hand object of two similar objects, one on the left, one of the right, is sited in the superior parietal lobe [36].

There is recent fMRI data indicating that there is also the experience of the motion after-effect (MAE) most strongly in BA 40 (the supramarginal gyrus), in the inferior parietal lobe. The experimental paradigm to observe this uses motion adaptation to a set of horizontal bars moving vertically downwards for 30 seconds, and then stopping. Subjects exposed to such a display usually experience the MAE for 9 or so seconds after cessation of the movement of the bars.

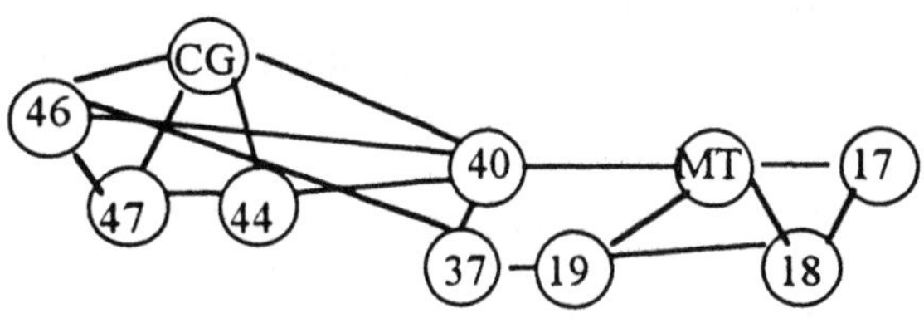

Figure 10. The network of areas active in the brain during the MAE experiment [39]. The lines joining the various modules denote those for which the correlation coefficient is at least 0.4. MT denotes the middle temporal area, and the other areas are numbered according to Brodmann's numeration.

Whole-head fMRI measurements during this paradigm [39] show a network of connected areas, as in figure 10. This network has a posterior group, involving especially the motion area MT, which was found to be responsive to all forms of motion as well as the MAE [40]. On the other hand there is a set of anterior modules, shown in figure 1, which are particularly active both during the MAE period and just after the cessation of oscillatory movement of the bars both up and down (after which there is no MAE experience reported). Finally there are inferior parietal regions which demonstrate activity almost solely in response to the MAE period.

In all, then, the experimental evidence points most strongly to the siting of awareness in the parietal lobe. It is possible that the superior parietal lobe is essential, as claimed in [33], since the patient DF lost the ability to make actions (involving the superior parietal lobe) in a conscious manner on oriented objects although was able to make the correct actions automatically. However it may be that such actions involve a 'motor' version of conscious experience, not usually regarded as consciousness. We will not overstep the experimental bounds if we conclude that at least the inferior parietal lobes are essential for awareness.

III.3. Characteristics of Consciousness

In this section we give a list of characteristics of neural activity needed for consciousness to be present, together with supporting evidence:
A. Temporal duration. Neural activity is needed to be present for at least 200 msecs for awareness to arise. This is supported by the data of Libet [41] and by the duration of activity in buffer working memory sites, suggested by a number of workers as being the sites of creation of consciousness [28].
B. Competition for consciousness. This is needed for a unique winner to occur, and for disambiguation of interpretations of inputs to be achieved [42].
C. Bodily inputs available. Such inputs are needed to give a perspectival character to inputs, following detailed investigations by psychologists [43].
D. Salience of inputs available. Coded in the limbic system, salience is needed to give a suitable level of importance to a given input. This salience can arise, for example, from the cingulate cortex, as involving motivational activation.
E. Past experience. This is to make use of previous relevant experience, and arises initially from hippocampus.

All of these inputs are available, it would seem almost uniquely, for the inferior parietal lobe. Thus the buffer working memory sites for space and time (and language) have been noted as being there [37], [38], as is a competition for consciousness associated with these sites [44]. Bodily inputs are also available there, as noted from effects of manipulation of the body in neglect, as well as from known neuroanatomy (connections with the vestibular apparatus and the cerebellum are well known). The limbic system is also well connected to the inferior parietal lobe [45] and so is episodic memory. For example, in rats the parietal lobe has been shown to be the final resting place for memories built during step-down inhibition learning [46].

Thus we conclude that the inferior parietal lobe is suitably connected and structured to satisfy all of the criteria A-E above. Moreover we have already noted that it contains a number of buffer working memory sites. These were modelled [47] as a set of dedicated self-recurrent nodes, holding activity in the fashion already indicated in section II.2. This activity then decays away through leakage from their membrane surface. We note here that the self-recurrence does not have infinite lifetime, as it did for the neurone of II.2; the neurones of the IPL only have a finite extension of the duration of their activity in this model. It was shown possible in this model to replicate results on recency in list learning, with a universal formula for the reaction time to recognition of an item as having been in the memorised set being

$$RT\ (n, N) = \tau \log \{a + (N-n)b\}$$

where a and b are constants and τ is the time constant of the leaky nodes of the model. A fit to the extant data gives a value of τ of about 1 second, in good agreement with that expected from the psychophysical data [5].

III.4. The Central Representation

Evidence from neglect studies and brain imaging on healthy subjects has been presented above to implicate the inferior parietal lobe as playing an important role in controlling attention and awareness. Both of these attributes can occur in a range of possible frames of reference: neglect can be observed tied to an object, or to a trunk-centred frame of reference or a variety of other reference frames [36]. This implies that the IPL is very likely composed of a set of modules, each carrying information from the environment as well as modulation by possible body input. Thus the IPL is eminently suited to carry what is termed the "Central Representation". This is defined as follows:
<u>The central representation is the combined set of multimodal activations involved in fusing sensory activity, body positions, salience and intentionality for future planning; it involves a competitive process between the various modules it contains to single one out to be conscious and be used for report to other working memory sites for further planning or action.</u>

There are several important features of the central representation (CR) defined above that need discussion, in relation to the criteria of section 3:
1) The CR must have access to sensory input, such as in vision, coded at a high level. Thus it must have good access to temporal lobe representations, so as to use the categorisation built there to guide action.
2) It also must have access to the bodily input needed to guide actions in terms of the intentionality coded in the superior parietal lobe. It was noted already that such intentionality is coded for various sorts of actions: of the limbs, eyes, head or

fingers [48]. This intentionality must be furnished with the parameters of the objects on which the actions must be taken; thus cerebellar and vestibular input is also needed to be accessible to the central representation, as it is in the parietal lobes.

3) Salience of the inputs in the sensory field is an important attribute for the guidance of actions; that arises from limbic input already activated to provide saliencies of inputs from the orbitofrontal cortex by way of the cingulate. This is compounded by activations in the retrosplenial (posterior cingulate) gyrus, encoded as parts of episodic memory [46], [49]. Such connections have been especially emphasised by Mesulam, who wrote, in discussing the important limbic connections to the parietal lobe involved in neglect, that "However, the cingulate and retrosplenial projections are much more selective and may be related to more complex and learned aspects of motivation". Both this and the aspects noted under points 1) and 2) above support the inferior parietal lobe as the site for the CR, since the IPL has good connections to the temporal lobe as well as to body inputs.

4) There are several modules involved in the CR in the IPL; the total activity must undergo an overall competition, possibly aided by thalamo-nucleus reticularis processing. A simulation of such a model has been given earlier [50]. The existence of such competition is supported by attentional deficits observed in subjects with pulvinar lesions [51].

5) Siting the emergence of awareness in the IPL, as the product of the competition ongoing there, is supported by a simulation of the data of Libet and colleagues [41]. This involved the creation of sensory experience (that of a gentle touch on the back of the patient's hand) by direct stimulation of cortex in patients being operated on for dyskinesia and related movement problems. The simulation [42] used a simplified model of the cortico-thalamo-nucleus reticularis circuit, and led to the observed dependence of the delay of awareness on the strength of the threshold current for experiencing the touch on the back of the patient's hand.

6) Such a competition has also been suggested [44] as occurring to explain experimental results of subliminal effects on lexical decision response times obtained by [52]. The experiment involved measurement of the reaction times of subjects to deciding if the first or third of three letter strings were words or not. Subliminal exposure to priming words occurred with the second letter string under one condition, with the presentation of polysemous words such as 'palm', on which the lexical decision had to be made to the third word. The prior exposure caused the decision to be speeded up or delayed in characteristic ways according to the semantic relations of the three words to each other; the simulation was able to explain these results by means of a competition assumed to occur on the phonological store, aided and abetted by activations from a semantic memory store.

In conclusion, we site the CR in the IPL as the confluence of information on salience, episodic memory, high level coding of inputs and body information.

III. 5. Discussion

Our main conclusion is that the IPL is the essential site in the brain for consciousness. We emphasise the crucial feature of the central representation, based there, in comparison to the feature integration theories [53], [54]: *the object file module has been fused with both the master map of spatial locations and the attentional module.* Awareness of an input arises by its representation on the object file (attentional map/central representation) winning the competition there [55], [28]. This leads in general to identification of initial awareness of an input with its 'token', as a specific object, not its 'type' or category. The identification of the type of an input will be expected to take longer, since there has to be conjoint activation of the object model module with the central representation, and then resulting activation of the relevant working memory, such as the phonological store for verbal report. Similar object-specific coupling to the sensory feature maps must also be enhanced. The differences in the speeds of processing to awareness for the different codes for vision mentioned earlier [31] may involve different access times to different codes.

IV. Overall Conclusions.

The analyses and related models and simulations presented above support the main thesis of the paper, which is that:

Higher mental processes (thinking, reasoning, consciousness) are based on simple reverberations in suitably designed neural networks.

This conclusion is thus a development of the original thesis of Eduardo Caianiello in [1]. It is this thesis which can be seen to be able to provide a broad framework from which these subtle processes of the brain can be appreciated. The simple reverberations in a network, that is its fixed points, have a controllable creation and annihilation by external inputs or by other already existent fixed points

We used, in Part II, cartoon versions of the architecture of the frontal lobes so as to be able to achieve the desired manipulative skills in an artificial system. In so doing we were able to recognise some of the functionality contributed to the frontal lobes by its architecture. Moreover many simulations, using different variations on the ACTION net architecture, did not lead to improvement or even to any systems with the capabilities of the benchmark-type of system arrived at earlier. Thus we conclude that the capabilities provided by the ACTION-style of architecture are highly sensitive to the details of this architecture.

In Part III we developed an analysis of the posterior sites of working memory and their relation to the nature of the information needed for conscious processing to be possible. This led us to equating the buffer working memory sites in posterior cortex as identical to the site of the Central Representation. This was introduced as a complex of fused information-carrying neural modules, centred in the inferior parietal lobes. These sites can be identified as those with reverberations of length of several seconds, able to handle the information fusion problem at the root of a central controlling system. Thus once again we find that reverberations are at the basis of higher cognition, in this case the decidedly elusive phenomenon of consciousness. We do not claim that we have given a reason as to why the central representation is conscious, but note here that it possesses all of the needed characteristics of that process.

Finally we suggest that the manner we can now begin to control the simplest reverberations of a network is a sign of an artificial technology ready to enter the field of artificial neural networks

References

[1] Caianiello ER Outline of a Theory of Thinking Machines and Thought Processes. Journal of Theoretical Biology 2:204

[2] Singer W & Gray CM (1995) Visual feature integration and the temporal lobe correlation hypothesis. Ann Rev Neuroscience 18:555-586.

[3] Goldman-Rakic P (1996) Regional and cellular fractionation of working memory. Proc Natl Acad Sci USA 93:13474-13480

[4] Callicott JH, Mattay VS, Bertolino A, Finn K, Coppola R, Frank JA, Gdlberg TE & Weinbergere DR (1999) Cerebral Cortex 9:20-26.

[5] Baddeley A (1986) Working Memory. Oxford: Oxford University Press.

[6] Taylor JG & Alavi FN (1996). In J. Sirosh, R. Miikulainen & Y. Choe (eds). Lateral Interactions in Cortex: Structure and Function.
At http:www.cs.utexas.edu/users/nn/web-pubs/htmlbook96.

[7] Nakahara H & Doya K (1998) Near-Saddle-Node Bifurcation Behaviour as Dynamics in Working Memory for Goal-Directed Behaviour. Neural Computation 10:113-132.

[8] Contreras-Vidal JL & Stelmach GE (1995). Biological Cybernetics 73: 467-476.

[9] Taylor JG (1995). Proc ICANN'95, Soulie F. and Gallinari P. (eds), Ec2 &Co, Paris, pp 543-548.

[10]Taylor JG & Taylor NR (1998a). Biological Cybernetics (submitted).

[11] Taylor NR & Taylor JG (1998b). Humphreys GW., Olson A,. Heinke D. (eds), Proc NCPW5 , Springer-Verlag, pp 92-101

[12] Taylor NR & Taylor JG (1999). To appear in Neural Networks.

[13] Taylor NR (1998). PhD thesis, King's College London, London.

[14] Monchi O & Taylor JG (1997). In: Bullinaria JA, Glasspool DW, Houghton G (eds) Connectionists Representations, Springer, London, pp 142-154.

[15] DeLong MR (1990). Trends in Neuroscience 13: 281-285

[16] Reiss M & Taylor JG (1991) Storing Temporal Sequences. Neural Networks 4:773-787.

[17] Tanji J & Shima K (1994). Nature 371: 413-416.

[18] Halsband U, Matsuzaka Y & Tanji J (1994) Neuroscience Research 20: 149-155

[19] Chappell GJ & Taylor JG (1993) The Temporal Kohonen Map. Neural Networks 6:441-446.

[20] Kohonen T (1988) Associative Memories. Berlin: Springer.

[21] Varsta M, Millan J del R. & Heikkonen (1997) A Recurrent Self-Organizing Map for Temporal Sequence Processing. pp 421-6 in Artifical Neural Networks-ICANN'97, W Gerstner, A Germond, M Hasler & J-D Nicoud (eds). Lecture Notes in Computer Science 1327. Berlin: Springer.

[22] Taylor JG (1990) On the Capacity of Neural Nets for Temporal Patterns. Int. J. Neural Systems 2:47-54.

[23] Shallice T (1988) From Neuropsychology to Mental Structure. Cambridge: Cambridge University Press.

[24] Mesulam M-M (1998). Brain 121: 1013-1052

[25] Passingham RE (1993) The frontal lobes and voluntary action. OUP, Oxford.

[26] Passingham RE (1987) In Motor Areas of the Cerebral Cortex, Ciba Foundation Symposium 132. John Wiley, Chichester, pp 151-164.

[27] Bergman T, Wichmann H, Karmin B & DeLong MR (1994). J. Neurophysiology 72: 507-520

[28] Taylor JG (1999) The Race for Consciousness. Cambridge: MIT Press.

[29] Pollen D (1999) On the Neural Correlates of Visual Perception. Cerebral Cortex 9:4-19

[30]Crick FHC & Koch C (1998) Consciousness and neuroscience. Cerebral Cortex 8:97-107

[31] Moutuoussis K and Zeki S (1997) Functional segregation and temporal hierarchy of the visual perceptive systems. Proc R Soc Lond B 264:1407-1414

[32] Tanaka K (1996) Inferotemporal Cortex and Object Vision. Ann. Rev. of Neurosci.19:109-139.

[33] Milner AD & Goodale MA (1995) The Visual Brain in Action. Oxford: Oxford University Press.

[34] Damasio A (199) Descartes Error. Cambridge: Cambridge University Press.

[35] Karnath H-O (1997) Neural encoding of space in egocentric coordinates? Evidence for and limits of a hypothesis derived from patients with parietal lesions and neglect. pp 497-520 in Parietal Lobe: Contributions to Orientation in 3D Space. P Thier & H-O Karnath (eds). Heidelberg: Springer

[36] Milner AD (1997) Neglect, Extinction and the Cortical Streams of Visual Processing. pp 3-22 in Parietal Lobe: Contributions to Orientation in 3D Space. P Thier & H-O Karnath (eds). Heidelberg: Springer

[37] Jonides J & Smith EE (1997) The architecture of working memory. ch 8, pp 243-276 in Cognitive Neuroscience, MD Rugg (ed). Hove, East Sussex: Taylor & Francis.

[38] Coull JT and Nobre AC (1998) Where and When to Pay Attention: The Neural Systems for Directing Attention to Spatial Locations and to Time Intervals as revealed by Both PET and fMRI. J Neuroscience 18:7426-7435

[39] Schmitz N, Taylor JG, Shah NJ, Ziemons K, Grosse-Ruykin M & Mueller-Gaertner H-W (1998) Brain Activation in the Motion After-Effect. Neuroscience abstracts 695.3, Soc for Neuroscience 24:1762

[40] Tootell RBH, Reppas JB, Dale AM, Look RB, Sereno MI, Malach R, Brady TJ & Rosen BR (1995) Visual motion aftereffect in human cortical area MT revealed by functional magnetic resonance imaging. Nature 375:139-141

[41] Libet B, Alberts WW, Wright EW, DeLattre LD, Levin G & Feinsein B (1964) Production of threshold levels of conscious sensation by electrical stimulation of human somatosensory cortex. J Neurophysiol. 27:546-578

[42] Taylor JG (1996) A competition for consciousness? Neurocomputing 11:271-296

[43] Bermduez JL,Marcel AJ & Eilan N (1995) The Body and the Self. Cambridge MA: MIT Press

[44] Taylor JG (1996) Breakthrough to awareness. Biol Cybern. 25:59-72

[45] Stein J (1992) The representation of egocentric space in the posterior parietal lobe. Behav. Brain sci. 15:691-700
[46] Izquierda I, Quillfield JA, Zanatta MS, Quevedo J, Schaeffer E, Schmitz PK & Medina JH (1997) Sequential Role of Hippocampus and Amygdala, Entorhinal Cortex and Parietal Cortex in Formation and Retrieval of memory for Inhibitory Avoidance in rats. Europ J Neuroscience 9:786-793 (1997)
[47] Hastings S & Taylor JG (1994) Modelling the Articulatory Loop. pp 1452-5 in vol 2 of Proc Int. Conf. Neural Networks ICANN'94, Marinaro M and Morasso P (eds). Berlin: Springer.
[48] Kalaska J, Scott SH, Cisek P & Sergio LE (1997) Cortical control of reaching movements. Current Opinion in Neurobiology. 7:849-859
[49] Mesulam M (1985) Attention, Confusional States amd Neglect. Ch 3, pp 125-168 in Principles of Neurobiology. MM Mesulam (ed). Philadelphia: FA Davis Co.
[50] Taylor JG & Alavi F (1995) A global competitive neural network. Biol. Cybern. 72:233-248
[51] Posner MI, Walker JA, Friedrich FA & Rafal RD (1987) How do the Parietal Lobes Direct Covert Attention? Neuropsychologia 25:135-145
[52] Marcel AJ (1980) Conscious and preconsious recognition on polysemous words: locating the selective effects of prior verbal contexts. In Attention and Performance VIII, RS Nickerson (ed). Hillsdale NJ:Lawrence Erlbaum
[53] Triesman A (1988) Features and Objects: The fourteenth Bartlett Memorial Lecture. Quart. Journ. Experiment. Psych. 40A:201-237
[54] Henderson JM (1994) Two Representational Systems in Dynamic Visual Identification. J Exper Psych.: Gen 20:410-426
[55] Taylor JG (1999) Neural Networks for Consciousness: The Central Representation. To appear in Proc Int.Joint Conference on Neural Networks (IJCNN'99). Hillsdale NJ: Erlbaum

Section 2
Review Papers

Computational Intelligence in Hydroinformatics: A Review

Gb. Cicioni(2) and F. Masulli(1,3)

(1) Istituto Nazionale per la Fisica della Materia
Via Dodecaneso 35, 16146 Genova, Italy
(2) Istituto di Ricerca sulle Acque del
Consiglio Nazionale delle Ricerche,
Via Reno 1, 00198 Roma, Italy
(3) DISI - Dipartimento di Informatica e
Scienze dell'Informazione - Università di Genova
Via Dodecaneso 35, 16146 Genova, Italy

Abstract

Hydroinformatics is the field of study of the flow of information and its processing by knowledge as applied to the flow of fluids and their interaction with the aquatic environment. Many new modeling techniques have been entered in Hydroinformatics successfully. Among them, the application Computational Intelligence methods in Hydroinformatics is a relatively new area of research, even if some successful results have been already obtained. In this review we present a general overview of the applications of Computational Intelligence methods to Hydroinformatics and analyze some promising cases study concerning, namely, estimation of sanitary flows, rainfall prediction, unit hydrograph estimation, groundwater monitoring, flood waves propagation, and pump scheduling.

1 Introduction

Computational Intelligence or *Soft Computing* area includes neural networks [30, 8], fuzzy systems [34], chaos theory [2], evolutionary computation [27, 7], and some others tools linked to inductive learning from data[1] [16, 56].

In recent years the applications of Computational Intelligence methods have been gradually increased in many areas including Physics, Chemistry and Biology, in particular when complex phenomena must be modeled, as, in principle, Computational Intelligence methods are data-driven methods and require a limited knowledge of the phenomena.

One particular challenging application field for Computational Intelligence is Hydroinformatics, that is the application of Computer Science to Water Resources Management.

[1]Sometimes the term *Soft Computing* is ascribed to the attempts to integrate all those computing paradigms, under the claim that when utilized together, the strengths of each of those techniques can be exploited in a synergistic manner for the creation of 'smart' systems.

application	MLP	SOM	TLRN	PRNN	TDNN	FL	EC	HM
river flow *MP*	[4, 39]							
lake water levels *MP*	[11]							
rainfall-runoff *MP*	[40] [32] [9]			[40]	[40]		[9]	[9]
water tide *MP*								[15]
sanitary flows *MP*	[22]							
flood *MP*	[33] [31]		[31]			[49]	[33] [49]	[33] [49]
contaminations in ground water *MP*	[53]					[53]	[53]	[53]
water distribution network *MP*							[42]	[51]
wastewater treatment plant *PC*	[29]							
river discharges *DA*		[11]						
ecological *DA*		[11]						
regional flood *DA*	[28]	[49]						

Table 1: Some applications of Computational Intelligence tools (columns) reported in the recent Hydroinformatics literature (rows). The acronyms are explained in the text.

Hydroinformatics deals with the modeling and control of non-linear systems with too many freedom degrees and uncertainties [29]. For example, Hydroinformatics faces the modeling of rainfall, floods, river flows or the control of wastewater treatment plants. All of those problems involve non-linear and possibly chaotic systems with many freedom degrees and uncertainties [29].

In the past decades, many new modeling techniques have been entered in Hydroinformatics successfully. Among them, the application of Computational Intelligence methods is a relatively new area of research, even if some successful results have been already obtained, e.g. in rainfall-runoff modeling, in predictions of water levels, currents and waves using hydrological and meteorological data, in the optimization of operational parameters of wastewater treatment plants, in groundwater management, and in modeling of ecological and water quality processes.

While classical approaches fail in presence of non-linear systems with many freedom degrees, Computational Intelligence approach offers a more flexible, less assumption dependent and self-adaptive approach to modeling, and moreover shows:

- the potential for improved performance, faster model development and execution times and therefore reduced costs;
- the capability to plug sits components directly into conventional models;
- the ability to provide a measure of prediction certainty via bootstrapping techniques and on-line retraining to adapt to rapidly changing future events.

Some relevant applications of Computational Intelligence tools to Hydrology reported in the recent literature are:

- **Modeling/Prediction** (*MP*): Multilayer Perceptron (MLP) [4, 5, 12, 26, 28, 31, 32, 39, 40], Auto-Regressive Neural Network (ARNN) [12], Time Lagged Recurrent Network (TLRN) [31], Partial Recurrent Neural Network (PRNN) [40], Time Delay Neural Network (TDNN) [40], Fuzzy Logic (FL) [42, 49, 53], Evolutionary Computation (EC) [33, 49, 51, 53].
- **Data Analysis** (*DA*): Self Organizing Feature Maps (SOM) [11].
- **Process Control** (*PC*): Multilayer Perceptron (MLP) [29].
- **Classification** (*CL*): Multilayer Perceptron (MLP) [15].

In Tab. 1 we report some Computational Intelligence tools successfully used in recent papers in Hydroinformatics.

It is worth noting that some papers try to find new strategies for applying Computational Intelligence methods in hybrid modeling (*HM*) approaches [15]. We can notice for example, combinations of neural networks and finite difference modeling methods [26], individual feedfoward networks linked using fuzzy logic controlling model and optimized with evolutionary computation algorithms [49], MLPs and Fuzzy Systems used to set up Rule Bases for Expert Systems [53], and Evolution Strategies used to optimize MLPs [5, 53].

In the next Sect.s we shall present some cases study of application of Computational Intelligence methods to Hydroinformatics concerning, namely, estimation of sanitary flows (Sect. 2), rainfall prediction (Sect. 3), unit hydrograph estimation (Sect. 4), groundwater monitoring (Sect. 5), flood waves propagation (Sect. 6), and pump scheduling (Sect. 7). In Sect. 8 conclusions are drawn.

2 Estimation of Sanitary Flows

In [22] Dejebbar and Alila applied Multilayer Perceptrons to the prediction of current and future wastewater flows in order to support the operation, maintenance, and upgrading of municipal wastewater collection systems.

Existing methods are based on parametric approach and are very imprecise. Given the variability of wastewater flow generation, Multilayer Perceptrons can constitute a better alternative to establish relationships between tributary area characteristics and wastewater flow rates. The work by Dejebbar and Alila [22] concerns the development of a MLP model to estimate wastewater production given the sewerage area characteristics.

Dry weather flow measurements at 40 monitoring sites in the Fraser Sewerage Area (FSA) of the greater Vancouver Regional District, Vancouver, British Columbia, were used in the MLP training. The training data base concerned a total area that varies between 120 ha and 1700 ha, and a population varying between 3700 and 58000. The validation data set has been obtained from a geographically separated area with significantly different demographics, land use, and hydraulic characteristics. It represents 16 flow monitors with tributary areas varying from 100 ha to 4400 ha, and populations varying from 1300 to 120000. The validation data set extended beyond the range of the development data set, providing a check of the MLP ability to extrapolate results.

As a first step, weights, optimization and architecture selection of the neural network model were performed. The selected input variables are: total surface area, residential population, dwelling units, commercial areas, industrial areas, institutional areas, other non residential areas.

Both for the development and validations data, the MLP model estimated dry weather flows with an average error of less than 16%, that is better than the results obtained with traditional estimation methods. Larger errors have been noticed for small areas with a large industrial component.

Such model can minimize the need for flow monitoring, and can fill the gap between flow monitoring sites. It can be used to track the wastewater increase as the land use characteristics change, and to forecast future flows more accurately. Also it can be used to check the reasonableness of measured flows and is a useful estimation tool for wastewater quality related issues.

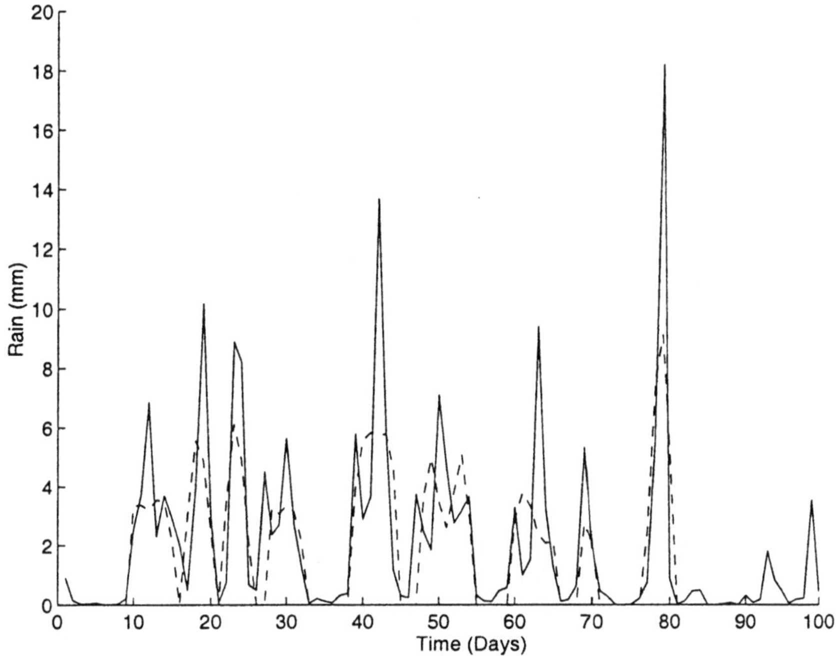

Figure 1: Mean Station: measured rainfall (continuous line) and 1 day ahead neural forecasting using SSA's components 1-76 (dashed line). Period 3/19/56 - 12/4/56.

3 Daily Rainfall Forecasting

In [44, 17, 18], Cicioni, Masulli, and Studer proposed a methodology for signal forecasting based on results and prescriptions related to the Takens-Mañé theorem [54, 41, 3] and its extension to the case of discontinuous signals prediction.

As an application, they faced the problem of forecasting of the rainfall intensities series collected by 135 stations distributed in the Tiber basin for a period of 10 years.

Classical regression techniques [13] are unable to obtain any suitable forecasting of the series of the hydrological variable, and also predictors based on neural or fuzzy systems give very poor results. The difficult of rainfall intensity forecasting can be ascribed mainly to the discontinuity of the hydrological variable, while Universal function approximation theorems for neural networks [20] and fuzzy systems [58] require the continuity of the function to be approximate.

In order to avoid the effect of the discontinues the authors applied to the hydrological series the Singular-Spectrum Analysis (SSA) [35, 45, 57, 38]. In SSA the state vector is a temporal window of the series of a given number of samples. A Principal Component Analysis [55] is performed in the space of state vectors, and the original temporal series can be decomposed in series components. The series components of the rainfall intensity are continuous and then we can use fuzzy or neural predictors for forecasting each of them.

A good property of SSA is that the original series can be recovered as the sum of all the individual series components. If we truncate this sum to an assigned number of component an estimation of the resulting reconstruction error is the sum of the eigenvalues corresponding the the remaining components.

In [18], the data analysis started by considering the virtual series of the Mean Station (MS), obtained by averaging all 135 rainfall intensity series. The application of the SSA to this signal produced an eigenvalue space that was separated in 10 sub-spaces (each of them representing a 10% of the reconstruction error) from where 10 signal components (waves) were obtained.

For each individual wave of the MS, the authors prepared a neural predictor based on a Multi-Layer Perceptron (MLP) [30] following the constructive approach described in [52, 44]. After the determination of the first minimum of the mutual information, the embedding dimension has been estimated, thorough the global false nearest neighbors technique. As a consequence, the selected MLP predictor has two hidden layers of 5 neurons, and an input layer of six neurons (equal to the embedding dimension of the signal).

The first waves, corresponding to the principal components of the SSA, are enough regular and easy to be predicted, while waves corresponding to sub-spaces with low eigenvalues are less regular and their predictions are unsatisfactory. For the first 6 waves, corresponding to the first 76 principal components, (i.e. 60% of the explained variance) the prediction results are of good quality (see Fig. 1). The sum of the prediction of the 6 waves at 1 and 2 days ahead give a signal well correlated with the original rainfall intensity of the MS.

In [18], the authors reported also good preliminary results of the application of the describe methodology to rainfall intensity series of individual stations using projections of the signal of the principal component space of the MS.

4 Unit hydrograph

One of the more effective means for the analysis of the formation of the outflows and their modulation, in correspondence of a section of a river basin, is given from the hydrograph, i.e., the diagram of the distribution of the flow in function of the time. The total outflow defined from the area of the hydrograph is equal to the effective rainfall to the river basin dependent from one rain of known duration and intensity.

The concept of the unit hydrograph was developed by Sherman in 1932 [50]. The unit hydrograph has many functions in hydrology, and it is used, e.g., in the characterization of the rainfall-runoff behavior of a catchment, in flood estimations, and, as an transfer-function, in existing conceptual models.

The First Differenced Transfer Function - Excess Rainfall and Unit Hydrograph by a Deconvolution Iterative Identification Technique (FDTF-ERUHDIT)

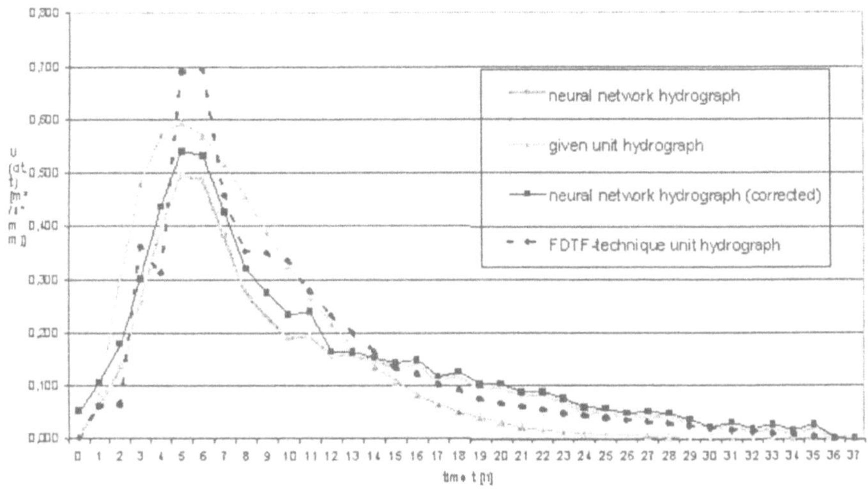

Figure 2: Computed hydrographs gauge *Bairawies* ($U(dt,t)[m^3/s^*mm]$ versus time $t[h]$). (Reprinted with permission from [6]).

derives numerically a unit hydrograph from rainfall and runoff records [24, 43]. The unit hydrograph is calculated in a iterative way by convolutions and deconvolutions. As the method uses measured rainfall and runoff records, it can estimate the effective rainfall and can cut off the base flow on its own. The user just has to give the size (area) of the catchment and the numbers of the ordinates of the expected unit hydrograph.

In [36], Lange proposed the derivation of shown the unit hydrograph using neural networks. By using the artificial neural network simulator SNNS (Stuttgart Neural Network Simulator) developed at University of Stuttgart [59], an optimal Multilayer Perceptron architecture was found after a great numbers of tests. The topology of the network with the best efficiency has 1 hidden layer. The input to the MLP are the measured rainfall data. The output represents the ordinates of the discharge at the gauge station. The number of hidden neurons is up to 15. The learning rule is the back-propagation algorithm [47] accelerated by using the flat-spot-elimination and the momentum-term (conjugated back-propagation method [14]).

After the training the neural network is able to produce a complete hydrograph depending on the given rainfall without simulation of any single physical process in the catchment. So the neural network is indeed a "black box" representing the catchment response.

In Fig. 2, a unit hydrograph of the catchment "Bairawies", is derived by using the neural network and the FDTF-technique. Both unit hydrographs are compared with the given unit hydrograph which was found in the documentation of the Bavarian state authority of water management. Both methods, the neural network and the FDTF-technique, are using the same input data so they can be compared likely. Data of rainfall and runoff from the year 1971 to 1981 for for this catchment were used. The time step of the measured data was five minutes for the rainfall data and fifteen minutes for the runoff data.

The FDTF-technique calculates a unit hydrograph directly. On the other hand the neural network presents the whole rainfall-runoff behavior of the catchment so the derivation of the unit hydrograph is just possible with a little trick. The trained network is able to calculate a hydrograph which is comparable with the unit hydrograph when the input value of the rainfall record is 1 mm. This methods works pretty well when the range of the training data is wide enough. That means that not only extreme events with high peak flows has to considered in the training set, but also events with low peak flows. This stands in contrast to the usual practice of unit hydrograph generation where just the largest events with spatially near-uniform rainfall are used. To receive a real unit hydrograph, the hydrograph of the neural network must be corrected to fulfill the condition of volume.

Compared with the given unit hydrograph the unit hydrograph of the neural network is in the first section to low and in the second part to high. The FDTF-technique produces a unit hydrograph which is similar to the neural network hydrograph except the maximum ordinate which is to high.

As reported in the paper by Lange [36], the derivation of a unit hydrograph by a neural network has some advantages than that using the FDTF-ERUHDIT approach. In particular, it needs less calculation time and, moreover, it is very robust concerning the training events. For example the calculation time of the catchment "Bairawies" was up to 15 hours on a regular PC with an Intel 200 Pentium Processor, while the calculation time of the training of the neural network costs about minutes.

The reduction of a neural network to a single hydrograph, i.e. the unit hydrograph , provides a tool which gives the user of neural network an idea about the quality of the events used in the training and how many events are necessary to describe the rainfall-runoff-relation in a perfect way. Another advantage is the possibility to use the unit hydrograph from the neural network in existing hydrologic procedure and models.

5 Groundwater monitoring

In [21], D'Agostino, Passarella and Vurro presented an approach to the optimization of networks of groundwater monitoring based on fuzzy geostatistics.

Recent literature reports a large number of methodologies for the design of monitoring networks and outline of sampling optimization procedures [1, 37]. Cokriging Estimation Variance (CEV) is a useful tool to determine the influence of the spatial configuration of monitoring networks on the estimations. CEV depends on the variograms [23]. Reduced and optimal configurations of a network based on the CEV suffers from the uncertainties caused by the variographer's choices on type and parameters of the variogram model when the

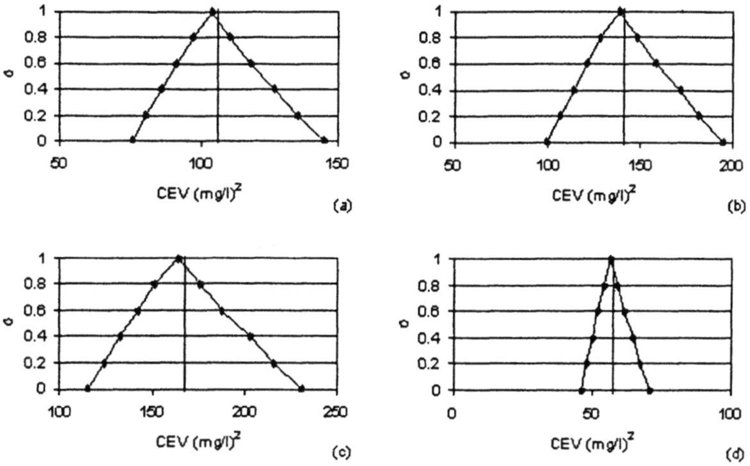

Figure 3: Membership functions of CEV and centroids in four critical points: (a) point n.2, (b) point n. 3, (c) point n. 8, (d) point n. 11. (Reprinted with permission from [46]).

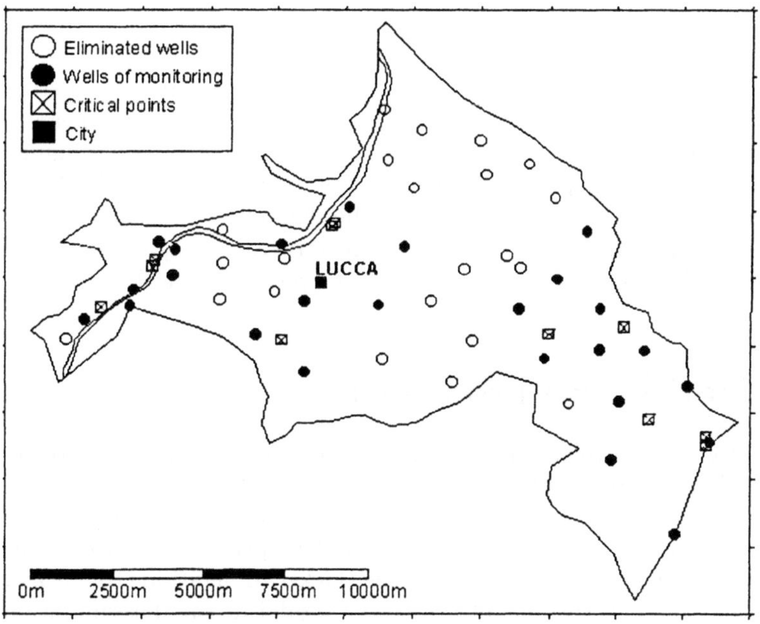

Figure 4: Reduced configuration (25 wells of monitoring) turning out from the case study. (Reprinted with permission from [46]).

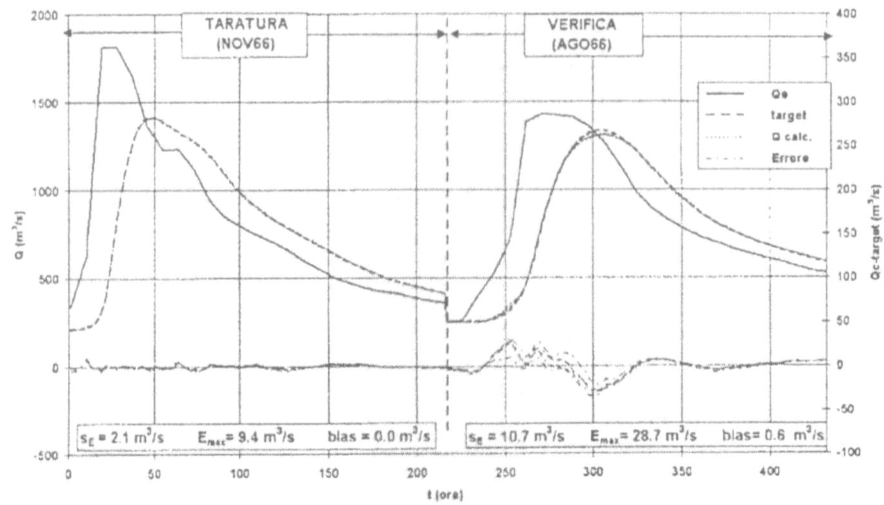

Figure 5: Comparison of the flood values estimated by the 46:4:1 MLP trained using the November wave and tested with the August wave (*novago.net*). (Reprinted with permission from [46]).

actual data are missing. Using a fuzzy approach to represent the uncertainties related to variogram parameters estimation, the variogram parameters can be described as fuzzy numbers and the parameter uncertainties by means of membership functions [10]. Using such fuzzy variograms [10], the cokriging method produces membership functions of both the estimation and the CEV. When an optimal network arrangement has to be defined, a defuzzification measure can be used. An application of the sequential research algorithm, based on the fuzzy CEVs, has been carried out in order to verify the methodology (see Fig.s 2, 3).

6 Flood waves propagation

The flood waves propagation is a typical non linear phenomenon of River Hydraulics. The propagation of the flood waves in the river hydraulics can be studied with hydraulic methods, conceptual models or black-box models.

In [25], Fiorillo presented a Multilayer Perceptron model for modeling the flood waves propagation trained using the Back-Propagation (BP) learning rule. They tested the effectiveness of the neural model in the determination of non linear transfer function between the relative waves of flood to the extreme sections of Adige river. For the training and verification, two events of flood (August 1996 and November of 1966) between the stations of "Marco" and "Boara Pisani", spaced about 180 km, have been in turn considered.

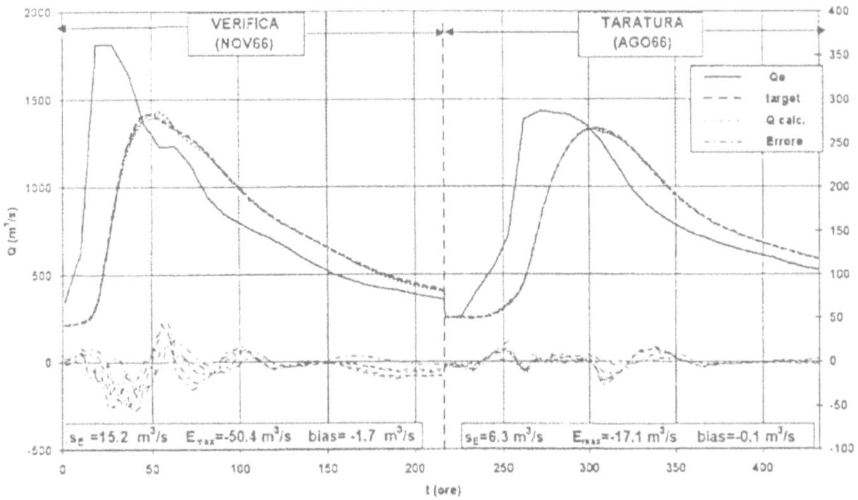

Figure 6: Comparison of the flood values estimated by the 46:4:1 MLP trained using the August wave and tested with the November wave (*agonov.net*). (Reprinted with permission from [46]).

The structure of the model of artificial neural network is based on the assumed functional dependence between input and output variables, on the basis of the propagation phenomenon, of the type $Q_t^{BP} = F(Q_t^m,\ Q_{t-1}^m, ...,$ $Q_{t-k}^m)$, in which Q_t^{BP} it is the outflow of gauge station "Boara Pisani" at time t, while $Q_t^m, Q_{t-1}^m, ..., Q_{t-k}^m$ are the outflows of the gauge station "Marco" at the time t and previous hourly intervals. k has been fixed to 45 hours on the basis of the evaluation of the propagation time in the section, of the $10 - 40$ order hours for variable outflow from 200 to 1800 m^3/s. The neural network is constituted by one input layer of 46 neurons, one hidden layer with 4 neurons and one output layer with one single neuron (i.e., 46:4:1). In the learning and test Q_t^{BP} target data have been corrected using the varying parameters Muskingum-Cunge model were used as target [19], taking into account the not linearity of the phenomenon.

As shown in Fig.s 5 and 6, the test with the floods not employed in training gave acceptable results, even if the task is difficult due to the hight diversity of the two waves of the mount section shown both in the overflow value and in the steepness of the growing part.

7 Pump Scheduling in Water Supply Systems

In [48], Schaetzen, Savic and Walters presented a method based on Evolution Programs for determining an optimal schedule of pumping on a 24-hour basis

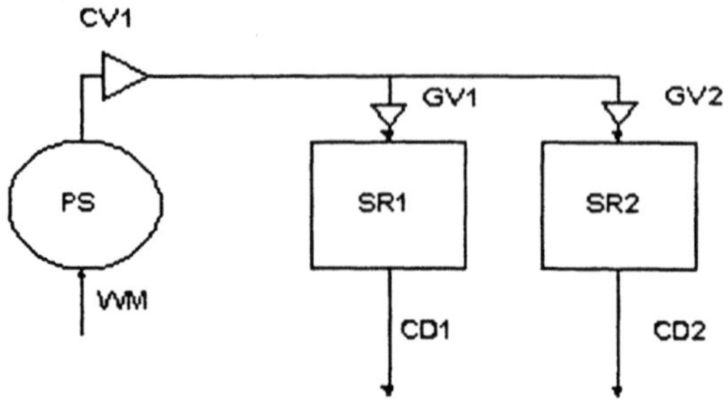

Figure 7: The model of the water network. (Reprinted with permission from [6]).

in a water supply system.

Pump scheduling is the process of choosing which of the available pumps within the system are to be used, the periods when they are to be run, and in the case of variable-speed pumps the pump speeds. The main objective of pump scheduling is to minimize operating costs whilst keeping within operating constraints.

A Genetic Algorithm (GA) [27] is applied in order to minimizing the total operating cost. Savings can be achieved by taking advantage of the lower unit cost of electricity at nights and weekends. An application of the method to the Sidmouth - Seaton water supply system in Devon (UK) is presented in the paper, showing that considerable savings are possible.

The required data for the GA are: The optimization period and the time interval, the demand plots during this optimization period, the initial minimum and maximum water levels of the reservoirs, the number of pumps comprising the pumping station, the different pump combinations and corresponding flows, the different valve combinations, and the electricity consumption for each pump combination and the pattern of the electricity tariffs used during this period.

A schematic representation of the Sidmouth - Seaton water network, as used in the optimization, is shown in Fig. 7. This simplified schematic is in direct correspondence to the proposed programming formulation of pump scheduling problems. The water supply network includes: two service reservoirs noted as SR1 and SR2, two lumped consumer demands noted as CD1 and CD2, one pumping station including three booster pumps noted as PS, two gate valves and one check valve noted as GV1, GV2 and CV1 respectively and a water main supplying the pumping station noted as WM.

The best GA-optimized pump schedule (Fig. 8) was determined in less than 8000 iterations. Tab. 2 compares the data resulting from the best chromosome found by the GA optimization process and the data which have been supplied by South West Water Services Ltd. (SWW). The hydraulic requirements of the GA-optimized pump scheduling are assumed to be satisfied.

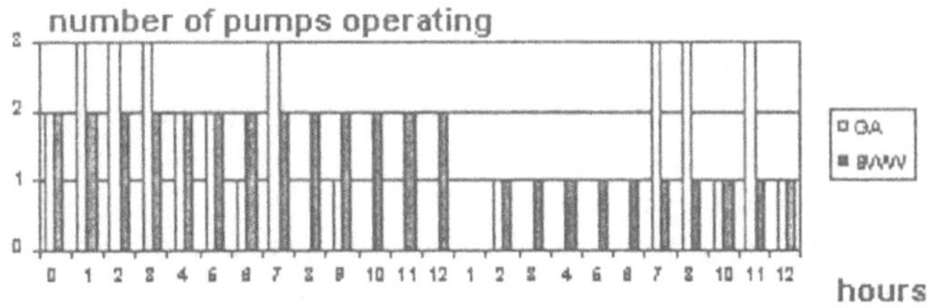

Figure 8: Comparison of Pump Schedules. (Reprinted with permission from [6]).

DATA DESCRIPTION	GA	SWW
Electricity Cost [£]	70.1	144.6
Total volume pumped [m^3]	5220	4964
Cumulative demand [m^3]	5159	

Table 2: Optimization results.

Comparing the GA-optimized pump scheduling with the original pump scheduling solution demonstrates theoretical savings of about 50% achieved by the GA optimization technique. The substantial theoretical savings are reduced when special operational costs are taken into account (i.e., costs associated with turning on or off pumps or valves).

As shown in the paper by Schaetzen and co-workers [48], Genetic Algorithms offer a valid way to achieve potential energy savings, and could be used in a wide range of water supply networks varying in size and complexity.

8 Conclusions

The information technology revolution of the last 30 years has altered the traditional planning, modeling and decision-making methodologies of the water-related sciences and technologies. Computer Science plays an essential role in the sustainable development of water resources and the responsible management of the aquatic environment. The general availability of sophisticated computers with ever-expanding capabilities has given rise to an increasing complexity in terms of computational ability in the storage, retrieval and manipulation of information flows.

Hydroinformatics is the field of study of the flow of information and its processing by knowledge as applied to the flow of fluids and their interaction with the aquatic environment. Many new modeling techniques have been entered in Hydroinformatics successfully. Among them, the application Computational Intelligence methods (such as neural networks, fuzzy systems, and evolutionary computation) in Hydroinformatics is a relatively new area of research, even if some successful results have been already obtained, e.g. in rainfall-runoff modeling, in predictions of water levels (or currents, or waves) from hydrological and meteorological inputs, in the optimization of operational parameters of wastewater treatment plants, in groundwater management, and in modeling of ecological or water quality processes.

In this review we have presented a general overview of the applications of Computational Intelligence methods to Hydroinformatics and we have analyzed some promising cases study concerning, namely, estimation of sanitary flows, rainfall prediction, unit hydrograph estimation, groundwater monitoring, flood waves propagation, and pump scheduling.

9 Acknowledgments

This works was partially supported by CNR, INFM, MURST.

References

[1] ASCE Task Committee on Geostatistical Techniques in Geohydrology of the Ground Water Hydrology. Review of Geostatistics in Geohydrology, ii: Applications. *Journal of Hydraulic Engineering*, 116(5):633–658, 1990.

[2] H.D.I. Abarbanel. The analysis of observed chaotic data in physical systems. *Reviews of Modern Physics*, 65(4):1331–1392, 1993.

[3] H.D.I. Abarbanel. *Analysis of Observed Chaotic Data.* Springer, New York, USA, 1996.

[4] R. Abrahart. Neural networks and the problem of accumulated error: an embedded solution that offers new opportunities for modelling and testing. In V. Babovic and L.C. Larsen, editors, *Hydroinformatics '98 - Proceedings of the third international conference*, Copenhagen, 1998. Balkema.

[5] V. Babovic and L.C. Larsen. A data mining approach to time series modelling and forecasting. In V. Babovic and L.C. Larsen, editors, *Hydroinformatics '98 - Proceedings of the third international conference*, Copenhagen, 1998. Balkema.

[6] V. Babovic and L.C. Larsen (Eds.). *Hydroinformatics '98 - Proceedings of the third international conference.* Balkema, Rotterdam, 1998.

[7] T. Baeck. *Evolutionary Algorithms in Theory and Practice: Evolution Strategies, Evolutionary Programming, Genetic Algorithm.* Oxford University Press, 1996.

[8] C. M. Bishop. *Neural Networks for Pattern Recognition.* Clarendon Press, Oxford, 1995.

[9] O. Blanpain, F. Boussemart, C. Lecoutre, and S. Merchez. Genetic algorithms to determine the profile of urban drainage networks from incomplete data. In V. Babovic and L.C. Larsen, editors, *Hydroinformatics '98 - Proceedings of the third international conference*, Copenhagen, 1998. Balkema.

[10] I Bogardi and A. Bardossy. Fuzzy sets geostatistics: Research needs. In *Proc. of Recent Developments and Applications in Surface and Subsurface Hydrology*, pages 80–90, Karlsruhe, 1990.

[11] H. van den Boogaard, S. Ali, and A.E. Mynett. Self organizing feature maps for the analysis of hydrological and ecological data sets". In V. Babovic and L.C. Larsen, editors, *Hydroinformatics '98 - Proceedings of the third international conference*, Copenhagen, 1998. Balkema.

[12] H. van den Boogaard, D.K. Gautam, and A.E. Mynett. Auto-regressive neural networks for the modelling of time series. In V. Babovic and L.C. Larsen, editors, *Hydroinformatics '98 - Proceedings of the third international conference*, Copenhagen, 1998. Balkema.

[13] G.E.P. Box and G.M. Jenkins. *Time Series Analysis Forecasting and Control.* Holden-Day, S. Francisco, 1970.

[14] R Brause. *Neurale Netze.* Teubner, Stuttgart, 1995.

[15] A. Breitscheidel, J. Cser, and P. van der Veer. Applicability of neural networks in hybrid water models. In V. Babovic and L.C. Larsen, editors, *Hydroinformatics '98 - Proceedings of the third international conference*, Copenhagen, 1998. Balkema.

[16] V. Cherkassky and F. Mulier. *Learning from Data: Concepts, Theory, and Methods.* J. Wiley and Cons, Inc., New York, 1988.

[17] Gb. Cicioni and F Masulli. Approccio alla previsione di altezze pluviometriche basato su analisi di componenti principali e sistemi fuzzy adattivi. In *Atti XXVI Convegno di Idraulica e Costruzioni Idrauliche - Catania)*, volume IV, pages 27–38. CUECM, Catania (Italy), 1998.

[18] Gb. Cicioni and F. Masulli. Forecasting discontinuous signals using singular spectrum analysis and neural networks. In *Proceedings of NEU99 - FINANCIAL APPLICATIONS OF NEURAL NETS AND FUZZY SYSTEMS*, 1999. *(invited paper - in press).*

[19] J.A. Cunge. On the subject of a flood propagation method (muskingum method). *Journal of Hydraulic Research, IAHR*, 7:205–230, 1969.

[20] G. Cybenko. Approximation by superpositions of a sigmoidal function. *Mathematics of Control Signals, and Systems*, 2:303–314, 1989.

[21] V. D'Agostino, G. Passarella, and M. Vurro. Supporto alla decisione basato sulla geostatistica fuzzy: ottimizzazione di reti di monitoraggio delle acque sotterranee. In *Atti del XXVI Convegno di Idraulica e Costruzioni Idrauliche - Catania*, volume IV, pages 39–51, Catania (Italy), 1998. CUECM Editore.

[22] Y. Djebbar and Y. Alila. Neural network estimation of sanitary flows. In V. Babovic and L.C. Larsen, editors, *Hydroinformatics '98 - Proceedings of the third international conference*, Copenhagen, 1998. Balkema.

[23] P.A. Dowd. The variogram and kriging: Robust and resistent estimators. In *Geostatistics for Natural Resources Characterization*, pages 91–106, Amsterdam, 1984. Reidel.

[24] D. Duband, C. Obled, and J.Y. Rodriguez. Unit hydrograph revisited: An alternate iterative approach to UH and effective precipitation identification. *Journal of Hydrology*, 150:115–149, 1993.

[25] G. Fiorillo. Modellazione della propagazione delle onde di piena con reti neurali artificiali. In *Atti del XXVI Convegno di Idraulica e Costruzioni Idrauliche - Catania*, volume IV, pages 65–76, Catania (Italy), 1998. CUECM Editore.

[26] I. Fulop, J. Jozsa, and T. Kramer. A neural network application in estimating wind induced shallow lake motions. In V. Babovic and L.C. Larsen, editors, *Hydroinformatics '98 - Proceedings of the third international conference*, Copenhagen, 1998.

[27] D.E. Goldberg. *Genetic Algorithms in Search, Optimization, and Machine Learning.* Addison-Wesley, Reading, 1989.

[28] M. J. Hall and A. Minns. Regional flood frequency analysis using artificial neural networks. In V. Babovic and L.C. Larsen, editors, *Hydroinformatics '98 - Proceedings of the third international conference*, Copenhagen, 1998. Balkema.

[29] W. S. Hanisch, E. C. Pires, and de Carvalho A. A neural network model for predicting operational parameters of a wastewater treatment plant. In V. Babovic and L.C. Larsen, editors, *Hydroinformatics '98 - Proceedings of the third international conference*, Copenhagen, 1998. Balkema.

[30] S. Haykin. *Neural Networks. A Comprehensive Foundation.* Macmillan College Publishing, New York, 1994.

[31] M. Khondker, G. Wilson, and A. Klinting. Application of artificial neural network in real time flash flood forecasting. In V. Babovic and L.C. Larsen, editors, *Hydroinformatics '98 - Proceedings of the third international conference*, Copenhagen, 1998. Balkema.

[32] J. Kim and C. H. Hahm. Hydrologic forecasting by artificial neural network with time series model structure. In V. Babovic and L.C. Larsen, editors, *Hydroinformatics '98 - Proceedings of the third international conference*, Copenhagen, 1998. Balkema.

[33] J. Kim, B. Park, and H. Cha. Parameters estimation of storage function model with genetic algorithms. In V. Babovic and L.C. Larsen, editors, *Hydroinformatics '98 - Proceedings of the third international conference*, Copenhagen, 1998. Balkema.

[34] G.J. Klir and B. Yuan. *Fuzzy Sets and Fuzzy Logic: Theory and Applications.* Prentice Hall, Upper Saddle River, New Jersey, 1995.

[35] R. Kumaresan and D.W. Tuffs. Data-adaptive principal component signal processing. In *IEEE Proc. Conf. on Decision and Control*, page 949, Albuquerque, USA, 1980. IEEE.

[36] N. Lange. Advantages of unit hydrograph derivation by neural networks. In V. Babovic and L.C. Larsen, editors, *Hydroinformatics '98 - Proceedings of the third international conference*, Copenhagen, 1998. Balkema.

[37] S.I. Lee and P.K. Kitadinis. Optimization of minitoring well installation time and location dusring aquifer decontamiation". *Water Resources Research*, 10:439–462, 1996.

[38] F. Lisi, O. Nicolis, and M. Sandri. Combining Singular-Spectrum Analysis and neural networks for time series forecasting. *Neural Computation*, 2:6–10, 1995.

[39] C. Liu and S. Lin. Artificial neural network forecasting for chaotic dynamical process in hydrology. In V. Babovic and L.C. Larsen, editors, *Hydroinformatics '98 - Proceedings of the third international conference*, Copenhagen, 1998. Balkema.

[40] K.C. Luk, J. E. Ball, and A. Sharma. Rainfall forecasting through artificial neural network. In V. Babovic and L.C. Larsen, editors, *Hydroinformatics '98 - Proceedings of the third international conference*, Copenhagen, 1998. Balkema.

[41] R. Mañé. On the dimension of the compact invariant sets of certain nonlinear maps. In D.A. Rand and L.-S. Young, editors, *Dynamical Systems and Turbulence*, volume 898 of *Lecture Notes in Mathematics*, pages 230–242, Warwick 1980, 1981. Springer-Verlag, Berlin.

[42] M. Manic and J. Muskatirovic. Selection of optimum solution for water supply in fuzzy decision environment. In V. Babovic and L.C. Larsen, editors, *Hydroinformatics '98 - Proceedings of the third international conference*, Copenhagen, 1998. Balkema.

[43] I. Nalbantis, C. Obled, and J.Y. RODRIGUES. Unit hydrograph and effective precipitation identification. *Journal of Hydrology*, 168:127–157, 1995.

[44] R. Parenti, F. Masulli, and L. Studer. Control of non-linear process by neural networks: Benefits using the Takens-Mañé theorem. In *Proceedings of the ICSC Symposium on Intelligent Industrial Automation, IIA'97*, pages 44–50, Millet, Canada, 1997. ICSC.

[45] E.R. Pike, J.G. MCWhirter, M. Bertero, and C. deMol. Generalized information theory for inverse problems in signal processing. *IEE Proceedings*, 59:660–667, 1984.

[46] *Proceedings of XXVI Convegno di Idraulica e Costruzioni Idrauliche - Catania.* CUECM Editore, Catania (Italy), 1998.

[47] D.E. Rumelhart, G.E. Hinton, and R.J. Williams. Learning internal representations by error propagation. In D.E. Rumelhart and J.L. McClelland, editors, *Parallel Distributed Processing*, volume 1, chapter 8, pages 318–362. MIT Press, Cambridge, 1986.

[48] de W. Schaetzen, D. Savic, and G.A. Walters. A genetic algorithm approach to pump scheduling in water supply systems. In V. Babovic and L.C. Larsen, editors, *Hydroinformatics '98 - Proceedings of the third international conference*, Copenhagen, 1998. Balkema.

[49] L. See and S. Openshaw. Using soft computing techniques to enhance flood forecasting on the river ouse. In V. Babovic and L.C. Larsen, editors, *Hydroinformatics '98 - Proceedings of the third international conference*, Copenhagen, 1998. Balkema.

[50] Sherman. In Ve Te Chow and L.W. Maidment, D.R. anmd Mays, editors, *Applied Hydrology*, pages 261–265, New York, 1988. McGraw-Hill.

[51] M. Stanic, Z. Kapelan, and D. Avakumovic. Evolutionary algorithm for determining optimal tree layout of water distribution networks. In V. Babovic and L.C. Larsen, editors, *Hydroinformatics '98 - Proceedings of the third international conference*, Copenhagen, 1998. Balkema.

[52] L. Studer and F. Masulli. Building a neuro-fuzzy system to efficiently forecast chaotic time series. *Nuclear Instruments and Methods in Physics Resesarch, Section A*, 389:264–667, 1997.

[53] M. Stundner and G. Zangl. Computational intelligence tools for groundwater management. In V. Babovic and L.C. Larsen, editors, *Hydroinformatics '98 - Proceedings of the third international conference*, Copenhagen, 1998. Balkema.

[54] F. Takens. Detecting strange attractors in turbulence. In D.A. Rand and L.-S. Young, editors, *Dynamical Systems and Turbulence*, volume 898 of *Lecture Notes in Mathematics*, pages 366–381, Warwick, 1981. Springer-Verlag, Berlin.

[55] C. W. Therrien. *Decision, Estimation, and Classification: An Introduction to Pattern Recognition and Related Topics.* Wiley, New York, 1989.

[56] V. N. Vapnik. *Statistical Learning Theory: Inference from Small Samples.* Springer, New York, 1998.

[57] R. Vautard, P. You, and M. Ghil. Singular-spectrum analysis: A toolkit for short, noisy chaotic signals. *Physica D*, 58:95–126, 1992.

[58] L. Wang and J.M. Mendel. Fuzzy basis functions, universal approximation, and orthogonal least-squares learning. *IEEE Trans. on Neural Networks*, 5:807–14, 1992.

[59] A. Zell, G. Mamier, N. Mache, R. Hubner, S. Doring, K.U. Herrman, T. Soyez, M. Schmalzl, T. Sommer, A. Hatzigeorgiou, D. Posselt, T. Schreiner, B. Kett, G. Clemente, and J. Wieland. SNNS Stuttgart Neural Networks Simulator, User Manual, Version 4.1. Technical Report Report No. 6/95, Stuttgart University, Stuttgart, 1995.

Theory, Implementation, and Applications of Support Vector Machines

Massimiliano Pittore
INFM - DISI, Università di Genova
Genova, Italy

Alessandro Verri
INFM - DISI, Università di Genova
Genova, Italy, and
Center for Biological and Computational Learning, MIT
Cambridge, MA (US)

Abstract

Support Vector Machines (SVMs) have been recently introduced as a new method for function estimation. In this survey we first review the main theoretical properties of SVMs, then present an implementation of SVMs able to work with training sets of very large size. Finally, we discuss two computer vision applications in which SVMs for both pattern recognition and regression estimation have been successfully employed.

1 Introduction

Support Vector Machines (SVMs) have been recently introduced as a new technique for solving pattern recognition [2] and regression estimation problems [12] in the framework of statistical learning theory. While traditional techniques are based on the so-called *Empirical Risk Minimization* (ERM) principle, statistical learning theory is based on the *Structural Risk Minimization* (SRM) principle. The main difference is that learning machines based on the ERM principle are trained by optimizing the performance on the training set. This training procedure is known to be asymptotically consistent [12], that is to converge to the optimal solution with infinite data sets. Learning machines based on SRM, instead, are built taking into explicit account the finiteness of the training set and the capacity, or VC dimension, of the hypothesis space (*i.e.*, the set of the possible solutions). The key idea is that learning machines trained according to the SRM principle look for the optimal performance on the training set while controlling the capacity of the hypothesis space.

SVMs minimize an approximation of the *structural risk* – that is, an upper bound on the generalization error for a fixed but unknown probability distribution of the data. What makes SVMs attractive is the ability to condense the information contained in the training set and find the optimal solution within the set of decision or approximating surfaces of relatively low VC-dimension

[13]. In the pattern recognition case, the key idea of a SVM can be explained in plain words. Given a training set S which contains points belonging to either of two classes, a linear SVM separates the classes with a hyperplane determined by certain points of S, termed *support vectors*. In the separable case, this hyperplane maximizes the *margin*, or twice the minimum distance of either class from the hyperplane, and all the support vectors lie at the same minimum distance from the hyperplane (and are thus termed *margin vectors*). In real cases, the two classes may not be separable, and both the hyperplane and the support vectors are obtained from the solution of a problem of constrained optimization. Since the support vectors are now either margin vectors or errors, the solution is a trade-off between the largest margin and the lowest number of errors, trade-off controlled by a regularization parameter.

The aim of this paper is to provide an elementary and concise introduction to SVMs from both the theoretical and the computational viewpoint. To stress the practical relevance of SVMs we also describe two computer vision applications developed in our laboratory which are essentially based on SVMs.

This survey is organized as follows. Section 2 is devoted to the presentation of the basic theory of SVMs for both pattern recognition and regression estimation as proposed by Vapnik [11]. An algorithm able to determine the support vectors of possibly very large training sets is reviewed in Section 3, while two computer vision applications based on SVMs are described in Section 4. Finally, a brief discussion of the main open issues is reported in Section 5.

2 The theory

An extensive and rigorous account of statistical learning theory can be found in [12]. Here, we give a concise summary of SVMs regarded as techniques for solving problems of pattern recognition and regression estimation. We close the section with a brief discussion on the connection between SVMs and structural risk minimization.

2.1 Pattern recognition

We start by studying the case of linearly separable data sets and then considering the more general nonlinear case.

2.1.1 *Problem setting*

Given a set S of points $\mathbf{x}_i \in \mathbb{R}^n$ with $i = 1, 2, \ldots, N$, each belonging to either class 1 or -1, we want to find the hyperplane that divides S leaving all the points of the same class on the same side while maximizing the distance between the two classes and the hyperplane. We denote the class of $\mathbf{x}_i$ with y_i (equal 1 or -1) and assume that S is *linearly separable*, or that there exist $\mathbf{w} \in \mathbb{R}^n$ and $b \in \mathbb{R}$ such that (for $i = 1, 2, \ldots, N$)

$$y_i(\mathbf{w} \cdot \mathbf{x}_i + b) \geq 1. \tag{1}$$

The hyperplane $\mathbf{w} \cdot \mathbf{x} + b = 0$ satisfying inequality (1) is named *separating hyperplane* and the signed distance d_i of a point $\mathbf{x}_i$ from the separating hyperplane $(\mathbf{w}, b)$ is thus given by

$$d_i = \frac{\mathbf{w} \cdot \mathbf{x}_i + b}{w}, \tag{2}$$

with w the norm of $\mathbf{w}$. Combining inequality (1) and equation (2), for all $x_i \in S$ we have

$$y_i d_i \geq \frac{1}{w}. \tag{3}$$

Therefore, $1/w$ is the lower bound on the distance between the points $\mathbf{x}_i$ and the separating hyperplane of equation $\mathbf{w} \cdot \mathbf{x} + b = 0$.

In order to establish a one-to-one correspondence between separating hyperplanes and their parametric representation, we define the *canonical representation* of the separating hyperplane as the representation which is obtained by rescaling the pair $(\mathbf{w}, b)$ into the pair $(\mathbf{w}', b')$ in such a way that the distance of the closest point, say $\mathbf{x}_j$, equals $1/w'$. Through this definition we have

$$\min_{\mathbf{x}_i \in S} \{y_i(\mathbf{w}' \cdot \mathbf{x}_i + b')\} = 1.$$

Consequently, for a separating hyperplane in the canonical representation, the bound in inequality (3) is tight. In what follows we will assume that a separating hyperplane is always given in the canonical representation and thus write $(\mathbf{w}, b)$ instead of $(\mathbf{w}', b')$. We are now in a position to define the *Optimal Separating Hyperplane* (OSH) as the separating hyperplane for which the distance of the closest point of S is maximum. Since the distance of the closest point equals $1/w$, the OSH can be regarded as the solution of the problem of minimizing $1/w$ subject to the constraint (1), or

Problem **P1**
Minimize $\frac{1}{2}\mathbf{w} \cdot \mathbf{w}$
subject to $y_i(\mathbf{w} \cdot \mathbf{x}_i + b) \geq 1, \quad i = 1, 2, \ldots, N$

The quantity $2/w$, the lower bound of the minimum distance between points of different classes, is named *margin*. Hence, the OSH can also be seen as the separating hyperplane which maximizes the margin (see Figure 1). We now study the properties of the solution of Problem **P1**.

2.1.2 Support vectors

Problem **P1** is usually solved by means of the classical method of Lagrange multipliers. For more details and a thorough review of the method see [1].

If we denote with $\boldsymbol{\alpha} = (\alpha_1, \alpha_2, \ldots, \alpha_N)$ the N nonnegative Lagrange multipliers associated with the constraints (1), finding the solution to Problem **P1** is equivalent to determining the *saddle point* of the function

$$L = \frac{1}{2}\mathbf{w} \cdot \mathbf{w} - \sum_{i=1}^{N} \alpha_i \{y_i(\mathbf{w} \cdot \mathbf{x}_i + b) - 1\}. \tag{4}$$

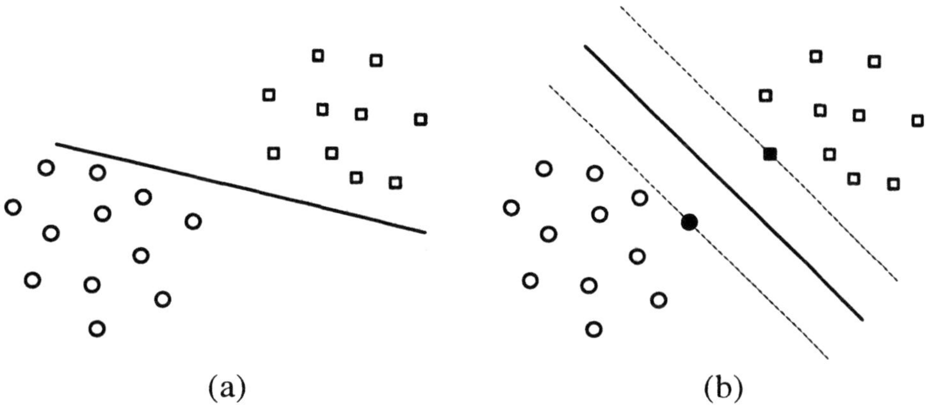

Figure 1: Separating hyperplane (a) and optimal separating hyperplane (b) for a training set consisting of open circles and squares. The filled circle and square in (b) are the support vectors.

with L = L($\mathbf{w}$, b, $\boldsymbol{\alpha}$). At the saddle point, L has a minimum for $\mathbf{w} = \bar{\mathbf{w}}$ and $b = \bar{b}$ and a maximum for $\boldsymbol{\alpha} = \bar{\boldsymbol{\alpha}}$, and thus we can write

$$\frac{\partial L}{\partial b} = \sum_{i=1}^{N} y_i \alpha_i = 0, \tag{5}$$

$$\frac{\partial L}{\partial \mathbf{w}} = \mathbf{w} - \sum_{i=1}^{N} \alpha_i y_i \mathbf{x}_i = 0 \tag{6}$$

with

$$\frac{\partial L}{\partial \mathbf{w}} = (\frac{\partial L}{\partial w_1}, \frac{\partial L}{\partial w_2}, \ldots, \frac{\partial L}{\partial w_N}).$$

Substituting equations (5) and (6) into the right-hand side of (4), we see that Problem **P1** reduces to the maximization of the function

$$\mathcal{L}(\boldsymbol{\alpha}) = \sum_{i=1}^{N} \alpha_i - \frac{1}{2} \sum_{i,j=1}^{N} \alpha_i \alpha_j y_i y_j \mathbf{x}_i \cdot \mathbf{x}_j, \tag{7}$$

subject to the constraint (5) with $\boldsymbol{\alpha} \geq 0$[1]. This new problem is called the *dual problem* and can be formulated as

Problem **P2**

Maximize	$-\frac{1}{2}\boldsymbol{\alpha}^{\top} D \boldsymbol{\alpha} + \sum \alpha_i$
subject to	$\sum y_i \alpha_i = 0$
	$\boldsymbol{\alpha} \geq 0,$

where both sums are for $i = 1, 2, \ldots, N$, and D is an $N \times N$ matrix such that

$$D_{ij} = y_i y_j \mathbf{x}_i \cdot \mathbf{x}_j. \tag{8}$$

[1]In what follows $\boldsymbol{\alpha} \geq 0$ means $\alpha_i \geq 0$ for every component α_i of any vector $\boldsymbol{\alpha}$.

As for the pair $(\bar{\mathbf{w}}, \bar{b})$, from equation (6) it follows that

$$\bar{\mathbf{w}} = \sum_{i=1}^{N} \bar{\alpha}_i y_i \mathbf{x}_i. \tag{9}$$

while $\bar{b}$ can be determined from $\bar{\boldsymbol{\alpha}}$, solution of the dual problem, and from the Kühn-Tucker conditions

$$\bar{\alpha}_i \left(y_i(\bar{\mathbf{w}} \cdot \mathbf{x}_i + \bar{b}) - 1 \right) = 0, \quad i = 1, 2, \ldots, N. \tag{10}$$

Note that the only $\bar{\alpha}_i$ that can be nonzero in equation (10) are those for which the constraints (1) are satisfied with the equality sign. This has an important consequence. Since most of the $\bar{\alpha}_i$ are usually null, the vector $\bar{\mathbf{w}}$ is a linear combination of a relatively small percentage of the points $\mathbf{x}_i$. These points are termed *Support Vectors* (SVs) because they are the closest points from the OSH and the only points of S needed to determine the OSH. Given a support vector $\mathbf{x}_j$, the parameter $\bar{b}$ can be obtained from the corresponding Kühn-Tucker condition as $\bar{b} = y_j - \bar{\mathbf{w}} \cdot \mathbf{x}_j$.

Once $\bar{\mathbf{w}}$ and $\bar{b}$ have been determined, a new data point $\mathbf{x}$ is assigned to class 1 or -1 according to

$$\operatorname{sign}(\bar{\mathbf{w}} \cdot \mathbf{x} + \bar{b}).$$

2.1.3 *Linearly nonseparable case*

If the set S is not linearly separable, the previous analysis can be generalized by introducing N nonnegative variables $\boldsymbol{\xi} = (\xi_1, \xi_2, \ldots, \xi_N)$ such that

$$y_i(\mathbf{w} \cdot \mathbf{x}_i + b) \geq 1 - \xi_i, \quad i = 1, 2, \ldots, N. \tag{11}$$

The purpose of the variables ξ_i is to allow for a small number of misclassified points. The generalized OSH is then regarded as the solution to

Problem **P3**
Minimize $\frac{1}{2}\mathbf{w} \cdot \mathbf{w} + C \sum \xi_i$
subject to $y_i(\mathbf{w} \cdot \mathbf{x}_i + b) \geq 1 - \xi_i \quad i = 1, 2, \ldots, N$
$\boldsymbol{\xi} \geq 0.$

The extra term $C \sum \xi_i$, where the sum is for $i = 1, 2, \ldots, N$, controls the number of misclassified points with the parameter C as a regularization parameter. The OSH tends to maximize the minimum distance $1/w$ for small C, and minimize the number of misclassified points for large C. For intermediate values of C the solution of problem **P3** trades errors for a larger margin.

In analogy with what done for the separable case, Problem **P3** can be transformed into the *dual*

Problem **P4**
Maximize $-\frac{1}{2}\boldsymbol{\alpha}^\top D \boldsymbol{\alpha} + \sum \alpha_i$
subject to $\sum y_i \alpha_i = 0$
$0 \leq \alpha_i \leq C, \quad i = 1, 2, \ldots, N.$

Note that the dimension of **P4** is given by the size of the training set, whereas the dimension of the input space gives the rank of D. From the constraints of Problem **P4** it follows that if C is sufficiently large and the set S linearly separable, Problem **P4** reduces to **P2**. The vector **w** is still given by equation (9), while $\bar{b}$ can again be determined from $\bar{\boldsymbol{\alpha}}$, solution of the dual problem **P4**, and from the new Kuhn-Tucker conditions

$$\begin{aligned} \bar{\alpha}_i \left(y_i(\bar{\mathbf{w}} \cdot \mathbf{x}_i + \bar{b}) - 1 + \bar{\xi}_i \right) &= 0 \quad (12) \\ (C - \bar{\alpha}_i)\bar{\xi}_i &= 0 \quad (13) \end{aligned}$$

where the $\bar{\xi}_i$ are the values of the ξ_i at the saddle point. Similarly to the separable case, the SVs are the points $\mathbf{x}_i$ for which $\bar{\alpha}_i > 0$. The main difference is that here we have to distinguish between the SVs for which $\bar{\alpha}_i < C$ and those for which $\bar{\alpha}_i = C$. In the first case, from condition (13) it follows that $\bar{\xi}_i = 0$, and hence, from condition (12), that the SVs lie at a distance $1/\bar{w}$ from the OSH. These SVs are termed *margin vectors*. The SVs for which $\bar{\alpha}_i = C$, instead, are: misclassified points if $\xi_i > 1$, points correctly classified but closer than $1/\bar{w}$ from the OSH if $0 < \xi \leq 1$, or margin vectors if $\xi_i = 0$. Neglecting this last rare (and degenerate) occurrence, we refer to all the SVs for which $\alpha_i = C$ as *errors*. All the points that are not SVs are correctly classified and lie outside the margin strip.

2.1.4 Nonlinear separating surfaces

The basic idea behind this important extension is to map the input points into feature points belonging to a space of very high dimension, and then look for the OSH in that space [2]. If $\mathbf{x} \in \mathbb{R}^n$ is an input point, we let $\varphi(\mathbf{x})$ be the corresponding feature point with φ a mapping from $\mathbb{R}^n$ to a certain feature space Z (typically a Hilbert space of finite or infinite dimension). Clearly, an OSH in Z corresponds to a nonlinear separating surface in input space.

From the dual formulation, it can be seen that a full knowledge of the mapping is not necessary as φ appears only in the dot product between feature points needed to compute the entry D_{ij}, as

$$D_{ij} = y_i y_j \varphi(\mathbf{x}_i) \cdot \varphi(\mathbf{x}_j),$$

and in the expression for the final classification as

$$\mathbf{w} \cdot \varphi(\mathbf{x}) + b = \sum \alpha_i y_i \varphi(\mathbf{x}_i) \cdot \varphi(\mathbf{x}) + b.$$

Consequently, it suffices to find a suitable function K, called *kernel*, expressing the dot product between feature points in terms of the corresponding points in input space, or

$$K(\mathbf{x}_i, \mathbf{x}_j) = \varphi(\mathbf{x}_i) \cdot \varphi(\mathbf{x}_j).$$

Examples of nonlinear kernels are the *polynomial* kernel

$$K(\mathbf{x}, \mathbf{y}) = (1 + \mathbf{x} \cdot \mathbf{y})^d, \qquad \mathbf{x}, \mathbf{y} \in [-a, a]^d,$$

which identifies a polynomial surface of degree d in input space, and the *Gaussian* kernel

$$K(\mathbf{x}, \mathbf{y}) = \exp\left(\frac{-\|\mathbf{x}-\mathbf{y}\|^2}{2\sigma^2}\right),$$

for some $\sigma \in \mathbb{R}^+$, which identifies a separating surface given by a weighted sum of Gaussians centered at the support vectors.

From the solution of the same dual problem of the previous case (but with $D_{ij} = y_i y_j K(\mathbf{x}_i, \mathbf{x}_j)$), the classification stage reduces to computing

$$\text{sign}\left(\sum \alpha_i y_i K(\mathbf{x}_i, \mathbf{x}) + b\right).$$

2.2 Regression estimation

We now extend the idea of SVM to regression estimation. In this case, the training set consists of N pairs $(\mathbf{x}_i, y_i)$ $(i = 1, 2, ..., N)$ with $\mathbf{x}_i \in \mathbb{R}^n$ and $y_i \in \mathbb{R}$. Similarly to pattern recognition, in the case of linear regression SVMs look for the linear function $\mathbf{w} \cdot \mathbf{x} + b$ solution to the problem

$$\begin{array}{lll} \text{Minimize} & \frac{1}{2}\boldsymbol{w} \cdot \boldsymbol{w} + C\sum(\xi_i + \xi_i^*) & \\ \text{subject to} & y_i - \boldsymbol{w} \cdot \boldsymbol{x}_i - b \leq \epsilon + \xi_i^* & \\ & \boldsymbol{w} \cdot \boldsymbol{x}_i + b - y_i \leq \epsilon + \xi_i & i = 1, 2, \ldots, N \\ & \boldsymbol{\xi} \geq 0, \boldsymbol{\xi}^* \geq 0. & \end{array}$$

The first two sets of constraints are derived from a function $|p|_\epsilon$, named *ϵ-insensitive loss function*, defined as

$$|p|_\epsilon = \begin{cases} 0 & \text{if } |p| < \epsilon \\ |p| - \epsilon & \text{otherwise} \end{cases}$$

with $p = y_i - \mathbf{w} \cdot \mathbf{x}_i - b$. Therefore, the term $\sum(\xi_i + \xi_i^*)$ is the ϵ-insensitive loss function evaluated over the training set for a given linear function. Note that for each i only one between ξ_i and ξ_i^* can be different from zero. The *ϵ-insensitive* loss function is robust with respect to both small numbers of outliers and large numbers of points affected by a small bias.

In the general case of nonlinear regression, and for a suitable kernel function K, the dual problem becomes

$$\begin{array}{ll} \text{Maximize} & -\frac{1}{2}(\boldsymbol{\alpha}^* - \boldsymbol{\alpha})^\top D(\boldsymbol{\alpha}^* - \boldsymbol{\alpha}) + \\ & \sum y_i(\alpha_i^* - \alpha_i) + \epsilon\sum(\alpha_i^* + \alpha_i) \\ \text{subject to} & \sum \alpha_i^* = \sum \alpha_i \\ & 0 \leq \boldsymbol{\alpha}^* \leq C, 0 \leq \boldsymbol{\alpha} \leq C. \end{array}$$

with $D_{ij} = K(\mathbf{x}_i, \mathbf{x}_j)$ and α_i^* and α_i the Lagrangian multipliers associated to the variables ξ_i^* and ξ_i respectively. The approximating function can then be expressed as

$$f(\mathbf{x}) = \sum_{i=1}^{N}(\alpha_i^* - \alpha_i)K(\mathbf{x}, \mathbf{x}_i) + b$$

where the only α_i^* or α_i strictly positive are those for which the constraints of the primal problem are satisfied with the equality sign. The corresponding $\mathbf{x}_i$ are again the support vectors, and the only points contributing to the sum for evaluating $f(\mathbf{x})$.

We conclude this section by listing the main properties of SVMs as large margin classifiers in the light of the connection between SVMs and structural risk minimization.

2.3 Main properties

Firstly, through the maximum margin property and some statistical properties of the OSH obtained by Vapnik [12], we observe that SVMs can be thought of as an approximate implementation of structural risk minimization. In essence, if the data points live in a ball of radius D, then the VC dimension h of the subset of hyperplanes separating a particular data set with margin A is

$$h \leq \min(D^2 A^2, n) + 1.$$

Therefore, it follows that the wider the margin, the smaller the VC dimension and, hence, the greater the capacity control of the classifier.

Secondly, SVMs condense all the information contained in the training set relevant to classification in the support vectors. This (a) reduces the size of the training set identifying the most important points, and (b) makes it possible to perform classification efficiently.

Thirdly, SVMs can be used to perform classification in high-dimensional spaces, even in the presence of a relatively small number of data points. SVMs effectively defeat the curse of dimensionality by maximizing the margin. From the quantitative viewpoint, the margin can also be used as a measure of the difficulty of the problem (the larger the margin the lower the probability of misclassifying a yet-to-be-seen point).

We now describe a computer implementation for determining the support vectors of a given training set.

3 Implementing SVMs

In the previous section we have seen that the problem of determining the OSH reduces to Problem **P4**, a typical problem of Quadratic Programming (QP) of size equal to the number of examples. This points to the need of an *ad hoc* method for dealing with large scale problems. In this section we review a decomposition method (proposed in [6]) which effectively finds the global solution by solving iteratively problems of relatively small size.

The method can be coupled to a generic QP solver. For example, the solver adopted in our implementation is based on the *Complementary Pivoting Algorithm* (CPA), a classical algorithm solving QP problems which, like problem **P4**, can be transformed into *Linear Complementary* Problems (LCPs) [1]. The decomposition method can be summarized as follows. Exploiting the particular form of the objective function, the method finds the solution of problem **P4** by solving iteratively subproblems which depend only on g variables (with g, typically, a small number). In each subproblem one maximizes the objective function of problem **P4** with respect to the selected g components of $\boldsymbol{\alpha}$, the other being kept constant. The obtained solution is then used to select a new set of g variables, with the components α_j violating the optimality conditions as possible candidate variables.

More precisely, we partition the set of indeces $I = \{1, \ldots, N\}$ in two sets: B, the *working set* and N, the *constant set*. Likewise, we decompose $\boldsymbol{\alpha}$ in the two vectors $\boldsymbol{\alpha}_B$ and $\boldsymbol{\alpha}_N$, and the matrix D in four blocks

$$D = \begin{pmatrix} D_{BB} & D_{BN} \\ D_{BN}^{\mathrm{T}} & D_{NN} \end{pmatrix}.$$

Then we consider the objective function of problem **P4** as a function of the $\boldsymbol{\alpha}_B$ only

$$W(\boldsymbol{\alpha}_B|\boldsymbol{\alpha}_N) = -\frac{1}{2}\boldsymbol{\alpha}_B^T D_{BB}\boldsymbol{\alpha}_B + \boldsymbol{\alpha}_B \cdot \mathbf{h} + a,$$

with

$$\mathbf{h} = \mathbf{1}_B - D_{BN}\boldsymbol{\alpha}_N,$$

and

$$a = -\frac{1}{2}\boldsymbol{\alpha}_N^T D_{NN}\boldsymbol{\alpha}_N + \boldsymbol{\alpha}_N^T \cdot \mathbf{1}_N.$$

The subproblem for the working set B induced by the constant vector $\boldsymbol{\alpha}_N$ is

$$\begin{array}{lll} \text{Maximize} & W(\boldsymbol{\alpha}_B|\boldsymbol{\alpha}_N) & \\ \text{subject to} & \sum_{i\in B} y_i\alpha_i = -\sum_{i\in N} y_i\alpha_i & \\ & 0 \leq \alpha_i \leq C, & i \in N. \end{array}$$

The method consists of three steps:

1. Randomly select g indeces in I to form the set B and set $\boldsymbol{\alpha}_N = 0$;

2. Solve the above problem (for example by means of CPA);

3. Look for an index $j \in N$ such that one of the following is true

    - $d(\mathbf{x}_j)y_j < 1$ and $\alpha_j = 0$;
    - $d(\mathbf{x}_j)y_j = 1$ and $\alpha_j = 0$ or $\alpha_j = C$;
    - $d(\mathbf{x}_j)y_j > 1$ and $\alpha_j > 0$

    where $d(\mathbf{x}_j)$ is the signed distance of point $\mathbf{x}_j$ from the current suboptimal separating hyperplane. If such an index j exists, then swap j with any index in B and go back to step 2. Otherwise stop: the vector $(\boldsymbol{\alpha}_B, \boldsymbol{\alpha}_N)$ is the solution of problem **P4**.

4 Applications

In the last few years there has been a growing interest in the development of SVMs for solving practical problems, including optical character recognition [11], object detection [8], particle identification [7], and automatic fish weighting from image measurements [5]. Here we discuss in some detail two computer vision applications based on SVMs. In the first application we describe the use of SVMs to perform aspect-based 3D object recognition directly from grey level images. In the second application we discuss a system for detecting, representing and recognizing visual dynamic events by means of SVMs for both regression estimation and pattern recognition.

4.1 3D object recognition

The recognition system presented in [10] consists of three stages: preprocessing, training set formation, and system testing, and was tested on the COIL database.

4.1.1 Preprocessing

The COIL database (which can be downloaded via anonymous ftp from the Columbia University web site) consists of 72 images of 100 objects positioned in the center of a turntable, observed from a fixed viewpoint. For each object, the turntable is rotated by 5° per image. Figures 2 shows a selection of the objects in the database. Figures 3 shows one every nine views of one particular object. As explained in detail in [4], the object region is re-sampled so that the larger of the two dimensions fits the image size. Consequently, the apparent size of an object may change considerably from image to image, especially for the objects which are not symmetric with respect to the turntable axis.

Figure 2: Images of 24 COIL objects.

The original images were color images (24 bits for each of the RGB channels) of 128×128 pixels. In the preprocessing stage each image was transformed into an 8-bit grey-level image by rescaling the obtained range between 0 and 255.

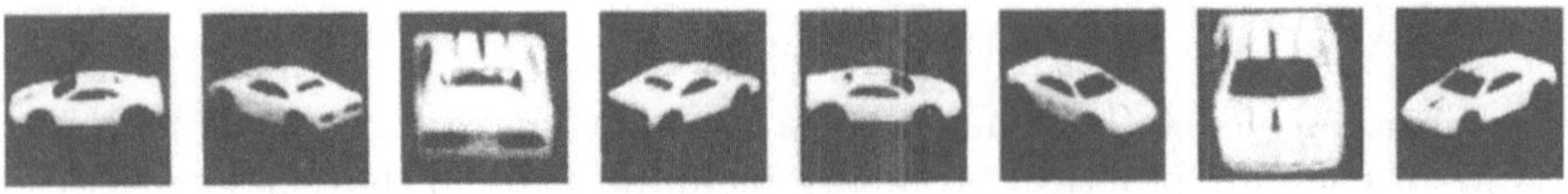

Figure 3: Eight views of a COIL object.

Finally, the image spatial resolution was reduced to 32×32 by averaging the grey values over 4×4 pixel patches.

4.1.2 Training set formation

The training set consists of 36 views (one every 10°) for each object. After the preprocessing stage, each image can be regarded as a vector $\mathbf{x}$ of $32 \times 32 = 1024$ components.

Depending on the classification task, a certain subset of the 100 objects (from 2 to 32) was considered. Then, the OSHs associated to each pair of objects i and j in the subset were computed, the SVs identified, and the obtained parameters, $\mathbf{w}(i,j)$ and $b(i,j)$, stored in a file. We have never come across errors in the classification of the training sets. The reason is essentially given by the high dimensionality of the object space compared to the small number of examples. The images corresponding to some of the SVs for a specific pair of objects are shown in Figure 4.

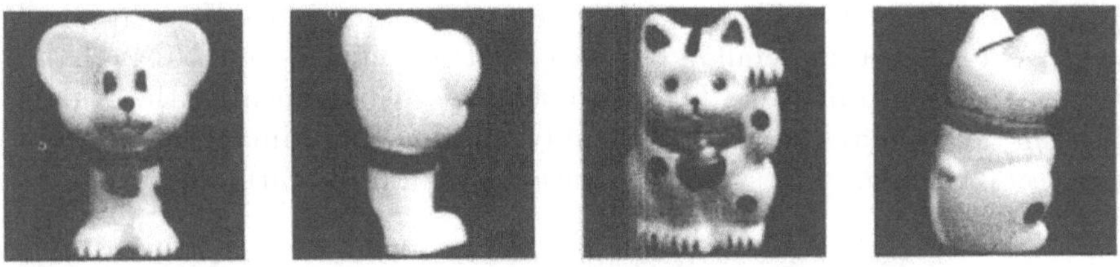

Figure 4: Four of the SVs for a specific object pair.

Typically, we have found a number of SVs ranging from 1/3 to 2/3 of the 72 training images for each object pair. This large fraction of SVs can be explained by the high dimensionality of the object space combined with the small number of examples.

4.1.3 System testing

Given a certain subset σ of the 100 objects and the associated training set of 36 images for each object in σ, the test set consists of the remaining 36 images per object in σ. Recognition was performed following the rules of a tennis tournament. Each object is regarded as a *player*, and in each *match* the system temporarily classifies an image of the test set according to the OSH relative to the pair of players involved in the match. If the players are objects i and j (see

Figure 5), the system classifies the viewed object of image $\mathbf{x}$ as object i or j depending on the sign of

$$\mathbf{w}(i,j) \cdot \mathbf{x} + b(i,j).$$

Note that the system determines the object identity without estimating pose.

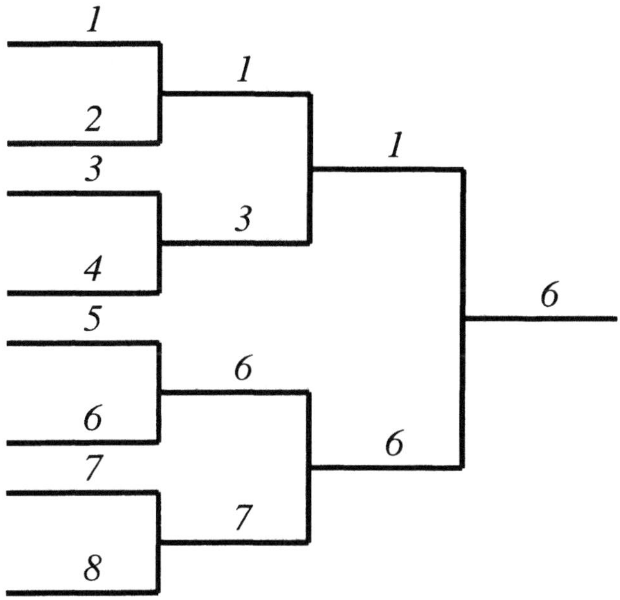

Figure 5: An unknown object is recognized as object "6" following the rules of a tennis tournament.

4.1.4 Experimental results

The proposed recognition system was tested on sets of 32 of the 100 COIL objects. The training sets consisted of 36 images for each of 32 objects and the test sets of the remaining 36 images for each object. For the 10 random choices of 32 of the 100 objects we tried, the system reached perfect score. Therefore, we decided to select by hand the 32 objects *most difficult* to recognize (*i.e.* the set of objects separated by the smallest margins). By doing so the system finally mistook a packet of chewing gum for another very similar packet of chewing gum in one case.

In order to verify the effectiveness and robustness of the recognition system, we performed experiments under increasingly difficult conditions: adding pixel-wise random noise, adding bias in the spatial registration, and in the presence of small amounts of occlusion. Restricting the analysis to the 32 objects most difficult to recognize, the system performed equally well for maximum noise up to ±100 grey levels, shifts of a few pixels, and occlusions of relatively small size (100 to 1000 pixels). As reported in [10] the performance degrades gracefully for increasing amounts of noise, but somewhat abruptly with respect to shifts and occlusions. Finally, the system compared very favorably with single-layer perceptrons. The poor performance of perceptrons can be easily explained in terms of the smaller margin associated with the separating hyperplane of each perceptron compared to the SVM margin. In this case, the perceptron margin is between 2 and 10 times smaller than the SVM margin. This means that both SVMs and perceptrons separate exactly the training set, but that the perceptron margin makes it difficult to classify correctly novel images in the presence of noise. Intuitively, this fact can be explained by thinking of noise perturbation as a motion in object space: if the margin is too small, even a slight perturbation can bring a point across the separating hyperplane.

4.2 Visual dynamic event recognition

We now move on to the description of *VIDERE*, a system for *VI*sual *D*ynamic *E*vent *RE*cognition introduced in [9]. We start by describing the scenario.

4.2.1 Scenario

We consider the task of *detecting* and *classifying* dynamic events from a sequence of images acquired by a camera looking down a corridor from a fixed position. Occasionally, one or more people are expected to walk up or down the corridor, possibly entering or leaving one of the adjacent rooms. The goal is to build a system able to learn from examples to

- detect the presence of moving people in the observed scene, and
- represent and classify the detected dynamic event.

The ability of learning to perform the given tasks from examples endows the system with the flexibility sufficient to effectively work in different environments and perform different tasks.

Let us give an operative definition of dynamic event and of the generic tasks to be performed. During the training stage a number of image sequences is acquired. In each sequence a person walks in the corridor and performs one of n possible *actions* labeled as action $1, 2, ...,$ or n. These actions are *dynamic events* and, in plain words, could be described as "somebody is crossing the corridor going from room A to B", or "somebody is walking up the corridor and entering room C". In the test stage the system acquires a new image sequence and is expected to properly detect and classify each of the observed actions, if any, as one of the n possible actions learned in the training stage.

In order to perform these tasks, in both the training and test stages, the system should be able to

(1) detect the beginning and end of a dynamic event, and

(2) compute a stereotyped representation of the observed event.

While in the training stage the event label is known *a priori*, in the test stage the system should also be able to

(3) classify the detected event on the basis of the computed representation.

To this purpose we built a system consisting of three modules, a *detection*, a *representation*, and a *recognition* module, which we now describe in some detail.

4.2.2 Event detection

This is the only module based on low-level visual processing. Since the camera is fixed and the background stationary, we employ a simple method which, while periodically updating the background image (kept fixed), starts off by locating all pixels of the current image at which the gray level difference with respect to the background image is larger than a fixed threshold m. If a *significant* change is detected the system

1. forms a minimal rectangular box of center $\mathbf{c}$ enclosing all nonisolated pixels where the change has been detected;

2. tracks the box over the image sequence by means of a linear Kalman filter until the change remains significant.

A change is considered *significant* if the number N of connected pixels over which the change is detected exceeds a given threshold t_1, and *remains* significant until N remains higher than a second (lower) threshold t_2. Events that last for less than a minimum number of frames, f, are discarded as noise. An example of this module at work is illustrated in Figure 6.

Apart from the linear Kalman filter, which controls $\mathbf{c}$ and N in a very simple way, this module depends on the four thresholds m, t_1, t_2, and f. Here t_1, t_2, and f were set manually observing a number of sequences beforehand; the threshold m is set to twice the standard deviation σ of the histogram of the pointwise signed image difference between two frames (both acquired while the camera looks at the stationary background).

A simple procedure has been employed to deal with a background that could be changing with time (some of the corridor lights switched on or off, or a door left open or closed). After the end of a dynamic event, or randomly if no event is detected for a while, the background image is updated with the current image if the standard deviation of the same histogram of above (this time computed between the current and the previous frame) is similar to σ. This procedure ensures that the background update does not take place during the onset of a new dynamic event.

4.2.3 Event representation

The choice of the representation to adopt is a crucial aspect of the entire system. The fact that the events to be represented last different amounts of time and have different spatio-temporal evolutions (true even of events corresponding to

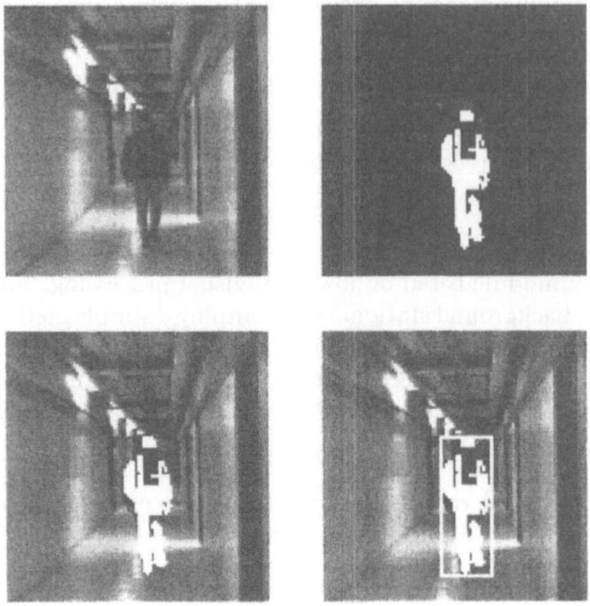

Figure 6: Event detection at a given frame. From upper left to lower right: the frame of a sequence, nonisolated pixels at which a change with respect to the background has been found, overlay of the previous two images, and minimal box enclosing the detected pixels.

the same class) makes this choice particularly difficult. For the time being, we adopted a very simple representation that satisfies the basic requirement of providing a stereotyped representation (in practice, a feature vector with a fixed number of components for each event, in which each component carries comparable information across all events).

The representation is built in three steps. In the first step, the detected event is described by three T-dimensional vectors $\mathbf{u}$, $\mathbf{v}$, and $\mathbf{a}$ (where T is the event duration in number of frames). The vectors $\mathbf{u} = (u_1, u_2, \ldots, u_T)$ and $\mathbf{v} = (v_1, v_2, \ldots, v_T)$ consist of the horizontal and vertical positions on the image plane of the box center $\mathbf{c}$, while the vector $\mathbf{a} = (a_1, a_2, \ldots, a_T)$ of the number of pixels N within the box at each frame.

In the second step, the vectors $\mathbf{u}$, $\mathbf{v}$, and $\mathbf{a}$ are normalized between $[0, 1]$ in both range and time and regarded as the sampling of three functions

$$\begin{aligned} U &: [0,1] \to [0,1] \\ V &: [0,1] \to [0,1] \\ A &: [0,1] \to [0,1]. \end{aligned}$$

In the third and final step an SVM for regression is run on the data $\mathbf{u}$, $\mathbf{v}$, and $\mathbf{a}$ in order to build a representation consisting of an ordered array of numbers of fixed size and homogeneous meaning across the events. This is obtained through the use of a kernel K corresponding to a piecewise linear spline with ℓ equally spaced fixed knots in the unit interval [12] (with $\ell = 11$ in the present implementation). The detected event is finally described by means of a $(3\ell + 7)$-dimensional vector $\mathbf{x}$ the components of which are obtained as follows. The first $\ell + 2$ components are the spline coefficients obtained from the regression estimation on the vector $\mathbf{u}$, the second $\ell + 2$ from $\mathbf{v}$, and the third $\ell + 2$ from $\mathbf{a}$. In an attempt tp make use of the information contained in the duration of the observed action (effectively lost in the previous normalization stages), the last component $x_{3\ell+7}$ is the duration T of the event.

4.2.4 *Event recognition*

This last module of *VIDERE* is only used in the test stage. During the training stage, instead, each event $\mathbf{x}$ in the training set is simply given by the supervisor one of the n possible labels $y_1, ..., y_n$ fixed beforehand.

The classification of a new event $\mathbf{x}$ is obtained by means of a linear SVM for pattern recognition trained to distinguish between all possible pairs of different event classes. As before, the multiclass classification follows the rules of a *tennis tournament*.

4.2.5 *Experimental results*

In its current version, *VIDERE* is able to detect the visual dynamic events in real time, but both the representation and the recognition stages are still computed off line. Therefore, in order to evaluate the performance of the system we acquired a number of image sequences, computed the representation for all the events detected in real time, and divided the obtained events to form the training and test sets.

VIDERE has been trained to recognize eight different actions taking place down a corridor leading to our laboratory (see Figure 7). Examples of these actions are *somebody is walking up the corridor and entering the last door on the left*, and *somebody is exiting from room A, walking up the corridor and entering room B* The shortest event lasted for 15 seconds, the longest for 42. Four different peoples were involved as actors.

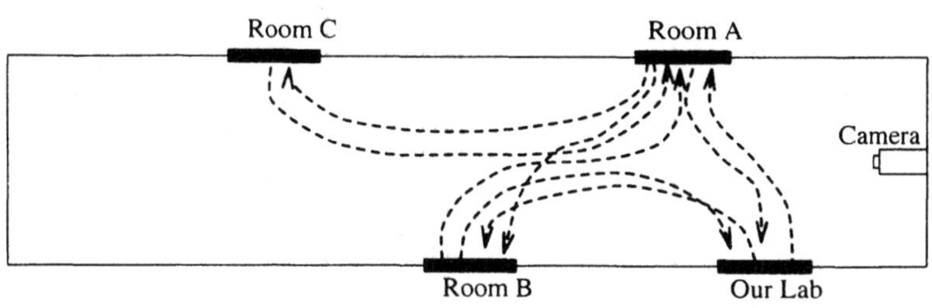

Figure 7: Typical trajectories of the 8 dynamic events that *VIDERE* is expected to detect and classify.

Given a total of 104 events (on average 13 examples for each class), each represented by a 40-dimensional vector, we checked the performance of *VIDERE* by means of the *leave-one-out* technique. This means that we trained the SVM for classification with 103 example and use the one left out for testing the system for all the 104 possible combinations. *VIDERE* classified correctly all but one event (with no event misclassified in training). Since this is a multiclass recognition task, the obtained result is bound to depend on the order in which the various classifications take place (or are organized). In the case of the *tennis tournament* this means that the final result depends on the initial positions on the table of the various players, or classes (see Figure 5, for example). *A posteriori* analysis of the data showed that in all cases the classification was order-independent, which means that the system accuracy in distinguishing between two classes is very high (99.8%). Interestingly, on about 76% of the testing, the classified point lied outside the margin. This suggests that the chosen representation, though exceedingly simple, is sufficient for the task at hand.

Current and future work concentrate on expanding and optimizing the three modules of *VIDERE*. For the *detection module* we are working on *(a)* adding a data association routine that will enable *VIDERE* to cope with the case of multiple actions taking place at the same time, and *(b)* looking at the possibility of introducing a stereo rig. More sophisticated kernels are under study for the SVM for regression employed in the *representation module*, possibly bypassing the various normalization steps required in the present version. The linear SVM for classification of the *recognition module*, instead, has been replaced by a Gaussian kernel SVM to be used in future tests involving larger numbers of examples and possible classes. Finally, we are working at a version of the system which can be both trained and tested online.

5 Conclusion

Support Vector Machines (SVMs) appear to be a very promising technique for learning the solution of pattern recognition and regression estimation problems. In this paper we reviewed the basic ideas behind SVMs and described in some detail two computer vision applications essentially based on SVMs.

While increasing attention is devoted to SVMs as a practical tool for solving problems which can be cast in the learning from examples paradigm, much work remains to be done especially at the theoretical level. In particular the connection between structural risk minimization and SVMs for regression remains to be explored. The interested reader may want to start from [3] in which the connection between SVMs and regularization networks is explored in depth.

Acknowledgments

Manuel Trucco read a draft of the paper and made useful remarks. The work described in this paper has been partially funded by a grant from the Ministero dell'Università e della Ricerca Scientifica e Tecnologica.

References

[1] Bazaraa, M. and Shetty, C.M. **Nonlinear programming** (John Wiley, New York, 1979).

[2] Cortes, C. and Vapnik, V.N "Support Vector Network", *Machine learning* **20**: 1-25 (1995).

[3] Evgenious, T., Pontil, M., and Poggio, T. "A Unified Framework for Regularization Networks and Support Vector Machines", A.I. Memo 1654, MIT A.I. Lab. (1999).

[4] Murase, H. and Nayar, S.K. "Visual Learning and Recognition of 3-D Object from Appearance", *Int. J. Comput. Vision*, Vol. 14, pp. 5–24 (1995).

[5] Odone, F., Trucco, E., and Verri, A. "Visual Learning of Weight from Shape Using Support Vector Machines", In *Proc. British Machine Vis. Conf.*, Southampton, 469–477 (1998).

[6] Osuna, E., Freund, R., and Girosi, F. "Support Vector Machines: training and applications", A.I. Memo 1602, MIT A.I. Lab. (1997).

[7] Pallavicini, M., Patrignani, C., Pontil, M., e Verri, A. "The Nearest-Neighbor Technique for Particle Identification", *Nuclear Instruments and Methods in Physics Research. A* **405**, 133–138 (1998).

[8] Papageorgiou, C., Oren, M., and Poggio, T. "A General Framework for Object Detection", In *Proc. Int. Conf. Computer Vision*, Bombay, 555–562 (1998).

[9] Pittore, M, Basso, C. e Verri, A. "Representing and Recognizing Visual Dynamic Events with Support Vector Machines", In *Proc. Int. Conf. Image Analysis and Proc.*, Venice (1999).

[10] Pontil, M. and Verri, A. "Support Vector Machines for 3-D Object Recognition", *IEEE Trans. Pattern Anal. Mach. Intell.* **20**, 637-646 (1998).

[11] Vapnik, V.N. **The Nature of Statistical Learning Theory**, Springer, 1995.

[12] Vapnik, V.N. **Statistical Learning Theory**, Wiley, 1998.

[13] Vapnik, V.N. and Chervonenkis, A.Ja. "On the uniform convergence of relative frequencies of events to their probabilities," *Theory Probab Appl.* **16** 264–280 (1971).

Section 3

Eduardo R. Caianiello Lecture

Sensitivity Analysis and Learning of Non-Linear Dynamic Systems by Two Dual Signal-Flow-Graph Approaches

Paolo Campolucci+

Dipartimento di Elettronica ed Automatica, Università di Ancona, Italy
E-mail : paolo@eealab.unian.it

Abstract

In this paper, two methods named Backward Computation (BC) and Forward Computation (FC) for both on-line and batch backward gradient computation of a system output (for sensitivity analysis) or cost function (for learning) with respect to system parameters are derived by the Signal-Flow-Graph representation theory and its known properties. The system can be any causal, in general non-linear and time-variant, dynamic system represented by a SFG, in particular any feedforward, time delay or recurrent neural network. In this work, we use discrete time notation, but the same theory holds for the continuous time case. The gradient is obtained in a straightforward way by the analysis of two SFGs, the original one and its adjoint (for the BC method) or its derivative (FC method) both obtained from the first by simple transformations without the complex chain rule expansions of derivatives usually employed.

The BC and FC methods are dual and the adjoint and derivative SFGs (of the same SFG) can be obtained one from the other by graph transformations. The BC method is local in space but not in time while the FC is local in time but not in space.

1. Introduction

In order to train Recurrent Neural Networks (RNNs) (Tsoi and Back 1994, Nerrand et al. 1993, Narendra and Parthasarathy 1991, Haykin 1994, Campolucci et al. 1995, Campolucci et al. 1999, Giles 1995) adaptively, an on-line learning algorithm must be derived for each specific architecture. In many learning algorithms, the gradient of a cost function must be estimated and then used in a parameter updating scheme, e.g. steepest descent or conjugate gradient. By gradient we mean the vector of derivatives of a given variable (here a system output or a function of it, e.g. a cost

---

+ Web page: http://nnsp.eealab.unian.it/Campolucci_P

function) with respect to a set of parameters that influence such a variable (here the parameters of a subset of network components).

When recurrent neural networks (or in general systems with feedback) are involved, the calculation of the gradient is a problem definitely more complex and difficult than in the case of feedforward neural networks (or systems without feedback). Due to feedback in fact, the current output depends upon the past outputs of the RNN, so that the present output depends not only on the present parameters but also on the past parameter values, and this dependence has to be considered in the calculation of the gradient. Quite often in the neural network literature this dependence has been totally or partially neglected to simplify the derivation of the gradient.

There are two different schemes to compute the gradient: the "forward computation" approach and the "backward computation" approach. These two names are motivated by the order in time in which the computations take place: forward means that both time and propagation of gradient information across the network flow as in the forward pass (i.e. the computation of the network outputs from its inputs), instead backward indicates the opposite flow. Real Time Recurrent Learning (RTRL) (Williams and Zipser 1989) and truncated Back Propagation Through Time (truncated BPTT) (Williams and Peng 1990) implement the forward and backward on-line computation, respectively, in the special case of fully recurrent neural networks.

An interesting approach to the computation of gradient information in recurrent NNs with arbitrary architectures was proposed by Wan and Beaufays (1996) who used a diagrammatic derivation to obtain the BPTT batch algorithm. The derived method can only be computed in batch mode since the defined reciprocal network (Wan and Beaufays 1996) is not causal due to the presence of the non-causal advance operators. The advance operators arise from the unfolding and transposing operations on which the method is based.

These authors also propose a proof of the equivalence of RTRL and BPTT in batch mode to show that the computed weight variations are the same, though the complexities of the two algorithms are different. The paper (Beaufays and Wan 1994) does not address the problem of graphical derivation of RTRL, as in (Wan and Beaufays 1996) for BPTT; nevertheless, the two approaches have been combined for that purpose in (Wan and Beaufays 1998) where a diagrammatic derivation of an on-line learning method is reported. However the resulting RTRL algorithm has to be derived in a two steps procedure, where the first step is the derivation of the batch backward method and the second is its transformation into the forward on-line algorithm.

Following the graph approach, now using the Signal Flow Graph (SFG) representation theory and its known properties (Mason 1953 and 1956, Lee 1974, Oppenheim and Schafer 1975), in this paper we propose two methods (Backward Computation or BC and Forward Computation or FC) which allow the gradient computation of a function (e.g. a cost function) of the system output with respect to some parameters both in batch and on-line modes. This system can be any causal, in general non-linear and time-variant, dynamic system represented by a SFG, in particular any feedforward (static), time delay or recurrent NN. In the following, we shall usually consider the NN framework but the method is fully general. In this work, we use discrete time notation, however the same theory holds for the continuous time case.

The presented approach generalizes and unifies the ordered derivatives (Werbos 1974) and automatic differentiation (Griewank 1988) approaches to gradient calculation both in forward and backward modes. The BC and FC methods are dual and the adjoint and derivative SFGs (of the same SFG) can be obtained one from the other by graph transformations. The BC method is local in space but not in time while the FC is local in time but not in space. Both the adjoint and derivative SFGs are obtained directly from the original SFG avoiding the two steps procedure previously mentioned (Wan and Beaufays 1998).

This work also extends a previous one by Osowski (1994) that proposed a SFG approach to neural networks learning. That work was basically developed for feedforward networks and the proposed extension to recurrent networks is valid only for networks that relax to a fixed point, as also in the work of Almeida (1987); moreover, it is not possible to accommodate delay branches in this framework. General recurrent networks able to process temporal sequences cannot be trained by the proposed adjoint equations.

Another related work is that by Martinelli and Perfetti (1991) which applies an adjoint transformation to an analog real circuit implementing a Multi-Layer-Perceptron (MLP) obtaining a circuit that implements backpropagation. Since this work is more related to hardware and the network is feedforward and static, it is less general and can be considered a particular case of the method presented here.

Finally, our formulation allows to deal both with the learning problem and with the sensitivity calculation problem, i.e. the problem of computing the derivatives of the system output with respect to some internal parameters. These derivatives can be used for designing robust electrical or numerical circuits.

The concepts detailed in this paper were briefly presented in (Campolucci et al. 1997, Campolucci et al. 1998) and extensively in (Campolucci 1998).

2. Signal Flow Graphs Definitions and Properties

A large class of complex systems, such as RNNs, can be represented by a signal-flow-graph (SFG). Therefore in this section, SFG notation and properties will be outlined following Lee's approach (1974). A different notation and properties of SFGs were also introduced by Oppenheim and Schafer (1975), while the SFG theory was first developed by Mason (1953, 1956).

A SFG is defined as a set of nodes and oriented branches. A branch is oriented from node a to node b if node a is the initial node and node b is the final node of the branch. An input node is one which has only one outgoing branch, the *input branch*, and no incoming branch. Associated with the input branch is an input variable u_j. An output node is one which has only one incoming branch, the *output branch*. Associated with the output branch is an output variable y_j. All other nodes besides the input and output nodes are called n-nodes. All other branches besides the input and output branches are called *f-branches*.

There are two variables related by a function associated with each of the f-branches. The initial variable x_j, at the tail of the branch, and the final variable v_j, at the head of the branch. The relationship between the initial and final variables for branch j is $v_j = f_j\left[x_j\right]$, where $f_j[.]$ is a general function describing the operator (or circuit component) of the j-th branch. By definition, for each n-node, the value of the x_j variables associated with its outcoming branches is the sum of all the v_j variables associated with its incoming branches, refer to Fig. 1. Let the SFG consist of $p+m+r$ nodes (m input nodes, r output nodes, and p n-nodes), and $q+m+r$ branches (m input branches, r output branches, and q f-branches). Please note that only the branches are indexed (and not the nodes), therefore all the reported indexes are to be considered as branch indexes.

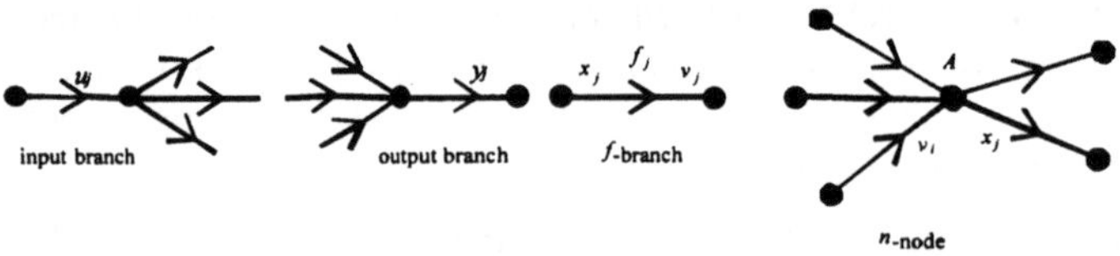

Fig. 1. Signal-flow-graph definitions. If I_A is the set of indexes of branches that comes in node A, and Z_A the set of indexes of branches that comes out from node A, then the following holds $x_j = \sum_{i \in I_A} v_i \ , \forall j \in Z_A$.

In the case of discrete time systems, the functional relationship between the variables x_j and v_j of the j-th f-branch (expressed as $v_j = f_j\left[x_j\right]$) is usually assumed to be:

$$\begin{cases} v_j(t) = g_j[x_j(t), \alpha_j(t), t] & \text{for a static branch (without memory)} \quad (2.1) \\ v_j(t) = \boldsymbol{q}^{-1} x_j(t) \overset{\Delta}{=} x_j(t-1) & \text{for a delay branch (one unit memory)} \quad (2.2) \end{cases}$$

where $\boldsymbol{q}^{-1}$ is the delay operator and g_j is a general differentiable function, which can depend on an adaptable parameter (or vector of parameters) α_j and also on time t to allow shape adaptation or variation in time of the non-linearity. For a static branch the relationship between x_j and v_j often particularizes as

$$\begin{cases} v_j(t) = w_j(t) x_j(t) & \text{for a weight branch} \\ v_j(t) = f_j(x_j(t)) & \text{for a fixed non-linear branch} \end{cases} \quad (2.3)$$

where $w_j(t)$ is the j-th (i.e. belonging to the j-th branch) weight parameter of the system at time t, and f_j is a differentiable function again of the j-th branch. In the neural networks context, $w_j(t)$ is a synaptic weight and f_j a sigmoidal activation function. However, equation (2.1) can represent more general non-linear adaptable functions controlled by a parameter (or vector of parameters) α_j, such as the recently introduced spline based activation functions (Uncini et al. 1999).

We are now able to state the following definitions:

DEFINITION 2.1 (SFG) The SFG G_N that represents the system is given by the connection (with or without feedback) of the branches described by (2.1), (2.2) and possibly (2.3).

The branches (components) are conceptually equivalent to the components of an electrical circuit and the relationship between the SFG and the implemented system is the same as that between the electrical circuit and the system it implements. Not all the SFG can represent real systems, it is known in fact that in order to be computed a SFG has to satisfy the constraint that each loop must include at least one delay branch.

DEFINITION 2.2 (Reversed SFG) A reversed SFG $\hat{G}_N$ of a given SFG G_N is a member of the set of SFGs obtained by reversing the orientation of all the branches in G_N, i.e. by replacing the summing junctions with branching points and viceversa.

Let $\hat{u}_j, \hat{y}_j, \hat{x}_j$ and $\hat{v}_j$, be the variables associated with the branches in $\hat{G}_N$ corresponding to u_j, y_j, x_j and v_j in G_N. In $\hat{G}_N$ the input variables are $\hat{y}_j$ $j = 1,...,r$, the output variables are $\hat{u}_j$ $j = 1,...,m$, $\hat{v}_j$ and $\hat{x}_j$ are the initial and the final variable of the j-th f-branch respectively. Let the indexes assigned to the branches of $\hat{G}_N$ be the same as the indexes assigned to the corresponding branches of G_N. As previously stated, node indexes are not required by this development. The functional relationships between the variables $\hat{x}_j$ and $\hat{v}_j$ in $\hat{G}_N$ are generic, therefore the graph $\hat{G}_N$ effectively belongs to a set of SFGs. For an example of SFG and reversed SFG, see Fig. 2 (a,b,c) where a locally recurrent MLP with Infinite Impulse Response (IIR) temporal filter synapses is shown.

DEFINITION 2.3 (Transposed SFG) The transposed SFG is a particular reversed SFG such that the relationships of the corresponding branches of G_N and $\hat{G}_N$ remain the same.

For a generic SFG, the following theorem, similar to Tellegen's theorem for electrical network (Tellegen 1952, Penfield et al. 1970), was derived by Lee (1974). It only relies on the topological properties of the original and reversed graphs and not on the functional relationships between the branch variables, that therefore can be described by any mathematical relation.

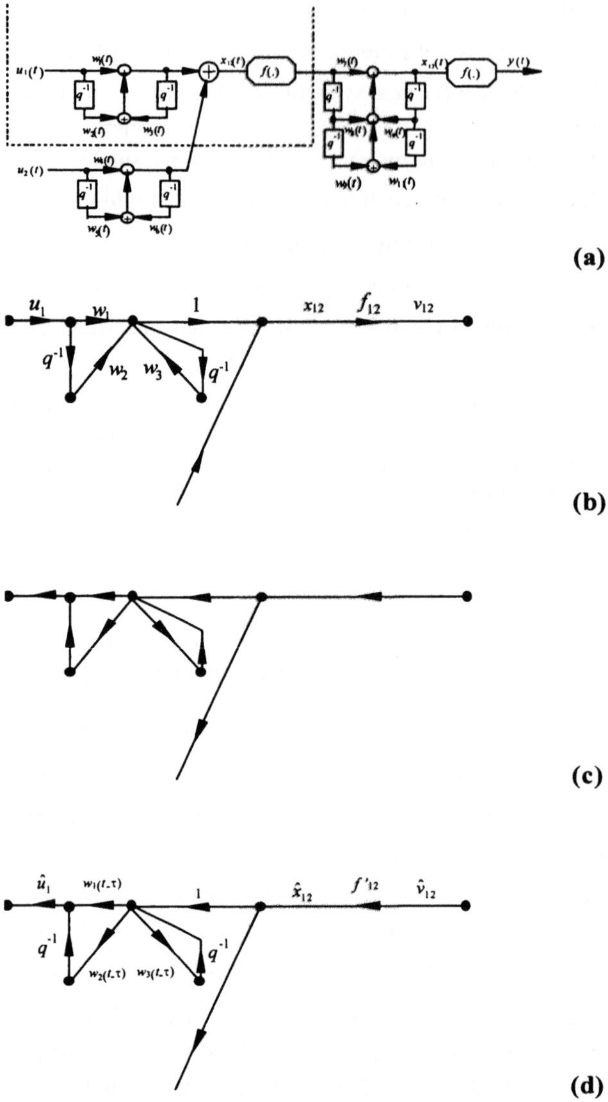

Fig. 2. (a) A simple IIR-MLP neural network with 2 layers, 2 inputs, 1 output, 1 neuron in the first layer with 2 IIR filter synapses (1 zero, 1 pole) and 1 neuron in the second layer with 1 IIR filter synapsis (2 zeros, 2 poles); (b) a section of its SFG G_N; (c) its reversed SFG $\hat{G}_N$; (d) its adjoint SFG $\hat{G}_N^{(a)}$.

THEOREM 2.1 (Lee 1974) Consider a causal dynamic system. Let G_N be the corresponding signal flow graph and $\hat{G}_N$ a generic reversed signal flow graph. If $\hat{u}_j, \hat{y}_j, \hat{x}_j$ and $\hat{v}_j$ are the variables in $\hat{G}_N$ corresponding to u_j, y_j, x_j and v_j in G_N, respectively, then

$$\sum_{j=1}^{r} \hat{y}_j(t) * y_j(t) + \sum_{j=1}^{q} \hat{x}_j(t) * x_j(t) = \sum_{j=1}^{m} \hat{u}_j(t) * u_j(t) + \sum_{j=1}^{q} \hat{v}_j(t) * v_j(t) \tag{2.4}$$

holds true, where "*" denotes the convolution operator, i.e.

$$y(t) * x(t) = \sum_{s=t_0}^{t} y(t-s)x(s) \tag{2.5}$$

where t_0 is the initial time.

For a static system the following equation holds

$$\sum_{j=1}^{r} \hat{y}_j y_j + \sum_{j=1}^{q} \hat{x}_j x_j = \sum_{j=1}^{m} \hat{u}_j u_j + \sum_{j=1}^{q} \hat{v}_j v_j \tag{2.6}$$

3. Sensitivity Analysis of Non-linear Dynamic Systems by Backward Computation

In network theory, a sensitivity is usually defined as the derivative of a certain output with respect to a given parameter. A general method to compute sensitivities of a dynamic continuous time system represented by a SFG was derived by Lee (1974) *for constant parameters*. Similarly an analogous result can be found for a discrete time system, however such a derivation can be used only for sensitivity computation in batch mode. Here we present a method to compute the sensitivity with respect to time-varying parameters, i.e. the derivative of a system output with respect to past or present parameters (the weights w_i and the non-linearity control parameters α_i). For an adaptive system they will depend on time, therefore the derivatives with respect to such parameters will also depend on time. This is a generalization of the works of Lee (1974) and Wan and Beaufays (1996).

The derivative of an output $y_k(t)$ of the system with respect to the parameter $w_i(t-\tau)$, where t is the time of the original SFG and τ a positive integer ($t \geq 0$ and $0 \leq \tau \leq t$), by the first equation in (2.3) is

$$\frac{\partial y_k(t)}{\partial w_i(t-\tau)} = \frac{\partial y_k(t)}{\partial v_i(t-\tau)} \frac{\partial v_i(t-\tau)}{\partial w_i(t-\tau)} = \frac{\partial y_k(t)}{\partial v_i(t-\tau)} x_i(t-\tau) \tag{3.1}$$

where $k=1,\ldots,r$ and i spans over the set of indexes of the weight branches. For the parameters of the non-linear function, by (2.1) the following more general expression holds:

$$\frac{\partial y_k(t)}{\partial \alpha_i(t-\tau)} = \frac{\partial y_k(t)}{\partial v_i(t-\tau)} \frac{\partial v_i(t-\tau)}{\partial \alpha_i(t-\tau)} = \frac{\partial y_k(t)}{\partial v_i(t-\tau)} \frac{\partial g_i}{\partial \alpha_i}\bigg|_{x_i(t-\tau),\alpha_i(t-\tau),t-\tau} \stackrel{\Delta}{=} \frac{\partial y_k(t)}{\partial v_i(t-\tau)} g_i'(t-\tau) \qquad (3.2)$$

where k=1,...,r and i spans over the set of indexes of the non-linear parametric branches, and a simplified notation for the derivative of g_i is implicitly defined.

In both cases we need to compute the term $\partial y_k(t) / \partial v_i(t-\tau)$, which is the derivative of the k-th output variable at time t with respect to a signal in the system, i.e. the final variable of the i-th f-branch at time $t-\tau$. As for the computation of the sensitivity in an electrical circuit, also in this case an adjoint network can be introduced to perform this task. The main idea is to apply Lee's theorem (equation 2.4) to the SFG G_N of the original system and to a particular reversed SFG $\hat{G}_N$, here called adjoint SFG or $\hat{G}_N^{(a)}$. It can be easily seen (see the proof in (Campolucci 1998)) that expression 2.4 holds true also when small variations of the variables are considered:

$$\sum_{j=1}^{r} \hat{y}_j(t) * \Delta y_j(t) + \sum_{j=1}^{q} \hat{x}_j(t) * \Delta x_j(t) = \sum_{j=1}^{m+1} \hat{u}_j(t) * \Delta u_j(t) + \sum_{j=1}^{q} \hat{v}_j(t) * \Delta v_j(t)$$

Since the functional relationships of the branches of $\hat{G}_N^{(a)}$ can be freely chosen, as equation (2.4) does not depend on them, these relationships are chosen in order to obtain an estimate of $\dfrac{\partial y_k(t)}{\partial v_i(t-\tau)}$ from this equation. Thus we can now state the following:

DEFINITION 3.1 (Adjoint SFG) The adjoint SFG $\hat{G}_N^{(a)}$ is the particular reversed graph whose f-branch relationships are related to the f-branch equations of the original graph G_N by the correspondence reported in Table 1.

Table 1 shows that for each delay branch in the original SFG, the corresponding branch of the adjoint SFG is a delay branch (with zero initial condition). The non-linearity without parameters f_j in the original SFG corresponds to a gain equal to the derivative of f_j evaluated at the value of the initial variable of the branch at time t-τ in the adjoint SFG, where t is the time of the original SFG and τ is that of the adjoint SFG. The weight at time t corresponds to the weight at time t-τ in the adjoint SFG. A similar transformation should be performed for the general non-linearity with control parameters. The outputs of the original SFG correspond to the inputs of the adjoint SFG. For an example of adjoint SFG construction for a simple MLP with Infinite Impulse Response (IIR) filter synapses see Fig. 2.

Branch Classification	Original Signal-Flow-Graph	Adjoint Signal-Flow-Graph for Parameter Sensitivity on y_k
Delay	q^{-1} $v_j(t) = q^{-1}x_j(t)\ ,\ v_j(0) = 0$	q^{-1} $\hat{x}_j(\tau) = q^{-1}\hat{v}_j(\tau)\ ,\ \hat{x}_j(0) = 0$
Static	f $v_j(t) = f_j(x_j(t))$	f_j' $\hat{x}_j(\tau) = \hat{h}_j(\tau)\hat{v}_j(\tau)$ $\hat{h}_j(\tau) = f_j'(x_j(t-\tau))$
	$v_j(t) = w_j(t)x_j(t)$	$\hat{x}_j(\tau) = \hat{w}_j(\tau)\hat{v}_j(\tau)$ $\hat{w}_j(\tau) = w_j(t-\tau)$
	[illegible]	g_j' $\hat{x}_j(\tau) = \hat{h}_j(\tau)\hat{v}_j(\tau)$ $\hat{h}_j(\tau) = \left.\frac{\partial g_j}{\partial x_j}\right\|_{x_j(t-\tau), \alpha_j(t-\tau), t-\tau}$
System Output		$\hat{y}_j(\tau) = \begin{cases} 0 \quad \forall\tau & \text{if } j \neq k \\ \begin{cases} 1\ ,\ \tau = 0 \\ 0\ ,\ \tau > 0 \end{cases} & \text{if } j = k \end{cases}$

Table 1. Adjoint SFG construction rules. For batch adaptation the adjoint branches must be defined remembering that $t=T$ in this case, where T is the final instant of the epoch.

Moreover the inputs of the adjoint SFG must be set to an impulse in correspondence with the output of G_N of which the sensitivity has to be computed, and to a constant null signal in correspondence with all the other outputs. In other words to compute the value of $\partial y_k(t) / \partial v_i(t-\tau)$ the input of the adjoint SFG should be:

$$\hat{y}_j(\tau)=\begin{cases}0 & \forall\tau \quad \text{if } j\neq k\\ \begin{cases}1\ , & \tau=0\\ 0\ , & \tau>0\end{cases} & \text{if } j=k\end{cases} \qquad j=1,\ldots,r \tag{3.3}$$

Let $\hat{u}_j, \hat{y}_j, \hat{x}_j$ and $\hat{v}_j$ be the variables associated with the branches in $\hat{G}_N^{(a)}$ corresponding to u_j, y_j, x_j and v_j in G_N ; it follows (see the proof in (Campolucci 1998)):

$$\frac{\partial y_k(t)}{\partial v_i(t-\tau)}=\hat{v}_i(\tau) \tag{3.4}$$

where k=1,…,r and i=1,…,q. Using this result in (3.1) we get

$$\frac{\partial y_k(t)}{\partial w_i(t-\tau)}=\hat{v}_i(\tau)x_i(t-\tau) \tag{3.5}$$

This result states that the derivative of an output variable of the SFG at time t with respect to its weight parameter $w_i(t-\tau)$ at time $t-\tau$ is the product of the initial variable of the i-th branch in the original SFG at time $t-\tau$ and the initial variable of the corresponding branch in the adjoint graph at τ time units. This result can be easily generalized for the non-linear parametric function as follows:

$$\frac{\partial y_k(t)}{\partial \alpha_i(t-\tau)}=\hat{v}_i(\tau)g_i'(t-\tau) \tag{3.6}$$

4. Learning of Non-linear Dynamic Systems by Backward Computation

In the previous section, we have shown how the SFG of the original network G_N and the SFG of the adjoint network $\hat{G}_N^{(a)}$ can be used to compute the sensitivity of an output with respect to a past or present system parameter (equations 3.5 and 3.6).

Now we show how to use this technique to adapt a dynamic network, minimizing a supervised cost function with respect to the network parameters.

The idea is to consider the cost function itself as a network connected in cascade to the dynamical network to be adapted. Therefore the whole SFG (named G_S in the following) has as inputs the inputs of the original network and the desired outputs, while the value of the cost function is its unique output. The gradient of this output (i.e. of the cost function) with respect to the system parameters (i.e. weights, etc.) now corresponds to the sensitivity of this output, which can be computed by using equation 3.5 and 3.6 again, applied now to the cascade SFG G_S .

4.1. Cost Functions and Parameter Updating Rules

Let us consider a discrete time non-linear dynamic system with inputs u_k, $k = 1,\dots,m$, outputs y_k, $k = 1,\dots,r$, and parameters w_i and α_i which have to be adapted with respect to an output error. Using gradient based learning and following the steepest-descent updating rule, it holds, e.g. for the weight w_i

$$\Delta w_i = -\mu \frac{\partial J}{\partial w_i} \quad , \quad \mu > 0 \tag{4.1}$$

where J is the cost function, Δw_i is the variation of the parameter w_i and μ is the learning rate. The major problem is given by the calculation of the derivative $\partial J / \partial w_i$.

Since for on-line learning the parameters can change at each time instant:

$$\frac{\partial J}{\partial w_i} = \sum_{\tau=0}^{t} \frac{\partial J}{\partial w_i(\tau)} = \sum_{\tau=0}^{t} \frac{\partial J}{\partial w_i(t-\tau)} \tag{4.2}$$

As the length of the summation linearly increases with the current time step t, the summation in (4.2) must be truncated in order to implement the algorithm

$$\Delta w_i(t) = -\mu \sum_{\tau=t-h+1}^{t} \frac{\partial J}{\partial w_i(\tau)} = -\mu \sum_{\tau=0}^{h-1} \frac{\partial J}{\partial w_i(t-\tau)} \tag{4.3}$$

where h is a fixed positive integer. In this way, not all the history of the system is considered but only the most recent part in the interval $[t\text{-}h\text{+}1,t]$. It is easy to show that a real truncation is necessary only for circuits with feedback, e.g. recurrent NN, while for feedforward networks with delays (e.g. Time Delay NN or TDNN) a finite h can be chosen, so that all the memory of the system is taken into account. An optimal selection of h requires an appropriate choice for each parameter; for layered TDNN h should depend on the layer and should be increased moving from the last to the first layer, since more memory is involved.

The above updating rule (equations 4.2 and 4.3) holds theoretically in the hypothesis of constant weights but practically it is only a good approximation. Equations (3.5) and (3.6) do not require that hypothesis, and can be used in a more general context. Equations equivalent to (4.1), (4.2) and (4.3) can also be written for the parameters α_i.

The most common choice for the cost function J is the MSE. The instantaneous squared error at time t is defined as

$$e^2(t) = \sum_{k=1}^{r} e_k^2(t) \qquad \text{with } e_k(t) = d_k(t) - y_k(t) \tag{4.4}$$

where $d_k(t)$ $k = 1,\dots,r$ are the desired outputs. So the mean squared error over a time interval $[t_0, t_1]$ is given by

$$E(t_0,t_1)=\sum_{t=t_0}^{t_1} e^2(t) \tag{4.5}$$

In the case of batch training, the cost function can be chosen as the mean squared error over the entire learning-epoch $E(0,T)$, where T is the final time of the epoch, whereas for on-line training only the most recent errors $e^2(t)$ must be considered, for example using $E(t-N_c+1,t)$ with the constant N_c properly chosen (Nerrand et al. 1993, Narendra and Parthasaraty 1991). Therefore it holds:

$$\begin{cases} E(t-N_c+1,t)=\sum_{s=t-N_c+1}^{t} e^2(s) & \text{on-line training} \\ E(0,T)=\sum_{s=0}^{T} e^2(s) & \text{batch training} \end{cases}$$

Assuming $J=E(t_0,t_1)$, it follows

$$\Delta w_i(t)=-\mu\sum_{\tau=0}^{h-1}\frac{\partial E(t-N_c+1,t)}{\partial w_i(t-\tau)} \quad \text{for on-line training} \tag{4.6a}$$

and

$$\Delta w_i=-\mu\sum_{\tau=0}^{T}\frac{\partial E(0,T)}{\partial w_i(T-\tau)} \quad \text{for batch training} \tag{4.6b}$$

4.2. Cost Function Gradient Computation by BC

Let us consider the SFG G_S obtained by a cascade connection of the SFG G_N of the system to be adapted and the SFG which implements the cost function J (named G_J in the following). The SFG G_J has as inputs the outputs of the adaptable system y_k and the targets d_k, while the value of J is its unique output. Obviously G_J can contain the same kind of operators as G_N, i.e. delays, non-linearities, parameters to be adapted (e.g. for regularization purposes) and feedback. The class of cost functions allowed by this approach is therefore enlarged with respect to other approaches since the cost expression can have memory and be recursive. Although some applications would require very complex cost functions, the SFG approach can easily handle them.

Using the adjoint network $\hat{G}_S^{(a)}$ of the new SFG G_S, it is easy to calculate the sensitivities of the output with respect to the system parameters (equations 3.5 and 3.6). Since the output y_k (k=1) of G_S is the cost function J, combining (4.2) with (3.5) or (3.6) and truncating the past history to h steps (on-line learning) it holds:

$$\begin{cases} \dfrac{\partial J(t)}{\partial w_i} = \displaystyle\sum_{\tau=0}^{h-1} \hat{v}_i(\tau) x_i(t-\tau) & \text{on - line} \qquad (4.7a) \\ \\ \dfrac{\partial J}{\partial w_i} = \displaystyle\sum_{\tau=0}^{T} \hat{v}_i(\tau) x_i(T-\tau) & \text{batch} \qquad (4.7b) \end{cases}$$

$$\begin{cases} \dfrac{\partial J(t)}{\partial \alpha_i} = \displaystyle\sum_{\tau=0}^{h-1} \hat{v}_i(\tau) g_i{}'(t-\tau) & \text{on - line} \qquad (4.8a) \\ \\ \dfrac{\partial J}{\partial \alpha_i} = \displaystyle\sum_{\tau=0}^{T} \hat{v}_i(\tau) g_i{}'(T-\tau) & \text{batch} \qquad (4.8b) \end{cases}$$

Equations (4.7) and (4.8) hold for any cost function J which can be described by a SFG; in the particular case when J is chosen to be a standard MSE cost function, we show that the explicit implementation of G_J can be avoided since the involved derivatives can be analytically derived.

For on-line learning, using the instantaneous squared error $J = e^2(t)$ the input-output relationship of G_J is given by (4.4), thus

$$\frac{\partial e^2(t)}{\partial y_k(t-\tau)} = \begin{cases} -2e_k(t) & ,\tau = 0 \\ 0 & ,\tau > 0 \end{cases} \quad , \quad k=1,\dots,r \qquad (4.9)$$

Using (3.4) and (4.9) and considering the output y_k of G_N as an internal signal v_i, we can explicitly compute the outputs of the adjoint network $\hat{G}_J^{(a)}$

$$\hat{y}_k(\tau) = \frac{\partial e^2(t)}{\partial y_k(t-\tau)} = \begin{cases} -2e_k(t) & ,\tau = 0 \\ 0 & ,\tau > 0 \end{cases} \quad , \quad \text{k=1,...,r} \qquad (4.10)$$

Such outputs can be used as the inputs of $\hat{G}_N^{(a)}$. Therefore if the signal (4.10) feeds the network $\hat{G}_N^{(a)}$ instead of (3.3), the explicit use of $\hat{G}_J^{(a)}$ can be avoided.

If $J = E(t - N_c + 1, t)$ is used, then equation (4.10) generalizes to the following:

$$\hat{y}_k(\tau) = \frac{\partial E(t - N_c + 1, t)}{\partial y_k(t-\tau)} = \begin{cases} -2e_k(t-\tau) & ,\tau \le N_c - 1 \\ 0 & ,\tau \ge N_c \end{cases} \quad , \quad k=1,\dots,r \qquad (4.11)$$

When batch learning is involved, i.e. when $J = E(0,T)$, the following equation has to be used instead of (4.10) or (4.11)

$$\hat{y}_k(\tau) = \frac{\partial E(0,T)}{\partial y_k(T-\tau)} = -2e_k(T-\tau) \ , \ k = 1,\dots,r \ , \ \tau = 0,1,\dots,T \qquad (4.12)$$

Note that the parameter variations computed by the summation of all gradient terms in $[0,T]$ are the same as those obtained by BPTT (Wan and Beaufays 1996).

Thus for standard MSE cost functions, the derivative calculation can be performed considering G_N and $\hat{G}_N^{(a)}$ only, but using either (4.10), (4.11) or (4.12) as inputs for $\hat{G}_N^{(a)}$ instead of (3.3).

Srinivasan et al. (1994) derive a theorem which substantially states what is expressed in (4.7) when $J = e^2(t)$; however such a theorem has been proved only for a neural network with a particular structure and by a completely different and very specific approach, whereas the method presented here is very general, including any NN as a special case.

In the following, we call the resulting algorithm BC(h) (Backward Computation), where h is the truncation parameter, or BC for a short notation or for the batch case.

Should be noted that not all of the variables of the adjoint SFG have to be computed, but only the initial variables of the branches containing adaptable parameters since they must be used in (4.7) and (4.8). It must be stressed that the delay operator of the adjoint SFG delays the τ and not the t index.

5. Sensitivity Analysis of Non-linear Dynamic Systems by Forward Computation

To compute the derivative of a certain output of the system considered $y(t)$ with respect to a parameter w_i (for α_i it is analogous), in a forward way, let's consider the sensitivity of each of the various operators of the SFG. Let's assume that the parameters are constant for a moment, at the end of this section such assumption will be removed.

a) For a summation node, $x_j(t) = v_k(t) + ... + v_p(t)$, it holds:

$$\frac{\partial x_j(t)}{\partial w_i} = \frac{\partial v_k(t)}{\partial w_i} + ... + \frac{\partial v_p(t)}{\partial w_i} \tag{5.1}$$

b) For a weight branch, $v_j(t) = w_j x_j(t)$, it holds:

$$\begin{cases} \dfrac{\partial v_j(t)}{\partial w_i} = w_j \dfrac{\partial x_j(t)}{\partial w_i} & , \quad j \neq i \\ \dfrac{\partial v_i(t)}{\partial w_i} = x_i(t) + w_i \dfrac{\partial x_i(t)}{\partial w_i} & , \quad j = i \end{cases} \tag{5.2}$$

c) For a non-linear branch, $v_j(t) = f_j(x_j(t))$, it holds:

$$\frac{\partial v_j(t)}{\partial w_i} = f_j'(x_j(t))\frac{\partial x_j(t)}{\partial w_i} \tag{5.3}$$

d) For a delay branch, $v_j(t) = q^{-1}x_j(t) = x_j(t-1)$, it holds:

$$\frac{\partial v_j(t)}{\partial w_i} = \frac{\partial x_j(t-1)}{\partial w_i} = q^{-1}\frac{\partial x_j(t)}{\partial w_i} \tag{5.4}$$

The initial states are not dependent on the parameters w_i, therefore null initial conditions can be used for the delay branch:

$$\frac{\partial v_j(0)}{\partial w_i} = 0 \tag{5.5}$$

e) The input variables are not dependent on the parameters w_i, therefore

$$\frac{\partial u_j(t)}{\partial w_i} = 0 \tag{5.6}$$

Based on the above equations, new SFGs can be defined, called *derivative SFGs* $G_{N,i}^{D}$ from the original network SFG G_N, by using the transformation rules in Table 2.

The systems associated with the derivative SFGs are linear and time-variant. Each of them has only one non-null input x_i and gives the desired derivative $\partial y(t)/\partial w_i$ as output.

Therefore, to compute $\partial y(t)/\partial w_i$ the initial variable of the *i*-th branch must be injected to the final node of the same branch in the derivative SFGs, as in Fig. 3.

Each of the derivative SFGs has a different temporal evolution (internal state) so the computation of *W* different SFGs is necessary for computing the output sensitivity with respect to all of the *W* weights of G_N.

At this point, we want to remove the initial assumption that weights are constant. Let's consider the following expansion of the gradient:

$$\frac{\partial y(t)}{\partial w_i} = \sum_{\phi=0}^{t}\frac{\partial y(t)}{\partial w_i(\phi)} \tag{5.7}$$

For computing each term inside the summation, i.e. $\partial y(t)/\partial w_i(\phi)$, the method of the derivative SFGs can be used with some corrections as detailed in the following.

Branch classification	Original signal flow graph	Derivative signal flow graph for computing $\partial y(t)/\partial w_i$
Delay	$x_j \xrightarrow{q^{-1}} v_j$	$\partial x_j(t)/\partial w_i \xrightarrow{q^{-1}} \partial v_j(t)/\partial w_i$
Static	$x_j \xrightarrow{f_j(.)} v_j$ $v_j(t) = f_j(x_j(t))$	$\partial x_j(t)/\partial w_i \xrightarrow{f'_j} \partial v_j(t)/\partial w_i$ $\partial v_j(t)/\partial w_i = f'_j(x_j(t))\,\partial x_j(t)/\partial w_i$
	$x_j \xrightarrow{w_j} v_j$	$\frac{\partial x_i(t)}{\partial w_i} \xrightarrow{w_i}$ (+ $x_i(t)$) $\quad j = i$ $\frac{\partial x_j(t)}{\partial w_i} \xrightarrow{w_j} \frac{\partial v_j(t)}{\partial w_i} \quad j \neq i$
Input	$\xrightarrow{u_j}$	$\xrightarrow{0}$

Table 2. Derivative SFGs construction rules.

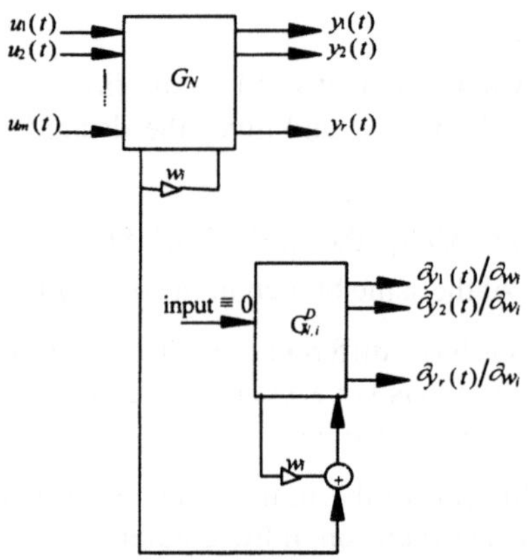

Fig. 3. Use of derivative SFGs for sensitivity computation.

The i-th static branch of G_N

$$v_i(\tau) = w_i(\tau)x_i(\tau) \tag{5.8}$$

should correspond to an operator of the derivative SFG $G^D_{N,i}$ described by the following

$$\begin{cases} \dfrac{\partial v_i(\tau)}{\partial w_i(\phi)} = w_i(\tau)\dfrac{\partial x_i(\tau)}{\partial w_i(\phi)} & , \ \tau > \phi \\ \\ \dfrac{\partial v_i(\phi)}{\partial w_i(\phi)} = x_i(\phi) & , \ \tau = \phi \end{cases} \tag{5.9}$$

instead of (5.2).

Therefore the derivative SFG can be constructed as before, but now its input must be a sequence with all zero but not at time ϕ when its value is $x_i(\phi)$. The output will be the derivative $\partial y(t)/\partial w_i(\phi)$. Therefore for computing each term in (5.7) t+1 forward propagations with ϕ from 0 to t could be used, but since the derivative SFGs are linear the summation in (5.7) can be obtained directly, without computing each term, using as input the sequence $x_i(\phi)$ $\phi = 0,1,...,t$ (without zeros) and so by the derivative SFGs defined by (5.2) and not (5.9).

For computing the gradient with respect to α_i parameters the same theory holds with the only difference that the input to be injected to the final node of the i-th branch (i.e. the non-linearity with parameters) is $\dfrac{\partial g_i(x_i(t), \alpha_i(t), t)}{\partial \alpha_i(t)}$ instead of $x_i(t)$.

This gradient computation method will be named Forward Computation due to the fact that the operations for the sensitivity computations follow the same order of the calculations of the original SFG G_N.

6. Learning of Non-linear Dynamic Systems by Forward Computation

In the previous section, a method to compute the gradient of the output of a SFG with respect to its parameters was presented, now we want to show how to extend this result for computing the gradient of a cost function of the outputs. The idea is the same as for Backward Computation: to describe the cost function as a SFG G_J that is cascaded after the SFG G_N of the adaptable system. Let's call the overall SFG G_S.

G_J is a single output SFG with output $y(t) = J$, and with inputs u_i, $i = 1,...,m$ (the trainable system inputs), and d_k, $k = 1,...,r$ (the desired outputs). The partial derivative of J with respect to a parameter w_i is obtained as output of the derivative SFG $G_{S,i}^D$, that is designed from G_S by the transformation rules in Table 2, see Fig. 4.

It is easy to show that the Forward Computation method applied to Fully Recurrent Neural Networks gives the Real Time Recurrent Learning (RTRL) algorithm (Williams and Zipser 1989). Applied to IIR filters it gives the Output-Error

Recursive Prediction Error (RPE) method, well known in the Signal Processing community (Shynk 1989).

The FC compared to the BC method has the advantage to be intrinsically on-line: there is no need for past history truncation strategies. The reason is that the derivative SFGs are designed from the original SFG keeping the same order of computation with respect to the branch operators and time, while in BC the order is reverted (backward order).

Therefore FC is particularly suited for on-line learning or sensitivity computation but still it can be used in batch mode by accumulating the instantaneous derivatives $\partial y(t)/\partial w_i$ on time and adapting the parameters with the accumulated variations. On the other hand, it should be noted that the complexity of BC and FC is not the same and therefore a choice should be made also considering complexity concerns, in most applications, please refer to the following section.

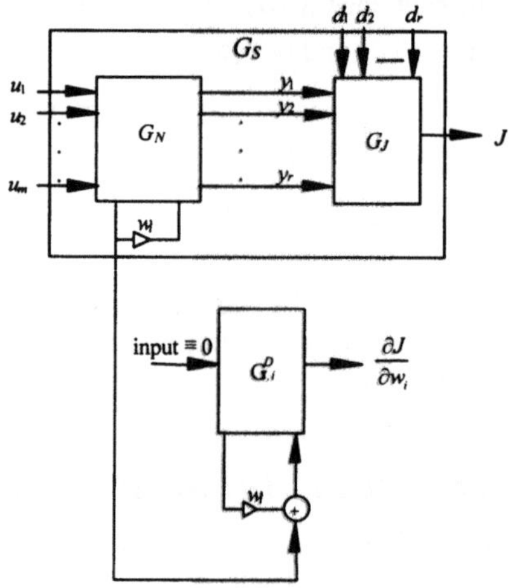

Fig. 4. Derivative SFGs for calculating the gradient of cost functions.

7. Complexity Analysis

The complexity of the BC gradient computation method linearly increases with the number of adaptable parameters. In particular, since the adjoint graph has the same topology of the original one, the complexity of its computation is about the same; in practice it is lower, since not all the variables of the adjoint graph have to be computed, but only the ones that correspond to a parameter to be adapted or of which the sensitivity has to be computed. For on-line learning the adjoint graph must be evaluated h times for each time step, therefore the number of operations for the computation of the gradient terms with respect to all the parameters is roughly (in practice it is lower) h times the number of operations of the forward phase, plus the computation needed by equ. (4.7) and (4.8) for each time step, whose complexity linearly increases with h and with the number of parameters W with respect to which, the derivative of the output or cost has to be computed.

On the contrary, by Forward Computation both in batch and on-line mode, W derivative SFGs must be computed, one for each parameter, for calculating all the gradient components. Since they are approximately of the same complexity of the original SFG, the complexity of FC is about W times the complexity of the original SFG.

Therefore the ratio of the complexities for the two approaches is in on-line mode

$$\frac{C_{BC}}{C_{FC}} \propto \frac{h}{W} \tag{7.1}$$

meaning that if h is small enough with respect to W then BC is a computationally simpler method than FC. Usually, for small networks FC can be simpler while for large SFGs with many branches the BC method should be preferred with respect to complexity concerns. Nevertheless, since the value of h that should be chosen, depends on the specific architecture and parameter values it should be considered when evaluating the complexities ratio.

On the contrary, in batch mode, the BC method is always simpler than the FC since the adjoint SFG has to be computed just once and therefore

$$\frac{C_{BC}}{C_{FC}} \propto \frac{1}{W} \tag{7.2}$$

For the simple case of a fully recurrent, single layer, single delay neural networks composed of n neurons, the computational complexity is $O(n^2)$ operations per time step for batch BC or $O(n^2h)$ for on-line BC compared with $O(n^4)$ for both batch and on-line FC (Williams and Zipser 1989). The memory requirement is $O(nT)$ for batch BC or $O(nh)$ for on-line BC, and $O(n^3)$ for FC. Therefore, as far as computational complexity is concerned, BC is significantly simpler than FC, in batch mode, whereas in on-line mode the complexity and also the memory requirement ratios depend on h/n^2. Considering that the number of parameters W is equal to n^2 for such neural architecture, this ratio can be obtained as a particular case of (7.1).

Conclusions

Two Signal Flow Graph approaches named Backward Computation and Forward Computation have been presented which allow an easy on-line or batch computation of the gradient terms needed in both sensitivity analysis and learning of non-linear dynamic adaptive systems represented by SFGs, even if their structure includes feedback or time delays. These two methods allow to avoid the complex chain rule derivative expansion, traditionally needed to derive the gradient equations. Both BC and FC methods generalize previous results in different fields and in particular the BC method provides an elegant analogy with the sensitivity analysis of analog networks by the "adjoint network" (Director and Rohrer 1969).

An interesting application of the FC method could be on VLSI design of neural networks or more general adaptive circuits, since it would not require a specific circuit for implementing the learning because it uses the derivative circuit that is a simple modification of the original circuit, preserving the evaluation order. This implementation can save chip area and computes the exact gradient, while other methods proposed in literature (Jabri and Flower 1992) for the same purpose are approximated.

The BC and FC methods are dual and the adjoint and derivative SFGs (of the same SFG) can be obtained one from the other by graph transformations. The BC method is local in space but not in time while the FC is local in time but not in space.

Acknowledgment

P. Campolucci wishes to thank Prof. F. Piazza for help revising part of this manuscript.

References

Almeida, L. B., 1987, "A learning rule for asynchronous perceptrons with feedback in combinatorial environment", Proc. Int. Conf. on Neural Networks, vol. 2, pp. 609-618.

Back A.D., Tsoi A.C., 1991, "FIR and IIR synapses, a new neural network architecture for time series modelling", Neural Computation 3: 375-385.

Back A.D., Tsoi A.C., 1993, "A simplified gradient algorithm for IIR synapse Multilayer Perceptron", Neural Computation 5: 456-462.

Beaufays F., Wan E., 1994, "Relating Real-Time Backpropagation and Backpropagation-Through-Time: An Application of Flow Graph Interreciprocity", Neural Computation 6: 296-306.

Campolucci P., 1998, "A circuit theory approach to recurrent neural networks architectures and learning methods", PhD dissertation, University of Bologna, Italy, 1998.

Campolucci P., Marchegiani A., Uncini A., Piazza F., 1997, "Signal-Flow-Graph derivation of on-line gradient learning algorithms" Proc. of ICNN-97, IEEE Int. Conference on Neural Networks, Houston (USA), June 1997.

Campolucci P., Piazza F., Uncini A., 1995, "On-line learning algorithms for neural networks with IIR synapses", Proc. of the IEEE International Conference of Neural Networks, Perth, November 1995.

Campolucci P., Uncini A., Piazza F., 1998, "Dynamical Systems Learning by a Circuit Theoretic Approach", Proc. of ISCAS-98, IEEE Int. Symposium on Circuits and Systems, 1998.

Campolucci P., Uncini A., Piazza F., Rao B.D., 1999, "On-line learning algorithms for locally recurrent neural networks", IEEE Transactions on Neural Networks, Vol. 10, No. 2, March 1999.

Director S.W., Rohrer R.A., 1969, "The generalized adjoint network and network sensitivities", IEEE Trans. on Circuit Theory, vol. CT-16, pp. 318-323, Aug. 1969.

Gherrity M., 1989, "A learning algorithm for analog, fully recurrent neural networks", Proc. Int. Joint Conference Neural Networks, I, 643-644.

Griewank A., 1988, "On Automatic Differentiation", Mathematics and Computer Science Division, Preprint ANL/MCS-P10-1088, November 1988.

Haykin S., 1994, "Neural Networks: a comprehensive foundation", IEEE Press-Macmillan.

Horne B.G. and Giles C.L., "An experimental comparison of recurrent neural networks", in Advances in Neural Information Processing Systems vol. 7, Cambridge, MA: MIT Press 1995 p. 697.

Jabri M. and Flower B., 1992, "Weight Perturbation: An Optimal Architecture and Learning Technique for Analog VLSI Feedforward and Recurrent Multilayer Networks", IEEE Trans. on Neural Networks, vol. 3, no. 1, pp. 154-157, January 1992.

Lee A.Y., 1974, "Signal Flow Graphs--Computer-Aided System Analysis and Sensitivity Calculations", IEEE Transactions on Circuits and Systems, vol. cas-21, no. 2, pp. 209-216, March 1974.

Martinelli, G. and Perfetti, R., 1991, "Circuit theoretic approach to the backpropagation learning algorithm", Proc. of the IEEE Int. Symposium on Circuits and Systems, 1991.

Mason S.J., 1953, "Feedback theory-Some properties of signal-flow graphs", Proc. of the Institute of Radio Engineers vol. 41, pp. 1144-1156.

Mason S.J., 1956, "Feedback theory-Further properties of signal-flow graphs", Proc. of the Institute of Radio Engineers vol. 44, pp. 920-926.

Narendra K.S., Parthasarathy K., 1991, "Gradient Methods for the Optimization of Dynamical Systems Containing Neural Networks", IEEE Trans. on Neural Networks, vol. 2, no. 2, March 1991.

Nerrand O., Roussel-Ragot P., Personnaz L., Dreyfus G., Marcos S., 1993, "Neural Networks and Nonlinear Adaptive Filtering: Unifying Concepts and New Algorithms", Neural Computation 5: 165-199.

Oppenheim, A. V. and Schafer, R. W., 1975, "Digital Signal Processing", Prentice-Hall.

Osowski, S. 1994, "Signal flow graphs and neural networks", Biological Cybernetics, 70, pp. 387-395.

Pearlmutter, B. A., 1995, "Gradient Calculations for Dynamic Recurrent Neural Networks: a Survey", IEEE Trans. on Neural Networks, vol. 6, no. 5.

Penfield, P., Spence, R. and Duiker, S., 1970, "Tellegen's Theorem and Electrical Networks", MIT Press, Cambridge, MA.

Shynk J.J., 1989, "Adaptive IIR filtering", IEEE ASSP Magazine, April 1989.

Srinivasan B., Prasad U.R. and Rao N.J., 1994, "Backpropagation through Adjoints for the identification of Non linear Dynamic Systems using Recurrent Neural Models", IEEE Trans. on Neural Networks pp. 213-228, March 1994.

Tellegen, B.D.H., 1952, "A general network theorem, with applications", Philips Res. Rep. 7, 259-269.

Tsoi A.C., Back A.D., 1994, "Locally recurrent globally feedforward networks: a critical review of architectures", IEEE Transactions on Neural Networks, vol. 5, no. 2, 229-239, March 1994.

Uncini A., Vecci L., Campolucci P., Piazza F., 1999, "Complex-valued Neural Networks with adaptive spline activation function for digital radio links nonlinear equalization", IEEE Transactions on Signal Processing, Vol. 47, No. 2, February 1999.

Wan E.A., Beaufays F., 1996, "Diagrammatic Derivation of Gradient Algorithms for Neural Networks", Neural Computation 8: 182-201.

Wan E.A., Beaufays F., 1998, "Diagrammatic methods for deriving and relating temporal neural networks algorithms" In adaptive Processing of Sequences and Data Structures, Lecture Notes in Artificial Intelligence – Springer Verlag 1998.

Werbos P.J., 1974, "Beyond regression: New tools for prediction and analysis in the behavioural sciences", Ph.D. dissertation, Committee on Appl. Math., Harvard Univ., Cambridge, MA, Nov. 1974.

Werbos, P. J., 1990, "Backpropagation through time: what it does and how to do it", Proc. of IEEE, Special issue on neural networks, vol. 78, no. 10, pp. 1550-1560.

Williams R.J., Peng J., 1990, "An efficient gradient-based algorithm for on line training of recurrent network trajectories", Neural Computation 2: 490-501.

Williams R.J., Zipser D., 1989, "A Learning Algorithm for Continually Running Fully Recurrent Neural Networks", Neural Computation 1: 270-280.

Williams, R. J. and Zipser, D., 1994, "Gradient-based learning algorithms for recurrent networks and their computational complexity", in Backpropagation: Theory, Architectures and Applications, Y. Chauvin and D.E. Rumelhart, Eds. Hillsdale, NJ: Lawrence Erlbaum Associates, pp. 433-486.

Section 4
Mathematical Models

Interval Arithmetic Multilayer Perceptron as Possibility-Necessity Pattern Classifier

Gian Paolo Drago
Istituto per i Circuiti Elettronici - CNR
via De Marini, 6 - 16149 Genova, Italy

Sandro Ridella
DIBE - Universita' di Genova
via Opera Pia, 11A - 16145 Genova, Italy

Abstract

In the work presented in this paper an Interval Arithmetic MLP (IAMLP) is used to detect the region in the input space to which an uncertainty decision should be appropriately associated. This region may be originated both by sub-regions which are not represented in the training set and by sub-regions where the probabilities of the two classes are very similar. To train the IAMLP, an algorithm will be presented which in particular is able detect the two certainty regions and the uncertainty one. The algorithm has been used for studying a simple artificial problem and one real-world application, the Breast Cancer data base.

Keywords: Learning, Interval Arithmetic, Fuzzy sets, Generalization, Neural Network Computing.

1 Introduction

In classification problems it may be important to recognize regions in the input space to which a very uncertain decision corresponds. Those uncertainty regions may originate from two different situations [1]:

A) low sampling regions, i.e. regions where the pattern probability density is low, and consequently are scarcely represented in the training set;

B) regions where the probabilities of the two classes are very similar.

In this work, the Interval Arithmetic (IA) will be used to detect the uncertainty region (UR) [2].

Our purpose is to identify an interval weight vector able to divide the pattern space into three regions [2]:

- one region containing only class 1 patterns (i.e. only patterns with negative targets);

- one region containing only class 2 patterns (i.e. only patterns with positive targets);
- one uncertainty region (i.e. containing patterns belonging to both classes).

This allows to detect, in the recognition phase, the uncertain situations, which should be rejected: this improves the generalization performance by reducing the classification error [1].

2 IAMLP training

In order to identify the UR with a reduced computational cost, the weight search has been split into two phases:

A) in the first phase, the weights of the MLP are found with well-known descending gradient techniques, minimizing the mean square error.

B) subsequently, evaluation of the interval weights is carried out through a direct search optimization routine [3], which minimizes a cost function proportional to the error in the UR with the constraint of an error function equal to zero within the two CRs.

The weights are initialized using SCAWI [4]. Then the MLP is trained using the AMBP algorithm [5] and the weights so obtained are used as starting values for the IAMLP interval weights search: as a first step, the weights are converted from real numbers into interval numbers, whose lower and upper values however are still coincident.

An interval variable (V) is defined through a pair of real values, the lower one (v^L) and the upper one (v^U), which delimit the range of values the variable itself is allowed to take.

Let us now describe how to evaluate the IAMLP interval output for identifying the three regions. The interval output of every neuron is given by:

$$Y = [y^L, y^U] = tanh(R) \quad ; \quad R = \sum W_i X_i \tag{1}$$

where W_i is the interval weight corresponding to X_i, i.e. the $i-th$ interval input of the neuron.

Two surfaces can be identified in the pattern space: one corresponding to $y^L = 0$ and the other to $y^U = 0$.

Let the following constraints be satisfied:

C1 In the region where $y^L < 0$ e $y^U < 0$, only patterns with negative target can be present.

C2 In the region where $y^L > 0$ e $y^U > 0$, only patterns with positive target can be present.

C3 The number of classification errors related to the y^L output must be minimum.

C4 The number of classification errors related to the y^U output must be minimum.

Classification error means the number of patterns in the training set which give, in the considered output, a different sign with respect to the sign of the target.

Through constraints C3 and C4, the number of training patterns assigned in the regions defined by constraints C1 and C2 is maximized. Once these four constraints are satisfied, two sets of degrees of freedom remain to be fixed. The first set regards the existence of an infinite number of of hypersurface couples between the patterns of the certain regions and those of the uncertain one. This situation is similar to the one which led Vapnik [1] to formulate the *maximum margin* search algorithm. The use of a maximum margin classifier should be convenient, since its generalization capability is considered to be the higher in the present state of art. Similarly to the Vapnik maximum margin formulation [1] and to the criterium used for the Interval Arithmetic Perceptron [3], besides satisfying the four constraints, the optimal surfaces couple features the property of maximizing the volume of the space corresponding to the UR. In terms of IA, this property can be described by the following condition

$$\max_{W_i} \sum_{ij} (w_{ij}^U - w_{ij}^L) \tag{2}$$

with the constraints

$$\sum_{ij} (w_{ij}^L)^2 = \sum_{ij} (w_{ij}^U)^2 = 1 \tag{3}$$

Where the sum on i, j means considering all the weights and the biases of the MLP.

The search for the solution to this problem is very complex and needs to be split into different phases. In the first phase, a weight vector is identified which is able to satisfy constraints C1 and C2. Subsequently a problem of constrained minimum is established that consists in finding an interval weight vector which minimizes the two errors described in constraints C3 and C4, while satisfying constraints C1 and C2. Once the two minimum errors are found, it is possible to maximize expression 2 by imposing the condition that the four constraints are satisfied, together with the requirement that expressions 3 are both equal to 1. Baba's algorithm [6] has been chosen as optimization routine for maximizing expression 2. This algorithm is not particularly efficient compared to algorithms like simulated annealing or genetic algorithms [3], but it has the advantage of being very simple and it proved very suitable, probably due in part to the very satisfying starting point supplied by the AMBP algorithm [5]. It should be remembered however that the Baba method is able to find the global minimum and avoid getting trapped within local minima, due to its random search feature. Finally, satisfying constraints with the Baba algorithm is very simple, since it is sufficient to define forbidden search regions for the interval weight search vector.

3 Computer Simulation

Two examples will be considered: an artificial problem and a real-world application whose dataset has been obtained from the UCI repository of machine learning databases (Murphy & Aha), available through the Internet. In the three examples all the input data have been normalised within the range $[-1, 1]$. In the figures the symbol I is used to indicate the separation line obtained at the IAMLP weight initialization (after AMBP), while, after the IAMLP weight training, the separation lines are indicated by the symbols L and U, which refer to the lower and upper values of interval weight vector. Moreover, the certainty regions (CR) are coloured grey, while the uncertainty one (UR) is white. Breast Cancer tests has been studied both using all the nine original inputs and only two ones. In the second case, the two inputs, which have the property of maximizing the recognition capability, have been identified using an algorithm described in [7]. In this paper, due to the lack of space, we present only the results obtained studying the problem with this reduced input number offers, with the advantage of supply the reader with a visual easily-interpretable representation. Moreover, it is well known as the reduction of the input number favors the decrease of the generalization error, even though an increase of the classification error on the training set occurs.

3.1 The artificial test

Samples of class 1 (filled circles) are randomly distributed around two centers (of coordinates $-1/\sqrt{3}$, 0.5 and $1/\sqrt{3}$, 0.5 rspectively) through a Gaussian random generator (standard dev.=0.3). Class 2 samples are similarly distributed around the center of coordinates 0, -0.5. (The three centers form an equilateral triangle). An IAMLP with 3 hidden units has been used. The result is shown in Fig.1. Within the UR (white-coloured), one can see both undersampled sub-regions, for instance on the left, and sub-regions where samples of the two classes are similarly distributed, expecially along the line I (obtained through the real weight MLP). The central prominence of the UR is due to the presence of isolated sampled whose class is different from that of the majority of their neighbours.

3.2 The Breast Cancer test

This real-world problem regards the data of the Breast Cancer file. This file includes 683 patterns, of which 444 are Benign and 239 Malignant (after removing 16 patterns with featuring missing inputs). Two of the 9 inputs maximize the recognition capability: the *Bare Nuclei* and the *Clump Thickness* [7]. An IAMLP with 2 hidden units has been used. The result is shown in Fig.2, where circles indicate those points where the number of malignant cases surpasses that of benign ones, crosses indicate the opposite occurrence, while the filled circles show points where the two numbers are equal. Within the UR, samples of both classes are present: among them, in particular, there are the all the (six) filled circles, where the two classes are equally represented. More details about this application may be found in [7]. A comparison with the result obtained with an Interval Arithmetic Perceptron can be made by looking at Fig.6 of [3]) (not shown here due to the lack of space): the IAP itself detects the UR with remarkable accuracy, due to the undersampled sub-region on the left of the UR.

4 Conclusions

Using an Interval Arithmetic MLP allows the detection of the uncertainty regions in the input space and consequently represents a powerful tool in order to estimate the quality and the reliability of any classifier decision.

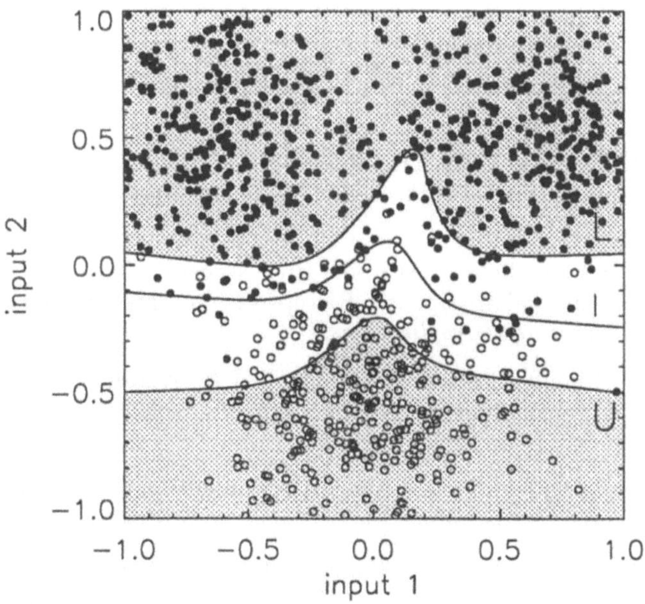

Figure 1: The artificial test. White and black circles represent class 1 and class 2 patterns respectively. The two certainty regions are grey-shaded, while the UR is white. I is the separation line after the weight initialization.

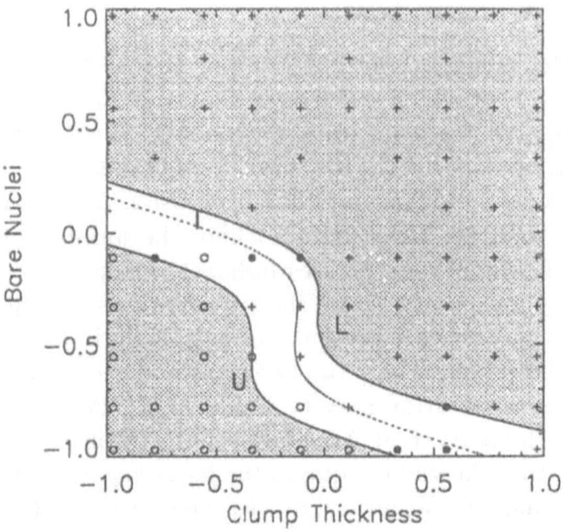

Figure 2: The Breast Cancer test taking account of only two inputs and modifying the the original targets. The two certainty regions are grey-shaded, while the UR is white. I is the separation line after the weight initialization. Circles indicate those points where the number of malignant cases surpasses that of benign ones, crosses indicate the opposite occurrence, while the filled circles show points where the two numbers are equal.

References

[1] Vapnik V.N., The Nature of Statistical Learning Theory, *New York: Springer-Verlag*, 1995.

[2] Ishibuchi H., Fujioka, Tanaka H., Possibility and necessity pattern classification using neural networks, *Fuzzy Nets and Systems*, 48, 1992, pp 331-340.

[3] Drago G.P. and Ridella S., Possibility and Necessity Pattern Classification using an Interval Arithmetic Perceptron, accepted for publication on *Neural Computing & Applications.*

[4] G.P.Drago, S.Ridella, Statistically Controlled Activation Weight Initialization, *IEEE Transactions on Neural Networks*, Vol.3, no. 4, July 1992, pp 627-631.

[5] Drago G.P. and Ridella S., An Adaptive Momentum Back Propagation (AMBP), *Neural Computing & Applications*, no.3, 1995, pp.213-221.

[6] Baba N., A new approach for founding the global minimum of error function of neural networks, *Neural Networks*, Vol.2, pp 367-373, 1989.

[7] Drago G.P. and Ridella S., Pruning with Interval Arithmetic Perceptron, *Neurocomputing*, Vol.18, no.1-3, 1998, pp. 229-246.

Polynomial Clusterons Exhibit Statistical Estimation Abilities

Simone Fiori and Pietro Burrascano *

Dipartimento di Ingegneria Industriale, Univesità di Perugia - Perugia, Italy
E-MAIL: SFR@UNIPG.IT, BURRASC@UNIPG.IT

Abstract. The aim of this paper is to investigate the behavior of neural clusterons that learn in an unsupervised fashion by means of an information-theoretic based rule. Particularly the aim is to investigate on the clusteron's statistical estimation abilities that naturally emerge from their learning behaviors.

1 Introduction

The concept of *clusteron* has been introduced by B.W. Mel in 1992 (see [12]) in order to take into account the non-linear computation performed by the dendritic trees in biological neurons. A clusteron consists of a cell-body where the global response is computed, and a dendritic tree that receives a set of excitatory weighted synaptic contacts from a set of afferent axons. The total activation value of a clusteron is computed according to the observation that it behaves as an approximative low-order polynomial function [12]. Other similar models have been proposed by different researchers, and the clusterons are also known as functional-link units [15,20] and σ-π neural units [3].

Various applications involve clusteron-based neural networks in engineering [4], ranging from the field of automatic control (see for instance [9]) to the area of communications (see e.g. [10]). Statistics estimation by non-parametric functions is a widely addressed problem in the Neural Networks literature [2,14,1,13,16–19,11]. Particularly, over the recent years much attention has been paid to *probability density function*, *cumulative distribution function* and *score function* approximation by neural networks applied to Blind Source Separation (see [2,18,5,8,6,7] and references therein).

In this paper we investigate the behavior of a clusteron that adapts by means of an *unsupervised* learning paradigm based on the minimization of the Kullback-Leibler Divergence (KLD) that is equivalent to maximum entropy criterion. Through computer simulations performed on both synthetic and natural data, the effectiveness of the proposed information-theoretic based learning theory is assessed, and the flexibility and approximation capability of the clusteron in absence of relevant information about the mapping to be learnt are illustrated.

* This research was financially supported by the Italian MURST.

2 Minimum KLD Connection Strengths Learning

As modeled here, the response of a polynomial clusteron is described by the relationship:

$$y = g(x) = \frac{1}{2} + \frac{1}{2}\text{th}[q(x)] \ , \tag{1}$$

where $x(t) \in \mathcal{X}$ represents the external excitation, $q(x)$ is a polynomial whose coefficients are adaptively changed through time according to an unsupervised learning rule, and $\text{th}(\cdot)$ denotes the hyperbolic tangent function. The excitation $x(t)$ is supposed to be a stationary random process characterized by a probability density function $f_x(x)$ and a cumulative distribuition function defined by $F_x(x) \stackrel{\text{def}}{=} \int_{-\infty}^{x} f_x(\xi)d\xi$.

For consistency purpose, here it is required that the polynomial $q(x)$ must be monotonically increasing almost everywhere within the range of $x(t)$, thus $q'(x) = dq/dx \geq 0$. This condition is surely fulfilled if $q'(x)$ has all positive coefficients and the variable x appears on it at powers of two only. Here the following expression for $q(x)$ is assumed:

$$q(x) = c_0 + \sum_{i=1}^{r} \phi(c_i) \left(\frac{x-\mu}{\nu}\right)^{2i-1} \ , \tag{2}$$

where $2r - 1$ is the order of the polynomial, $(c_0, c_1, \ldots, c_r)$ are variable parameters, $\phi(\cdot)$ is whatever (almost everywhere) positive and differentiable function, and (μ, ν) are fixed constants used for contracting the range $\mathcal{X}$ to $[-1, +1]$, as recommended in [15]. A schematic of such clusteron with proper connection-strength variables is depicted in Figure 1, where each block marked with a Π evaluates the product of its inputs, while blocks marked with a Σ perform summations.

The target is to approximate $F_x(x)$ by $g(x)$, that is the same of approximating $f_x(x)$ by $\psi(x) \stackrel{\text{def}}{=} dg/dx$. The function $\psi(x)$ depends upon coefficients $(c_0, c_1, \ldots, c_r)$, that may be adaptively changed in order to make $\psi(x)$ as close as possible to $f_x(x)$. To this aim, the following cost function may be considered as closeness measure:

$$\mathcal{D}(f_x||\psi) \stackrel{\text{def}}{=} \int_{\mathcal{X}} f_x(\xi) \log \frac{f_x(\xi)}{\psi(\xi)} d\xi = -\text{h}_x - E_x[\log \psi] \ , \tag{3}$$

where h_x is the differential entropy of x, that does not depend upon coefficients $(c_0, c_1, \ldots, c_r)$, and the symbol $E_x[\cdot]$ denotes mathematical expectation w.r.t. x under the hypothesis $(c_0, c_1, \ldots, c_r)$. The functional $\mathcal{D}(u||v)$ is known as the Kullback-Leibler Divergence (KLD) between regular distributions u and v. Note that by construction $\psi(x) > 0$ almost everywhere, thus $\mathcal{D}$ is well-defined. It is known that $\mathcal{D}(u||v) \geq 0$ with the equality if and only if $u = v$.

To minimize $\mathcal{D}$ with respect to $(c_0, c_1, \ldots, c_r)$ by means of the Gradient Steepest Descent rule $\Delta c_i = -\rho \partial\mathcal{D}/\partial c_i$, where ρ is a positive factor and $i =$

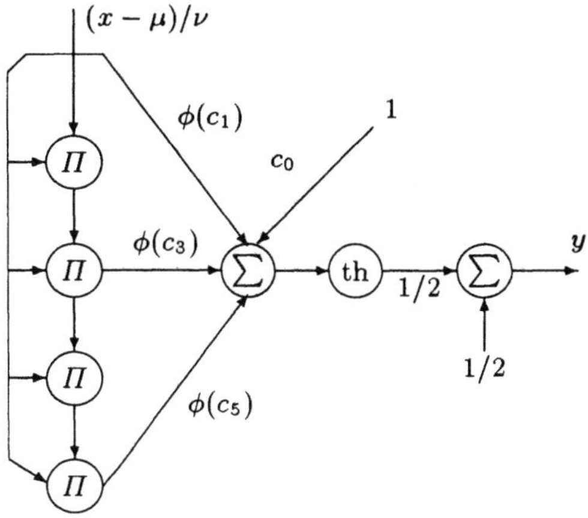

Fig. 1. The clusteron associated to equation (1) (shown for $r = 3$).

$0, 1, \ldots, r$, requires derivatives:

$$\frac{\partial \mathcal{D}}{\partial c_i} = -E_x \left[\frac{1}{\psi} \frac{\partial \psi}{\partial c_i} \right] .$$

Direct calculations allow finding the required quantities as:

$$\psi(x) = \frac{1}{2}\{1 - \text{th}^2[q(x)]\}q'(x) \ , \ \nu q'(x) = \sigma(x) \ ,$$
$$\sigma(x) \stackrel{\text{def}}{=} \sum_{i=1}^{r} (2i-1)\phi(c_i) \left(\frac{x-\mu}{\nu}\right)^{2i-2} ,$$

that lead to the following expression:

$$\frac{1}{\psi}\frac{\partial \psi}{\partial c_i} = -2\text{th}(q)\frac{\partial q}{\partial c_i} + \frac{1}{\sigma}\frac{\partial \sigma}{\partial c_i} .$$

Partial derivatives are then found to be:

$$\frac{\partial q}{\partial c_0} = 1 \ ; \ \frac{\partial q}{\partial c_i} = \phi'(c_i)\left(\frac{x-\mu}{\nu}\right)^{2i-1} \ , \ i \geq 1 \ ,$$
$$\frac{\partial \sigma}{\partial c_0} = 0 \ ; \ \frac{\partial \sigma}{\partial c_i} = (2i-1)\phi'(c_i)\left(\frac{x-\mu}{\nu}\right)^{2i-2} \ , \ i \geq 1 \ ,$$

where $\phi'(t) = d\phi/dt$, hence it is easily seen that:

$$\frac{1}{\psi}\frac{\partial \psi}{\partial c_0} = -2(2y-1) \ , \ \text{while for } i \geq 1 \ : \tag{4}$$

$$\frac{1}{\psi}\frac{\partial \psi}{\partial c_i} = \left(\frac{x-\mu}{\nu}\right)^{2i-2} \phi'(c_i) \left[\frac{2i-1}{\sigma} - 2(2y-1)\left(\frac{x-\mu}{\nu}\right)\right] . \tag{5}$$

In order to implement learning equations $\Delta c_i = \rho E_x[\psi^{-1}(\partial\psi/\partial c_i)]$, the expectation is dropped, thus equations (4)-(5) directly represent the required learning terms.

3 Discussion on the Proposed Theory

It is worth noticing that also the quantity $\sigma(x)$ may be computed by the same neural architecture depicted in Figure 1 where some additional elements have to be considered, as shown in Figure 2.

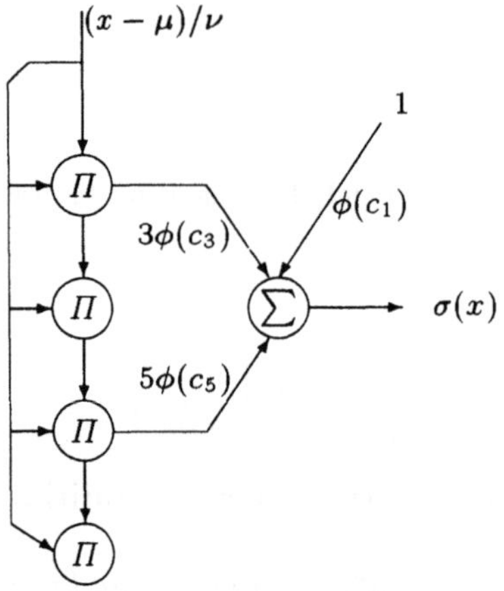

Fig. 2. Adjoint elements for computing function $\sigma(x)$ (shown for $r = 3$).

It is known [17] that if a cumulative transformation corresponding to a random process $x(t)$ is applied to $x(t)$ itself, that is $F_x[x(t)]$, then the resulting random process has a flat (uniform) density distribution. This result may be considered as an interesting effect in communications [10], where uniformly distributed signals are extensively used, and in other fields, as in data-bases applications [5]. Note that if coefficients $(c_0, c_1, \ldots, c_r)$ may change continuously through time, the hypothesis that $x(t)$ is stationary may be relaxed, and a slow variability of its probability density function may be allowed.

4 Numerical simulations

The aim of this part is to simulate the proposed clusteron's structure and learning rule in order to test for its statistics estimation capability.

For the following simulations, common choices were: $r = 2$ (that is a low-order polynomial, according to Mel), 0.01 as initial value for the coefficients c_i, and $\phi(t) = t^2$.

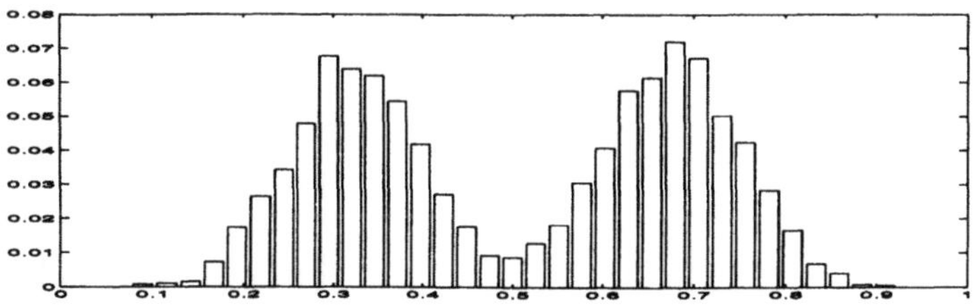

Fig. 3. Bi-Gaussian stimulus.

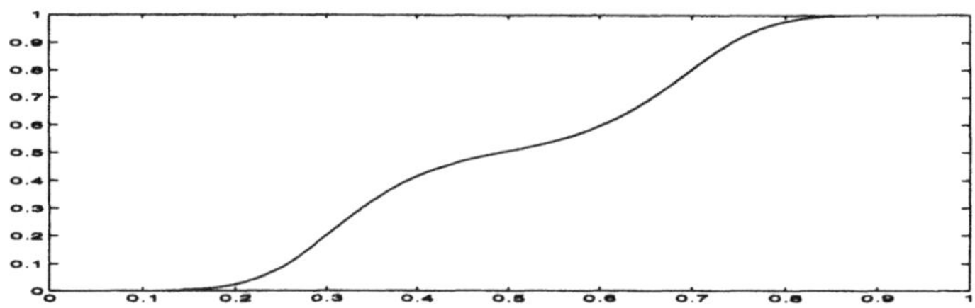

Fig. 4. Function $g(x)$ in the bi-Gaussian case.

First, the bi-Gaussian excitation, whose distribution histogram is shown in Figure 3, was presented to the functional-link neuron. For this simulation parameters were chosen as: $\mu = \nu = 0.5$ and $\rho = 0.04$. In Figure 4 the approximating function $g(\cdot)$ is drawn.

Figures 5 and 6 depict instead the bi-constant input density histogram and the obtained $g(x)$ for $\mu = \nu = 0.5$ and $\rho = 0.04$. The bi-constant excitation is the most difficult one to be treated due to the large amount of lacking samples within the central 'valley'.

Figures 7 and 8 depict the result obtained for a natural excitation, that is a sampled human voice stream, with $\mu = 0$, $\nu = 0.5$ and $\rho = 0.02$.

5 Conclusion

In this paper we studied neural clusterons that learn in accordance to an information-theoretic based optimization principle. In forthcoming works it will be interesting to investigate the property that emerges from Section 3: As equations in [2] are a particularization of the more general paradigm (4)-(5), Bell-Sejnowski's Blind Source Separation algorithm could be a particularization of a more general one based on the approach explained here. Preliminary results along this way may be found in [6–8].

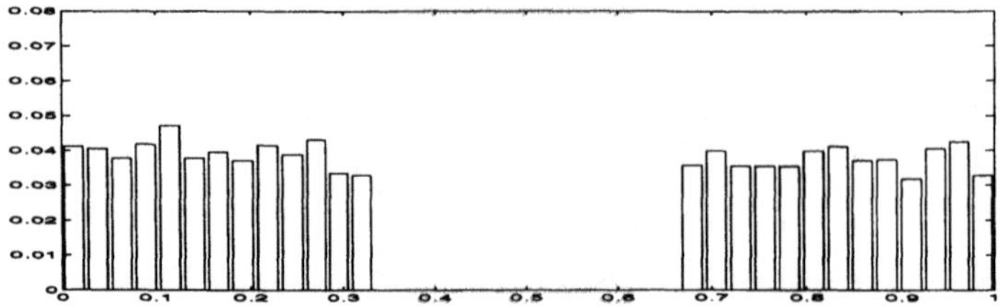

Fig. 5. Biconstant stimulus.

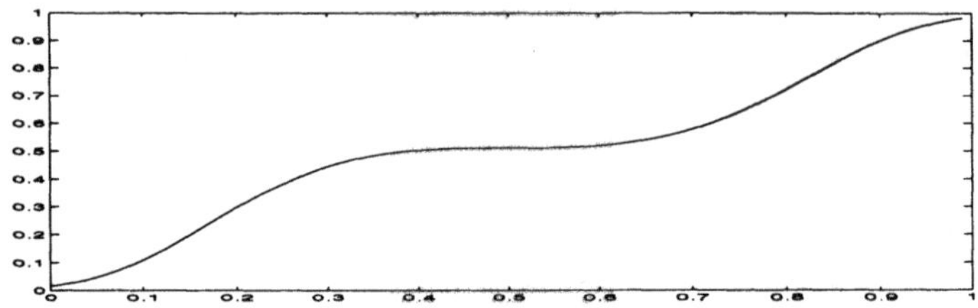

Fig. 6. Function $g(x)$ in the biconstant case.

References

1. M. Battisti, P. Burrascano, and D. Pirollo, *Efficient minimisation of the KL distance for the approximation of posterior conditional probabilities*, Neural Processing Letters, Vol. 5, pp. 47 – 55, 1997
2. A.J. Bell and T.J. Sejnowski, *An information maximisation approach to blind separation and blind deconvolution*, Neural Computation, Vol. 7, No. 6, pp. 1129 – 1159, 1996
3. J.L. McClelland and D.E. Rumelhart (Eds.), *Parallel distributed processing*, Vol. I – Foundations, The MIT Press, 1986
4. J.A. Feldman and D.H. Ballard, *Connectionist models and their properties*, Cognitive Science, Vol. 6, pp. 205 – 254, 1982
5. S. Fiori, P. Bucciarelli, and F. Piazza, *Blind signal flatting using warping neural modules*, Proc. of International Joint Conference on Neural Networks (IJCNN), pp. 2312 – 2317, 1998
6. S. Fiori, *Blind Source Separation by New M–WARP Algorithm*, Electronics Letters, Vol. 35, No. 4, pp. 269 – 270, Feb. 1999
7. S. Fiori, *Entropy Optimization by the PFANN Network: Application to Blind Source Separation*, Network: Computation in Neural Systems, Vol. 10, No. 2, pp. 1 – 16, Apr. 1999
8. S. Fiori, P. Baldassarri, and F. Piazza, *An efficient architecture for independent component analysis*, To appear on the Proc. of Int. Symposium on Circuits and Systems (ISCAS), May 30 - June 2, 199, Orlando (Florida - USA)
9. E.B. Kosmatopoulos, A. Chassiakos, H. Boussalis, M. Mirmirani, and P.A. Ioannou, *Neural network control of unknown systems*, Proc. of International Joint Conference on Neural Networks (IJCNN), pp. 943 – 948, 1998

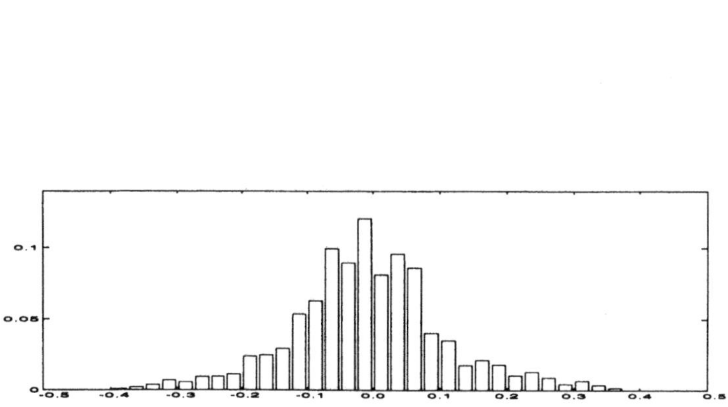

Fig. 7. Human voice excitation.

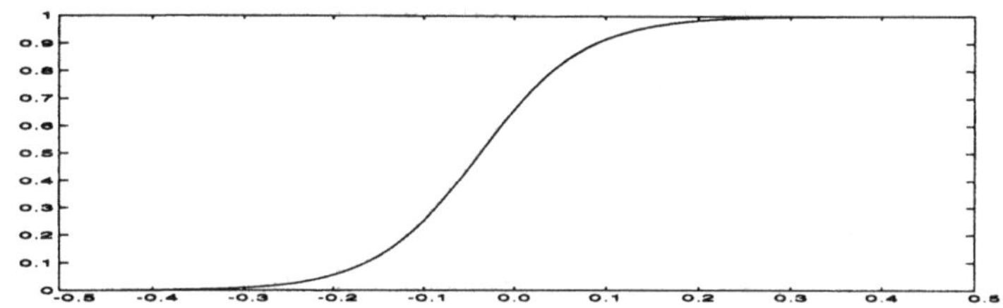

Fig. 8. Function $g(x)$ in the real-world case.

10. P. Kulczycki and H. Schioler, *Estimating conditional distributions by neural networks*, Proc. of International Joint Conference on Neural Networks (IJCNN), pp. 1344 - 1349, 1998
11. S. Laughlin, *A Simple Coding Procedure Enhances a Neuron's Information Capacity*, Z. Naturforsch, Vol. 36, pp. 910 - 912, 1981
12. B.W. Mel, *Information processing in dendritic trees*, Neural Computation, Vol. 6, pp. 1031 - 1085, 1994
13. G. Miller and D. Horn, *Probability density estimation using entropy maximization*, Neural Computation, Vol. 10, pp. 1925 - 1938, 1998
14. P. Moreland, *Mixture of experts estimate a-posteriori probabilities*, Artificial Neural Networks, pp. 499 - 505, Springer-Verlag, 1997
15. Y.-H. Pao, *Adaptive Pattern Recognition and Neural Networks*, Addison-Wesley Publishing Company, 1989 (Chpt. 8)
16. Z. Roth and Y. Baram, *Multidimensional density shaping by sigmoids*, IEEE Trans. on Neural Networks, Vol. 7, No. 5, pp. 1291 - 1298, Sept. 1996
17. A. Sudjianto and M.H. Hassoun, *Nonlinear Hebbian rule: a statistical interpretation*, Proc. of Int. Conf. on Neural Networks (ICNN), Vol. 2, pp. 1247 - 1252, 1994
18. A. Taleb and C. Jutten, *Entropy optimization - Application to source separation*, Artificial Neural Networks, pp. 529 - 534, Springer-Verlag, 1997
19. Y. Yang and A.R. Barron, *An asymptotic property of model selection criteria*, IEEE Trans. on Information Theory, Vol. 44, No. 1, Jan. 1998
20. S.M. Zurada, *Introduction to Neural Artificial Systems*, West Publishing Company, 1992

The N-SOBoS model

F. Frisone, P. G. Morasso
DIST - Department of Informatics, Systems and Telecommunicatio
University of Genova
Via Opera Pia 13, 16145 Genova, Italy
V: +39 10 3532749; F: +39 10 3532154; E:{friso,piero}@dist.unige.it

Abstract

The purpose of this work is to outline a computational architecture for the intelligent processing of sensorimotor patterns. The focus is on the nature of the internal representations of the outside world which are necessary for planning and other goal-oriented functions. A model named *N-SOBoS* (new self-organizing body-schema), based on the SOBoS model [10] and on the dual Extended Topology Representing Network architecture is proposed, which integrates a number of concepts and methods partly explored in the field [15, 11, 12]. The novelty and the biological plausibility is related to the global architecture which allows to deal with sensorimotor patterns in a coordinate-free way, using population codes as internal representations and communication channels among different cortical maps.

1 Introduction

Since the times of N. Bernstein, two key problems for a theory of motor control are, respectively, the formation of coordinate transformations between different motor spaces and the reduction of the excess degrees of freedom implied by them.

The problem of sensory-motor transformations has been addressed from the point of view of learning, in the context of neural network models. Several contributions [6, 4] were inspired by the principle that, during active movements, there is a natural association between the efferent motor patterns and the (re)-afferent patterns that measure their sensory consequences.

This efference-reafference cycle, known as *circular-reaction* and colorfully called *sensory-motor dialogue*, allows humans, particularly during early development, to learn sensory-motor transformations via an active exploration of the environment. From the computational point of view, circular-reaction can be seen as a self-organizing or self-supervised strategy for learning the association between sensory stimuli, taken as targets, and the movements which are able to acquire that targets: the basic idea is self-teaching by means of un-targeted (or pseudo-random) active movements that provide the brain with coherent sensory and motor patterns.

Based on the previous suggestion of [13] that motor commands may be generated by performing a 'mental simulation' on an internal model of the body[1], this work explores a comprehensive architecture which is organized as a computational map [5] and combines a paradigm of *self-organization* (for building robust and coherent associative maps of the different motor spaces, representing the internal model) with *attractor-based dynamics*, or gradient-descent in a potential field (for the run-time integration of task constraints) which includes a dynamic mechanism for the generation of target signals.

2 N-SOBoS architecture

We now briefly outline a possible computational mechanism which exploits the lateral connectivity of the TRN [8] model, thus behaving as a Cellular Automaton (CA) or, more precisely, as a *coupled map lattices* because space and time are discrete but the state variables vary in a continuous way. In other words, we can view the topology-preserving cortical map not merely as a mechanism for a "passive" topographic representation of space but as a dynamic computational engine which operates on the population code $\mathcal{P}$, shifting it on the field as a *soliton*, according to local interaction rules. This concept is synthesized in the ETRN (Extended Topology Representing Network) [3] model.

Let us address the problem of how the cortical spatial map might plausibly solve the problem of generating the internal representation of a virtual trajectory, as a response to the presentation of a target $\vec{s}_T$. This stimulus point, perceived through the different sensory channels, is reflected in the map as a cluster of activity $\mathcal{P}_T$, centered around a neuron of receptive field center $\vec{\pi}_T^x$. Moreover, we may think that the initial position of the end-effector is represented by a similar cluster $\mathcal{P}_0$, centered around another neuron of receptive field center $\vec{\pi}_0^x$. The problem, then, is to smoothly shift $\mathcal{P}_0$ until it reaches $\mathcal{P}_T$ and this can be achieved by means of the attraction dynamics determined by a convergent field, such as the gradient of a *distance field* from the target $\delta = \delta(\vec{x}; \vec{x}_T)$, which measures the Euclidean distance of the end-effector from the target and operates as a Liapunov function for the map of neurons. This distance is only indirectly accessible, via the connectivity of the map, which is a topological or gestaltian knowledge. Therefore, we need a mechanism for translating gestaltian knowledge, relevant at the population level, into local knowledge, exploitable at the neuron level.

The natural solution is to establish an analogy between Euclidean distances in the external space and distances in the graph of neurons which represents that space. Given the target $\vec{x}_T$ (represented by its population code), the distance field can be expressed by the map using a codebook of distance values: $\Pi^\delta = \{\pi_i^\delta,\ i \in \mathcal{I}\}$, but how can we assign each neuron ν_j in the map such a distance estimate? This function can carried out in a parallel and distributed way, which only involves local computations, by exploiting the topographical connections of the network, i.e. as a CA which carries out a diffusion process of

[1] Mental analogic simulations, as general methods of reasoning and planning, have found support in different areas [16, 1, 2].

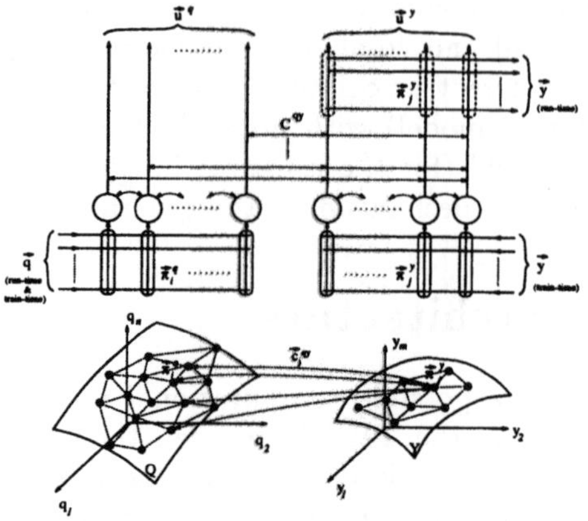

Figure 1: Dual-ETRN architecture: representation of the forward model $\vec{y} = \mathcal{K}(\vec{q})$. Q: input manifold; Y: output manifold. $\vec{q}$: input vector; $\vec{u}^q$: input population code; $\vec{y}$: output vector; $\vec{u}^y$: output population code; $\vec{\pi}_i^q$: prototype input vector; $\vec{\pi}_j^y$: prototype output vector; $\vec{c}_j^{qy}$: inter-connection vector to neuron ν_j^y from $\mathcal{N}^q$. On-line flow: $\vec{q} \rightarrow \vec{u}^q \rightarrow \vec{u}^y \rightarrow \vec{y}$. Continuous ovoids: soft-max input transfer function. Dashed ovoids: linear output transfer function. Dotted vectors: intra and inter-connections.

local distance information. In particular, in a preliminary simulation we used a parallel version of the well known Dijkstra algorithm, which propagates to the nodes of a graph of neurons local estimates of distance:

$$\pi_j^\delta(t+1) = \min\{\pi_j^\delta(t), \min_{k \in \mathcal{P}(j)} (\pi_k^\delta(t) + \delta_{jk})\} \tag{1}$$

where $\mathcal{P}(j)$ is the set of neurons which are neighbors of neuron ν_i and the initial state is characterized by the fact that π_T^δ is set to 0 and the distance values of all the others are set to a very large number. After diffusion, the shortest path from any neuron to the target, similarly to the Dijkstra algorithm, is simply found by jumping from neuron to neuron in order of decreasing $\vec{\pi}_j^\delta$ values.

We show now, how the proximal and distal manifolds, together with the corresponding fields, can be represented by means of a pair of ETRNs: one for the proximal manifold Q and the other for the distal manifold Y, which are trained according to the ETRN equations (see [3]):

$$\Delta\vec{\pi}_i^x = \eta(\vec{x} - \vec{\pi}_i^x)U_i^x \tag{2}$$

and the lateral connections among neurons

$$\Delta\vec{c}_i^{\,x} \propto U_i U_j \quad i, j \in \mathcal{I}^x \tag{3}$$

thus, as shown in figure 1, we have two population codes, U^q and U^y, and two codebooks of prototype vectors, $\Pi^q = \{\vec{\pi}_i^q, i \in \mathcal{I}^q\}$ and $\Pi^y = \{\vec{\pi}_j^y, j \in \mathcal{I}^y\}$. In addition, we propose to apply the same Hebbian rule 3, used for learning the

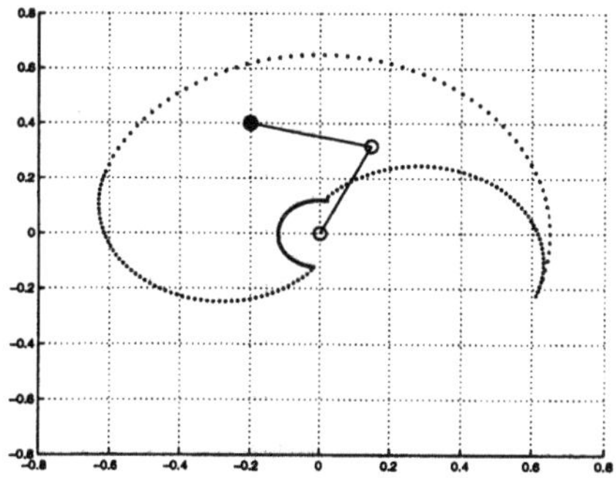

Figure 2: Work space for a 2 degrees of freedom planar arm. The angles range for the two joints are respectively: $(-\frac{\pi}{9}; \frac{8}{9}\pi)$ and $(0; \frac{8}{9}\pi)$.

intra-connections, $C^y = \{\vec{c}_i^{\,y},\ i \in \mathcal{I}^y\}$ and $C^q = \{\vec{c}_j^{\,q},\ i \in \mathcal{I}^q\}$, also for learning a set of cross or inter-connections C^{qy} among the two networks[2]. The set of proximal neurons converging on a given output neuron ν_j can be characterized by a inter-connectivity vector $\vec{c}_j^{\,qy}$: they are a *quantization of the no-motion manifold* of the forward model for the corresponding value of the distal output vector $\vec{y}$. The population code U^y of this vector is computed by projecting, for each output neuron ν_j^y, the proximal population code U^q onto the cross-connectivity vector $\vec{c}_j^{\,qy}$:

$$U_j^y = \vec{c}_j^{\,qy} \cdot U^q = \sum_{i \in \mathcal{P}^q} c_{ij}^{qy} U_i^q \tag{4}$$

which preserves the normalization of the output population code.

3 Simulation Results

Let us consider, for example, the planar arm with two joints of figure 2. Figure 3 shows the ETRN representing the proximal space.

Figure 4 shows the result of the combined relaxation in the two maps to a multiple target. The whole process was started by selecting three targets, indicated by the crosses, representing a closed triangular trajectory. The planned trajectory, reconstructed by means of the time-varying population code, is slightly bended, staying away from the boundary of the workspace, which would be reached by attempting to impose a straight path from the initial to the final position. Figure 5 shows the population code dynamics of the dual-map in three different instants.

[2]The two maps are trained simultaneously, with pairs of input vectors. For each sample pair there is a pair of "winners" in the two maps, respectively, whose connection c_{ij} is increased proportionally to the product of the activation values. All the other lateral connections are decreased according to an exponential decay term. The surviving connections, at the end of the training, quantize the no-motion manifold.

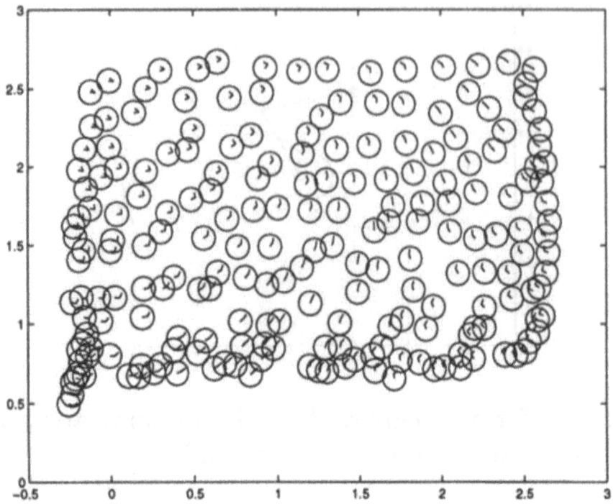

Figure 3: Body-icons learned by the cortical map of a planar arm with 2 degrees of freedom

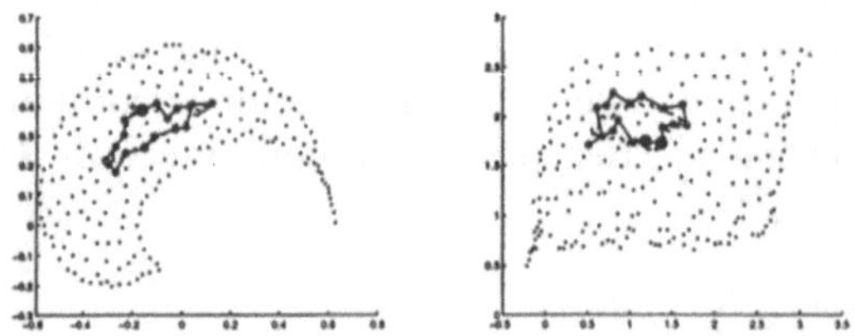

Figure 4: Trajectory formation in the dual-ETRN architecture: proximal (right) and distal (left) space . Solid line represents the trajectory carry out by the population code peak; dashed line represents the trajectory recovered by a interpolation formula. Crosses are the three targets representing a closed triangular trajectory.

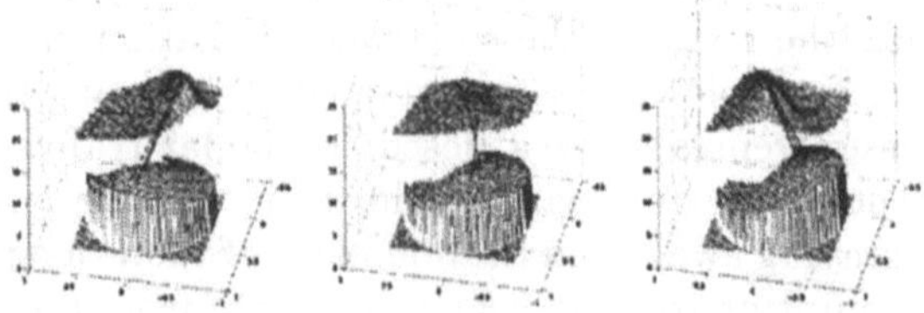

Figure 5: The population code dynamics, of the dual-map in three different instants. A sample of the inter-connections is also shown, which links an arbitrary node in the proximal ETRN (top) with the corresponding nodes in the distal ETRN (bottom).

This is a significant example, even in such elementary model with a very limited number of neurons, of the way in which the dynamic system can automatically take into account the mechanical constraints, combining them with the task-related attractor field to the target.

4 Conclusion

The N-SOBoS model has a similar representing power of others models but is more compact and has an explicit representation of the topologies of both manifolds, Q and Y. With respect to several neural network models of sensorimotor planning, which use multi-layer perceptron for the forward model, we can observe the following: the perceptron is an efficient approximator of non-linear mappings but is not compatible with a field computation based approach which, on the contrary, is naturally suited for an architecture based on computational maps. In a sense, this is another version of the computation/memory trade-off.

References

[1] J. Decety and D.H. Ingvar. Brain structures partecipating in mental simulation of motor behavior: a neuropsychological interpretation. *Acta Psychologica*, 73:13–34, 1990.

[2] J. Decuyper and D. Keymeulen. A reactive robot navigation system based on a fluid dynamics metaphor. In Hans-Paul Schwefel and Reinhard Maenner, editors, *Parallel Problem Solving from Nature*, pages 348–355. Springer, 1991.

[3] F. Frisone and P.G. Morasso. Extending the TRN model in a biologically plausible way. In *ICANN97- Int. Conf. on Artificial Neural Networks*, pages 201–206, Lousanne, October 1997. Springer.

[4] P. Gaudiano and S. S. Grossberg. Vector associative maps: unsupervised real-time error-based learning and control of movement trajectories. *Neural Networks*, 4:147–183, 1991.

[5] E. I. Knudsen, S. du Lac, and S.D. Esterly. Computational maps in the brain. *Annual Review of Neuroscience*, 10:41–65, 1987.

[6] M. Kuperstein and J. Rubinstein. Implementation of an adaptive neural controller for sensory-motor coordination. In R. Pfeifer, Z. Schreter, F. Fogelman-Soulie, and L. Steels, editors, *Connectionism in Perspective*, pages 49–61. Elsevier Science Publishers, Amsterdam, 1989.

[7] T. Martinetz and K. Schulten. A 'Neural-Gas' network learns topologies. In T. Kohonen, K. Makisara, O. Simula, and J. Kangas, editors, *Artificial Neural Networks*, Amsterdam, 1991. North-Holland.

[8] T. Martinetz and K. Schulten. Topology Representing Networks. *Neural Networks*, 3:507–522, 1994.

[9] P. Morasso and V. Sanguineti. Neurocomputing concepts in motor control. In J. Paillard, editor, *Brain and Space*, pages 404–432. University Press, Oxford, UK, 1991.

[10] P. Morasso and V. Sanguineti. Self-organizing body-schema for motor planning. *Journal of Motor Behavior*, 26:131–148, 1993.

[11] P. Morasso and V. Sanguineti. Kinematic invariances and body schema. *Behavioral and Brain Sciences*, 18:769–770, 1995.

[12] P. Morasso and V. Sanguineti. How the brain can discover the existence of external egocentric space. *Neurocomputing*, 12:289–310, 1996.

[13] F. A. Mussa Ivaldi, P. Morasso, and R. Zaccaria. Kinematic networks – A distributed model for representing and regularizing motor redundancy. *Biological Cybernetics*, 60:1–16, 1988.

[14] H. Ritter, T. Martinetz, and K. Schulten. Topology conserving maps for learning visuo-motor coordination. *Neural Networks*, 2:159–168, 1989.

[15] V. Sanguineti, T. Tsuji, and P. Morasso. A dynamical model for the generation of curved trajectories. In S. Gielen and B. Kappen, editors, *International Conference on Artificial Neural Networks*, pages 115–118, London, 1993. Springer.

[16] R.N. Shepard and L.A. Cooper. *Mental Images and their transformations.* MIT Press, Cambridge, MA, 1982.

A Neural Network approach to detect functional MRI signal

F. Frisone, P. G. Morasso
DIST - Department of Informatics, Systems and Telecommunicatio
University of Genova
Via Opera Pia 13, 16145 Genova, Italy
V: +39 10 3532749; F: +39 10 3532154; E:friso@dist.dist.unige.it

P. Vitali, G. Rodriguez
DIMI - Department of Internal Medicine, Service of Clinical Neurophysiology
University of Genova
Ospedale S. Martino, L.go R. Benzi 10, 16132 Genova, Italy

A. Pilot
Department of Medical Physics
Ospedale S. Martino, L.go R. Benzi 10, 16132 Genova, Italy

F. Sardanelli
Radiology Institute
Ospedale S. Martino, L.go R. Benzi 10, 16132 Genova, Italy

M. Rosa
Service of Neuroradiology
Ospedale S. Martino, L.go R. Benzi 10, 16132 Genova, Italy

Abstract

In fMRI the key problem of data analysis is to detect the weak BOLD signal component (about 2-5%) in the MR signal. Standard approaches, that typically use cross-correlation analysis or statistical parametric mapping, imply a presumptive knowledge of the expected stimulus-response pattern, which is not available in spontaneous events like hallucinations, sleep, or epileptic seizures. To evidence the possibility of analyzing these events by means of fMRI, we investigated a computational approach based on a self-organizing neural network (Neural Gas) that detects time-dependent alterations in the regional intensity of the functional signal.

1 Introduction

The recent development of event-related fMRI technique [1], allows now to investigate in a single trial the hemodynamic response in each activation area. However this technique does not exploit the intrinsic temporal resolution of a single EPI slice acquisition. At present, even if the fMRI temporal resolution is limited by the intrinsic characteristic of the BOLD signal (delayed 2-4 sec with respect to neural firing) it is theoretically possible to single out peaks of activation with a temporal resolution in the order of 1 sec (depending on the desired spatial resolution). Standard approaches typically use cross-correlation analysis or statistical parametric mapping and thus reduce a whole sequence of acquisition to one static map of activation. Moreover, since the typical method for extracting the weak BOLD signal component (about 2-5%) of the MR noisy signal is based on the explicit knowledge of stimulus-response patterns, it is not applicable for obvious reasons to the analysis of spontaneous events like hallucinations, sleep, or epileptic seizures.

This paper describes a neural network based algorithm that attempts to overcome both problems, i.e. has the purpose of reconstructing the spatio-temporal patterns of brain activation without any stimulus trigger. The algorithm uses a self-organizing neural network (Neural Gas) [3] that is trained to detect time-dependent alterations in the regional intensity of the functional signal. The trained network can then be used as a sort of adapted filter for extracting the dynamic spatio-temporal patterns from the overall MR signal during a variety of complex tasks, with the full temporal resolution of the EPI technique. The performance of the application of this approach is compared with the results obtained by a standard statistical package (SPM96).

2 Methods

Let us consider an fMRI experiment in which n consecutive frames have been acquired and let us focus our attention on a single slice. Each pixel of the slice will be characterized by a time series that can be represented as a vector $\vec{v} \in \mathcal{R}^n$. The whole set of vectors is a spatio-temporal pattern that is a function of the anatomical structure of the slice as well as the functional activity captured during the observation time. The anatomical component of the vector is much stronger than the functional one but there is no reason to believe that they are correlated and thus we may expect that a suitable clustering algorithm can isolate the small anatomical component from the rest of the pattern.

We used the Neural Gas model for carrying out such clustering task. Empirically we found that a formulation in the frequency domain is much more effective than a direct approach in the time domain. Each n-dimensional time-vector $\vec{v}$ was transformed into a d-dimensional frequency-vector $\vec{x}$ with $d = n/2 - m$ according to the following procedure:

- The images of the sequence were smoothed by means of a Gaussian spatial filter;
- For each pixel of each image, the power spectrum of the time-series $\vec{v}$ is computed by means of the FFT algorithm, yielding a sequence of $n/2$ real values;
- The first m elements of the sequence (typically $m = 2$) are discarded, because the low-frequency part of the spectrum is much more likely to re-

flect anatomical structure than functional activity, producing the output feature $\vec{x}$-vector.

These feature vectors were used for training a Neural Gas (NG)network with N neurons, each of which is characterized by a prototype spectral vector or receptive field center $\vec{\pi}$. The NG algorithm applies an iterative update rule each time a data vector $\vec{x}$ is presented:

$$\Delta\vec{\pi}_g = \varepsilon \cdot h_\lambda(k_g(\vec{x}))\cdot \parallel \vec{x} - \vec{\pi}_g \parallel \qquad g = 1, \ldots, N \tag{1}$$

where $\varepsilon \in [0, 1]$ modulates the speed of adaptation and $h_\lambda(k_g(\vec{x}))$ is the activity level of neuron g that is a decaying function of the distance of its receptive field center $\vec{\pi}_g$ from $\vec{x}$: $h_\lambda(k_g(\vec{x})) = e^{\frac{-k_g(\vec{x})}{\lambda}})$ and $k_g(\vec{x})$ is the number of vector prototypes $\vec{\pi}$ for which the relation $\parallel \vec{x} - \vec{\pi} \parallel < \parallel \vec{x} - \vec{\pi}_g \parallel$ holds. The parameter λ identifies the "size" of the receptive field; ε and λ must be decreased during the learning process by a suitable "freezing" curve. It has been shown that this learning rule is approximately equivalent to the stochastic gradient descent of the following cost function:

$$E(\vec{\pi}, \lambda) = \frac{1}{2C(\lambda)} \sum_{g=1}^{N} \int d\vec{x} P(\vec{x}) h_\lambda(k_g(\vec{x})) \parallel \vec{x} - \vec{\pi}_g \parallel^2 \tag{2}$$

with $C(\lambda) = \sum_1^N h_\lambda(k_g)$ ($P(\vec{x})$ is the probability density function of the data).

3 Results

The experiments were carried out at the Service of Neuroradiology of the S. Martino Hospital in Genova, using the available Siemens 1,5 T scanner. Structural T1-weighted and functional EPI images (TE 66, TR 0.96, Flip-Angle 90°) were acquired on the same slice and movement artifacts were compensated by automatic image alignment (SPM96 software).

In particular, we analyzed four different experimental paradigms: the classic finger-tapping task, two language tasks (phonetic fluency and, verbal judgment), and a word recall task.

Subjects were healthy young volunteers, who gave an informed consent: a 28 years old right-handed woman, performed the finger-tapping and word recall tasks; four subjects (2 male and 2 female, 20-32 years, 2 right and 2 left-handers) performed the two language tasks.

We used the finger-tapping task to train the NG network. Each sequence had $n = 40$ images. We used networks of different sizes and we obtained the best results with $N = 30$ neurons, but this number is not critical at all. Each neuron was characterized by an 18-dimensional feture vector (40/2-2). The training set of 19000 samples was presented to the network four times.

At the end of the training procedure we analyzed separately each neuron $\vec{\pi}_g$ of the network by generating an image in the following way (a sort of non-linear neural filter): the intensity of each pixel of the image is proportional to

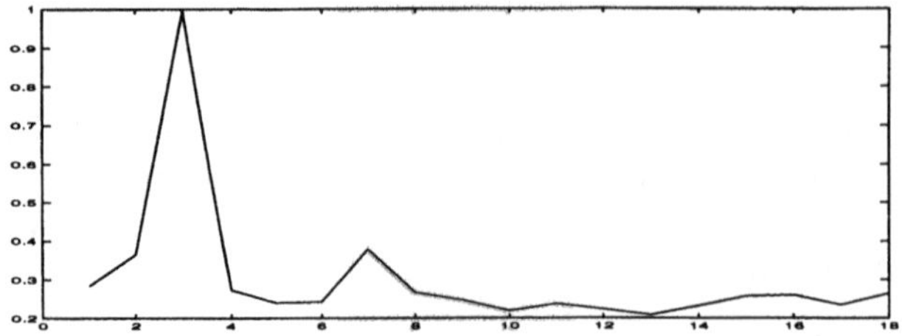

Figure 1: The features values (at the end of the training) of the weight prototype vector $\vec{\pi}$ relative to the neuron that classifies the functional signal in the finger-tapping task.

the correlation between $\vec{\pi}_g$ and the spectral vector $\vec{x}$ of that pixel. For ease of interpretation we superimposed that image with the standard MR anatomical image. According to the judgement the neuroradiologist only one neuron, in all the experiments, captured functional activity. All the other neurons were specialized to single out anatomical features. Figure 1 shows the vector prototype of the "functional neuron". Figure 2 displays the result of the neural filter applied to different slices of the brain.

As we can see in figure 2 our approach is able to identify all major clusters of activation also determined by the standard SPM96 package. In addition it can find some cortical and sub-cortical areas of activation that are not captured by SPM96. This is true also in cognitive experiments, where the intensity of activation is smaller compared to motor studies. The power of the neural network approach is the independence from the stimulation paradigm and the ability to detect the functional rate of the fMRI signal without presumptive knowledge of the stimulus-response model: the network trained with the finger-tapping task is able to detect activation in other tasks.

In general the approach can be applied to sequences of arbitrary duration. The critical parameter is n that identifies the size of the time window observed by the neurons of the netrwork: this windows must be large enough, in order to capture characteristic spectral features of the relevant transients, but short enough, in order to have a suitable time resolution. In this way, the novelty of our approach to fMRI analysis by self-organizing neural networks is the production of a time-varying map of activation.

We applied this method to the analysis of the word-recall data (40 sec experiments), using the neural network trained with the finger tapping data that is characterized by a 4 sec time window. Figure 3 shows the time-varying pattern of brain activity: it can be seen that the "peaks of activation" were shifted in time in the different areas of the same slice.

References

[1] R. L. Buckner, P. A. Bandettini, R. M. O'Craven, R. L. Savoy, S. E. Petersen, Raichle M. E., and Rosen. B. R. Detection of cortical activation

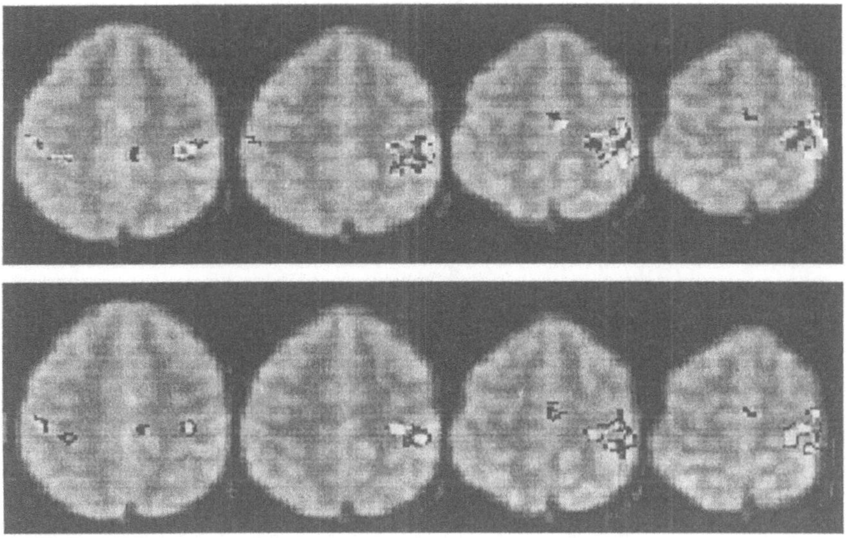

Figure 2: Activation map in the finger-tapping task. Top panel: NG approach; bottom panel: SPM96 approach

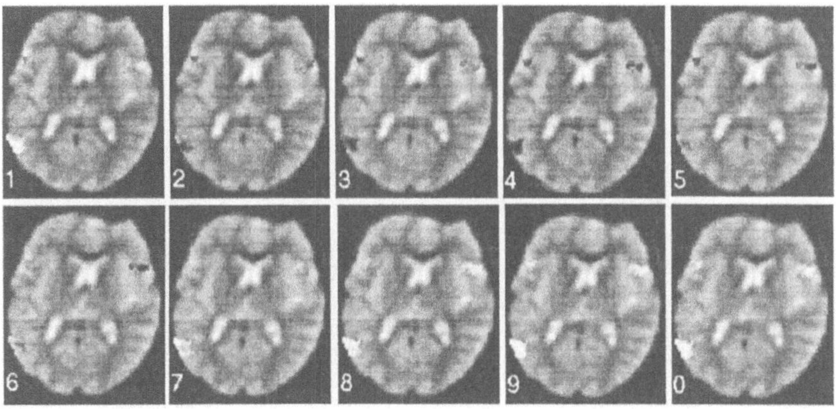

Figure 3: In this left-handers, the activation in right sided Wernicke's and Broca's areas presents a peak almost 4 sec before the left-sided Broca's area. The five images in the top panel are acquired during the activation phase; the others five (bottom panel) during control phase. Activation areas are white in all slices. The activation intensity is in grey scale.

during averaged single trials of a cognitive task using functional magnetic resonance imaging. *Proc. Nat. Acad. Sci. USA*, 93:1478–1483, 1996.

[2] T. Martinetz. Neural gas network for vector quantization and its application to time-series prediction. *IEEE Transactions on Neural Networks*, 4(4):558–569, 1993.

Continual Prediction using LSTM with Forget Gates

Felix A. Gers — felix@idsia.ch
Jürgen Schmidhuber — juergen@idsia.ch
Fred Cummins — fred@idsia.ch

IDSIA, Corso Elvezia 36 , 6900 Lugano, Switzerland , http://www.idsia.ch/

Abstract

Long Short-Term Memory (LSTM,[1]) can solve many tasks not solvable by previous learning algorithms for recurrent neural networks (RNNs). We identify a weakness of LSTM networks processing continual input streams without explicitly marked sequence ends. Without resets, the internal state values may grow indefinitely and eventually cause the network to break down. Our remedy is an adaptive "forget gate" that enables an LSTM cell to learn to reset itself at appropriate times, thus releasing internal resources. We review an illustrative benchmark problem on which standard LSTM outperforms other RNN algorithms. All algorithms (including LSTM) fail to solve a continual version of that problem. LSTM with forget gates, however, easily solves it in an elegant way.

1 Introduction

Recurrent neural networks (RNNs) constitute a very powerful class of computational models, capable of instantiating almost arbitrary dynamics. The extent to which this potential can be exploited, is however limited by the effectiveness of the training procedure applied. Gradient-based methods ("Back-Propagation Through Time" (BPTT) or "Real-Time Recurrent Learning" (RTRL)) share an important limitation. The backpropagated error quickly either vanishes or blows up [1, 2]. Hence standard RNNs fail to learn in the presence of time lags greater than 5 - 10 discrete time steps between relevant input events and target signals.

A recent model, *"Long Short-Term Memory"* (LSTM), is not affected by this problem [1]. LSTM solves complex long time lag tasks that have never been solved by previous RNN algorithms.

In this paper, however, we will show that even LSTM fails to learn to correctly process certain continual time series that are not *a priori* segmented into training subsequences with clearly defined ends. We present a remedy.

While we present a specific solution to the problem of forgetting in LSTM networks, we recognize that *any* training procedure for RNNs which is powerful enough to span long time lags must also address the issue of forgetting in short-term memory (unit activations). We know of no other current training method for RNNs which is sufficiently powerful to have encountered this problem.

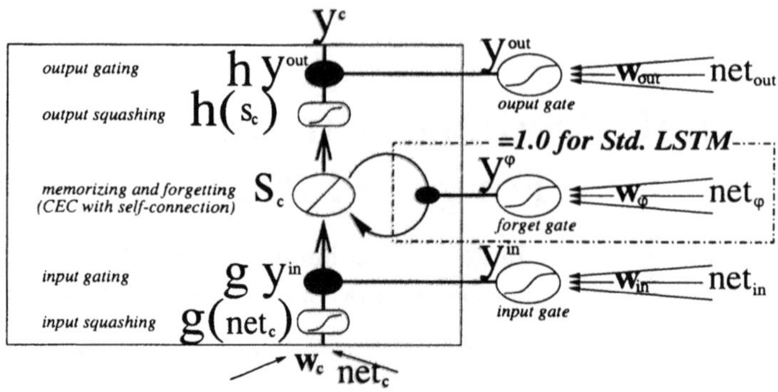

Figure 1: LSTM memory block with a single cell.

2 Standard LSTM

The basic unit in the hidden layer of an LSTM network is the *memory block*, which contains one or more *memory cells* and a pair of adaptive, multiplicative gating units which gate input and output to all cells in the block. Each memory cell has at its core a recurrently self-connected linear unit called the "Constant Error Carousel" (CEC), whose activation we call the cell *state*. The CECs solve the vanishing error problem: in the absence of new input or error signals to the cell, the CEC's local error back flow remains constant, neither growing nor decaying. The CEC is protected from both forward flowing activation and backward flowing error by the input and output gates respectively. When gates are closed (activation around zero), irrelevant inputs and noise do not enter the cell, and the cell state does not perturb the remainder of the network. Fig. 1 shows a memory block with a single cell. The cell state, s_c, is updated based on its current state and three sources of input: net_c is input to the cell itself; net_{in} and net_{out} are inputs to the input and output gates.

We consider discrete time steps t, j indexes memory blocks; v indexes memory cells in block j, such that c_j^v denotes the v-th cell of the j-th memory block; w_{lm} is the weight from unit m to unit l. Index m ranges over all source units, as specified by the network topology. f, g and h are logistic sigmoids with ranges $[0,1]$, $[-2,2]$ and $[-1,1]$ respectively.

Input gate activation y^{in} and output gate activation y^{out} are computed as follows:

$$net_{out_j}(t) = \sum_m w_{out_j m}\, y^m(t-1) \; ; \quad y^{out_j}(t) = f_{out_j}(net_{out_j}(t)) \; ,$$

$$net_{in_j}(t) = \sum_m w_{in_j m}\, y^m(t-1) \; ; \quad y^{in_j}(t) = f_{in_j}(net_{in_j}(t)) \; .$$

The internal state of cell $s_c(t)$ (with $s_{c_j^v}(0) = 0$) and the cell's output y^c are calculated by:

$$s_{c_j^v}(t) = s_{c_j^v}(t-1) + y^{in_j}(t)\, g(net_{c_j^v}(t)) \; ,$$

$$y^{c_j^v}(t) = y^{out_j}(t)\, h(s_{c_j^v}(t)) \; .$$

Learning. See [1] for details of standard LSTM's backward pass. Essentially, as in truncated BPTT, errors arriving at net inputs of memory blocks and their gates do not get propagated back further in time, although they *do* serve to change the incoming weights. When an error signal arrives at a memory cell output, it gets scaled by the output gate and the output nonlinearity h; it then enters the memory cell's linear CEC, where it can flow back indefinitely without change (thus LSTM can bridge arbitrary time lags between input events and target signals). Only when the error escapes from the memory cell through an open input gate and the additional input nonlinearity g, does it get scaled once more and then serves to change incoming weights before being truncated.

2.1 Limits of standard LSTM

LSTM allows information to be stored across arbitrary time lags, and error signals to be carried far back in time. This potential strength, however, can contribute to a weakness in some situations: the cell states s_c often tend to grow linearly during the presentation of a time series (the nonlinear aspects of sequence processing are left to the squashing functions and the highly nonlinear gates). If we present a continuous input stream, the cell states may grow in unbounded fashion, causing saturation of the output squashing function, h.

3 Solution: Forget Gates

Our solution to the above problem is to use adaptive "forget gates" which learn to reset memory blocks once their contents are out of date and hence useless. We replace standard LSTM's constant CEC weight 1.0 (Fig. 1) by the multiplicative forget gate activation y^{φ}, which is calculated like the activations of the other gates and squashed using a logistic sigmoid, $[0, 1]$. y^{φ_j} functions as the weight of the self recurrent connection of the internal state s_c in equation (1). The revised update equation for s_c in the extended LSTM algorithm is (for $t > 0$):

$$s_{c_j^v}(t) = y^{\varphi_j}(t)\ s_{c_j^v}(t-1)\ +\ y^{in_j}(t)\ g(net_{c_j^v}(t)) \quad ,$$

with $s_{c_j^v}(0) = 0$. Bias weights for LSTM gates are initialized with negative values for input and output gates (see [1] for details), positive values for forget gates. This implies that in the beginning of the training phase the forget gate activation will be almost 1.0, and the entire cell will behave like a standard LSTM cell. It will not explicitly forget anything until it has learned to forget.

3.1 Backward Pass of LSTM with Forget Gates

Space precludes a full description of the equations used in updating network weights (see [3, 1] for details). In essence, the novel forget gate is treated just as the input gate of standard LSTM: weights feeding the gate are updated using a slightly modified version of RTRL.

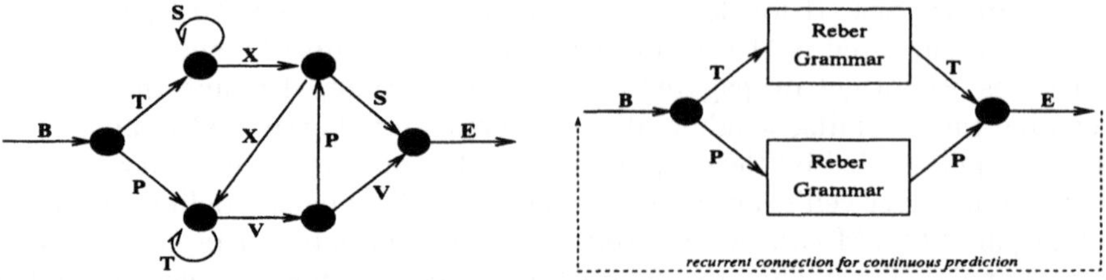

Figure 2: Transition diagrams for standard (left) and embedded (right) Reber grammars. The dashed line indicates the continual variant.

The storage complexity for the backward pass does not depend on the length of the input sequence. Like standard LSTM, extended LSTM is local in space and time.

4 Experiments

To generate an infinite input stream we extend the well-known "embedded Reber grammar" (ERG) benchmark problem [4, 5, 6, 1] to a continuous task (CERG). See Fig. 2. An ERG string is generated from left to right (beginning with an empty string) by following randomly chosen edges ($p = 0.5$), and appending the associated symbols to the current string. Input and target patterns are represented using a localist encoding. The task is to read strings, one symbol at a time, and to continually predict the next possible symbol(s). A prediction is considered correct if the mean squared error at each of the 7 output units is below 0.49 (error signals occur at every time step).

The minimal time lag is 7 (at the limit of what standard recurrent networks can manage); time lags have no upper bound though (for our test sets the expected value of the longest string is greater than 50).

A performance comparison of previous RNNs (RTRL [4], "Elman net trained by Elman's procedure" [5], "Recurrent Cascade-Correlation" [6] and LSTM [1]) on the standard ERG problem showed that only LSTM always solves the task. It learns more than 20 times as fast as the second best approach (RCC).

CERG. Our more difficult continual variant of the ERG problem (CERG) does not provide information about the beginnings and ends of symbol strings. Without intermediate resets, the network is required to learn, in an on-line fashion, from input streams consisting of concatenated ERG strings. Input streams are stopped as soon as the network makes an incorrect prediction or the 10^5-th successive symbol has occurred. Learning and testing alternate: after each training stream we freeze the weights and feed 10 test streams. Our performance measure is the average test stream size; 100,000 corresponds to a so-called "perfect" solution (10^6 successive correct predictions).

4.1 Network Topology and Parameters

The 7 input units are fully connected to a hidden layer consisting of 4 memory blocks with 2 cells each (8 cells and 12 gates in total). The cell outputs are fully connected to the cell inputs, to all gates, and to the 7 output units (which have

Algorithm	%sol	train seqs
Std LSTM with external reset	74	7441
Std LSTM	0	> 30000
LSTM with Forget Gates	18	18889
LSTM with Forget Gates and sequential α-decay	62	14087

Table 1: Continual Embedded Reber Grammar (CERG): Percentage of "perfect" solutions (correct prediction of 10 streams of 100,000 symbols each), and the number of sequences presented until solution was reached.

additional "shortcut" connections to the input units). Gates and output units have a bias connection (see [3] for details). Weights are initialized randomly in the range $[-0.2, 0.2]$. Weight changes are made after each input symbol presentation. The learning rate α is initialized with 0.5. It either remains fixed or decays by a fraction of 0.99 per time step.

4.2 Results

Training was stopped after at most 30000 training streams, each of which was ended when the first prediction error or the 100000th successive input symbol occurred. Table 1 compares extended LSTM with and without learning rate decay to standard LSTM. External resets (non-continual) case allow LSTM to find excellent solutions in 74% of the trials, according to our stringent testing criterion. Standard LSTM fails, however, in the continual case. Extended LSTM with forget gates can solve the continual problem. The best results were obtained by combining forget gates with learning rate decay.

4.3 Analysis of the Results

How does extended LSTM solve the task on which standard LSTM fails? Section 2.1 already mentioned LSTM's problem of uncontrolled growth of the internal states. Without forget gates, the internal states tend to grow linearly. Fig. 3 shows the evolution of the internal states s_c during the presentation of a test stream. At the starts of successive ERG strings, the network is in an increasingly active state. At some point, the high level of state activation leads to saturation of the cell outputs, and performance breaks down. Extended LSTM, however, learns to use the forget gates for resetting its state when necessary. Fig. 4 (top) shows a typical internal state evolution after learning.

5 Conclusion

Continuous input streams generally require occasional resets of the network. Partial resets are also desirable for tasks with hierarchical decomposition. For instance, re-occurring subtasks should be solved by the same network module, which should be reset once the subtask is solved. Since typical real-world input streams are not *a priori* decomposed into training subsequences, and since

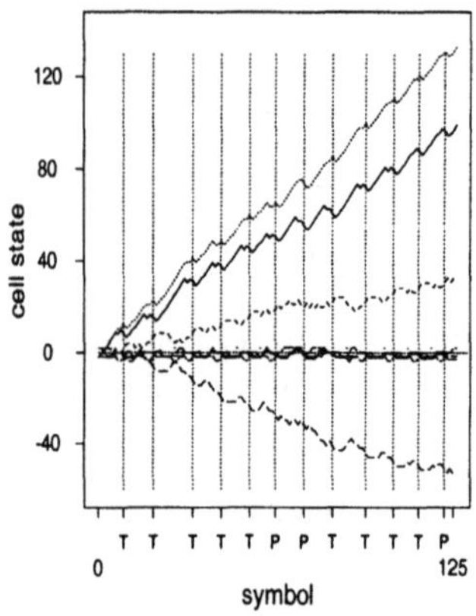

Figure 3: Evolution of standard LSTM's internal states s_c during presentation of a test stream stopped at first prediction failure. Starts of new ERG strings are indicated by vertical lines labeled by the symbols (**P** or **T**) to be stored until the next string start.

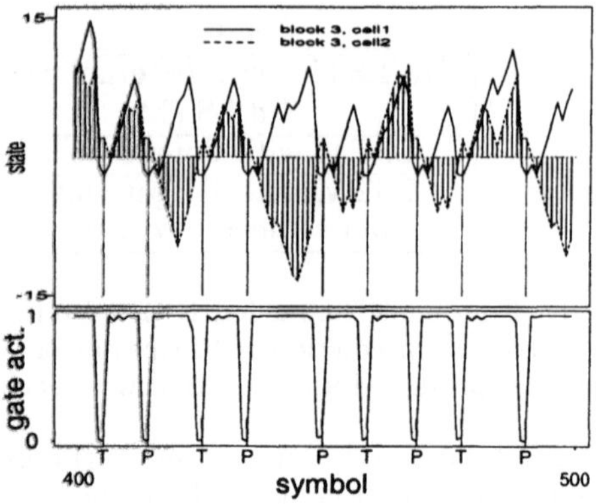

Figure 4: Top: Internal states s_c of the two cells of the self-resetting third memory block in an extended LSTM network during a test stream presentation. The figure shows 100 successive symbols taken from the longer sequence presented to a network that learned the CERG. Starts of new ERG strings are indicated by vertical lines labeled by the symbols (**P** or **T**) to be stored until the next string start. Bottom: simultaneous forget gate activations of the same memory block.

typical sequential tasks are not *a priori* decomposed into appropriate subproblems, RNNs should be able to *learn* to achieve appropriate decompositions. Our novel forget gates naturally permit LSTM to learn local self-resets of memory contents that have become irrelevant.

Acknowledgment. This work was supported by SNF grant 2100-49'144.96 .

References

[1] S. Hochreiter and J. Schmidhuber, "Long short-term memory", *Neural Computation*, **9**(8):1735–1780, 1997.

[2] Y. Bengio, P. Simard, and P. Frasconi, "Learning long-term dependencies with gradient descent is difficult", *IEEE Transactions on Neural Networks*, **5**(2):157–166, 1994.

[3] F. A. Gers, J. Schmidhuber, and F. Cummins, "Learning to forget: Continual prediction with LSTM", Tech. Rep. IDSIA-01-99, IDSIA, Lugano, CH, 1999.

[4] A. W. Smith and D. Zipser, "Learning sequential structures with the real-time recurrent learning algorithm", *International Journal of Neural Systems*, **1**(2):125–131, 1989.

[5] A. Cleeremans, D. Servan-Schreiber, and J. L. McClelland, "Finite-state automata and simple recurrent networks", *Neural Computation*, **1**:372–381, 1989.

[6] S. E. Fahlman, "The recurrent cascade-correlation learning algorithm", in *NIPS 3*, R. P. Lippmann, J. E. Moody, and D. S. Touretzky, Eds. 1991, pp. 190–196, San Mateo, CA: Morgan Kaufmann.

Dynamics of On-Line Learning in Radial Basis Function Neural Networks

Maria Marinaro[a,b,c] and Silvia Scarpetta[a,b]
[a] INFM, Unita' di Salerno, Salerno, Italia
[b] Universita' di Salerno, Dipartimento di Fisica Teorica ' E.R.Caianiello",
Via S. Allende, 84081 Baronissi (SA), Italia
[c] International Institute for Advanced Science Studies,
Via G. Pellegrino 19, 84019 Vietri sul Mare, Salerno, Italia

Abstract

We present a method for analyzing the behavior of RBFs in an on-line scenario which provides a description of the learning dynamics without invoking the thermodynamic limit. Our analysis is based on a master equation that describes the dynamics of the weight space probability density for any value of the input space dimension. Because the transition probability appearing in the master equation cannot be written in closed form, some approximate form of the dynamics is developed. We assume a arbitrary small learning rate (small noise) and we derive in this limit the dynamic evolution of the means and the variances of the net weights. The analytic results are then confirmed by simulations.

1 Introduction

Learning from examples in layered neural networks is usually formulated as an optimization problem based on the minimization of a learning error, defined as an additive error over a set of examples (training set). In the on-line learning scenario the weights of the synapses are updated sequentially according to the error computed on a single new example. The average that accounts for the disorder introduced by the independent random selection of an example at each time step can be performed directly, without invoking the replica method. In this scenario the multilayer perceptron networks have been successfully analyzed [5]. The analysis makes use of the thermodynamic limit that is obtained by assuming infinite input space dimension N. When the thermodynamic limit is introduced the error is expressed in terms of a finite set of mean-field observables whose dynamic is governed by a closed set of deterministic differential equations. The application of the on-line scenario to Radial Basis Function Network (RBFN) presents some difficulties, in fact for RBFN the thermodynamic limit is totally meaningless, as the activation functions are localized. We present a method for analyzing the learning dynamics of RBFNs in an on-line scenario for an arbitrary number of hidden and input nodes. Earlier work on on-line RBFs from a different perspective can be found in [4]. In our approach the learning process is viewed as a stochastic process governed by a continuous-time

Master Equation (ME). We use statistical tools that have been often successfully applied in neural networks analysis [1, 2]. Since the ME cannot be solved exactly, stochastic approximation methods are developed. Typically, as here, a small fluctuation expansion provides a description of the dynamics in terms of suitably scaled fluctuations around a deterministic flow. Making a small fluctuation Ansatz for small learning parameters we derive a set of ordinary differential equations for the expectation value of the net parameters and their variances. The performance of the network during the learning process is then measured by an estimation of the generalization error.

2 The Framework

The RBF architecture consists on a feed-forward two-layer network in which the argument of activation function $\Psi_i(\boldsymbol{\xi})$ of each hidden unit is given by the distance between the input vector $\boldsymbol{\xi}$ and a prototype vector. We will focus our attention on Gaussian activation functions, since they are the most commonly used basis functions, and they have a number of useful analytical properties. Let the activation function of each hidden unit be $\Psi_i(\boldsymbol{\xi}; \bar{\boldsymbol{x}}_i, \gamma_i) = exp(-\gamma_i ||\boldsymbol{\xi} - \bar{\boldsymbol{x}}_i||^2)$, where $||.||$ is the Euclidean distance, and γ_i is the parameter that controls the spread of the function around the center $\bar{\boldsymbol{x}}_i \in S_{input}$. Consider that the net have to approximate the mapping $T : \boldsymbol{\xi} \rightarrow f_t$, from N-dimensional input space, $\boldsymbol{\xi} \in S_{input}$, to one dimensional output space, $f_t \in S_{out}$. So the network has N input nodes, one linear output node and an hidden layer with an arbitrary number K of hidden nodes. The output nodes compute a linear combination of the activation of the hidden nodes, parameterized by the weights $\boldsymbol{u}$ between hidden and output layers. The output of the RBF network is:

$$net(\boldsymbol{\xi}) = \sum_{i=1}^{K} u_i exp(-\gamma_i ||\boldsymbol{\xi} - \bar{\boldsymbol{x}}_i||^2) \tag{1}$$

The net is supposed to learn *on line* from a sequence of training examples $\{(\boldsymbol{\xi}^\mu, {f_t}^\mu),\ \mu = 1 \ldots\}$. At each iteration a new example is given and the network parameters are updated on line according to the error computed on that last example. In the training scenario examined here, the components of the training input vectors $\boldsymbol{\xi}^\mu$ are chosen as uncorrelated Gaussian random variables of mean 0 and spread parameter σ; i.e. their probability distribution is $P_\sigma(\boldsymbol{\xi}) = \frac{1}{(\pi/\sigma)^{N/2}} exp(-\sigma \boldsymbol{\xi}^2)$. It is useful to suppose that the training data are produced by an unknown RBF *teacher* network which may be corrupted by noise. The vector of M hidden-to-output weights of the *teacher* net are denoted by an M dimensional vector $\mathbf{v}$, while the centers of teacher hidden nodes are denoted by $\bar{\boldsymbol{x}}_j^0$, $j = 1 \ldots M$, and the spread parameters by β_j, $j = 1 \ldots M$. The network that have to learn from the *teacher* is called *student* network. As far as the student is concerned, the teacher is an 'oracle' or 'black box'; the only information available about the inner parameters of the teacher is contained in the various targets of examples it provides. Although the framework enables us to consider a wide range of cases we will consider here the noiseless case and moreover the number K of student basis functions is supposed equals to the number M of teacher basis functions. The function computed by the teacher is therefore: $f_t(\boldsymbol{\xi}) = \sum_{j=1}^{K} v_j \Psi_j(\boldsymbol{\xi}; \bar{\boldsymbol{x}}_j^0, \beta_j)$. The learning dynamics, we are interested in, follows the on-line gradient descendent rule, that is parameters are adjusted according to the gradient of the error after the presentation of each new example. The error made by student with parameters $\{\gamma_i, \bar{\boldsymbol{x}}_i, u_i\}, i = 1 \ldots K$ on a given input $\boldsymbol{\xi}^\mu$ is denoted by $\epsilon^\mu = \epsilon(\boldsymbol{\xi}^\mu; \{\gamma_i, \bar{\boldsymbol{x}}_i, u_i\}) = (net(\boldsymbol{\xi}^\mu) - f_t(\boldsymbol{\xi}^\mu))^2$. We can write the update rule in a compact form by introducing the vector of parameters $\boldsymbol{J} = \{u_i, \gamma_i, \bar{\boldsymbol{x}}_i\}, i = 1 \ldots K$. The update rule can be written in the form:

$$\boldsymbol{J}^{\mu+1} = \boldsymbol{J}^\mu - \eta \boldsymbol{F}_J(\boldsymbol{J}^\mu, \boldsymbol{\xi}^\mu) \tag{2}$$

where $\boldsymbol{F}_J = \{\boldsymbol{F}_u, \boldsymbol{F}_\gamma, \boldsymbol{F}_{\bar{\boldsymbol{x}}}\}$ is the gradient of the error :

$$F_{\gamma_i} = \frac{\partial \epsilon}{\partial \gamma_i}, \quad F_{ui} = \frac{\partial \epsilon}{\partial u_i}, \quad F_{\bar{\boldsymbol{x}}i} = \frac{\partial \epsilon}{\partial \bar{\boldsymbol{x}}_i} \tag{3}$$

The training-error is an average on all training examples of the quadratic deviation $\epsilon(\boldsymbol{\xi}; \{\gamma_i, \bar{\boldsymbol{x}}_i, u_i\})$, while the performance on a typical input defines the generalization error,

$$\epsilon_g(\boldsymbol{J}) \equiv \epsilon_g(\{\gamma_i, \bar{\boldsymbol{x}}_i, u_i\}) = \langle \epsilon(\boldsymbol{\xi}; \{\gamma_i, \bar{\boldsymbol{x}}_i, u_i\}) \rangle_{\boldsymbol{\xi}} \tag{4}$$

where the average is over the distribution $P_\sigma(\boldsymbol{\xi})$ of input vectors.

The most common situation for RBF architecture is a structure where there are enough hidden neurons to cover the input domain and therefore the number of hidden neuron K is typically much greater then the input dimension N. Therefore we will assume an arbitrary finite input dimension N and an arbitrary number of hidden units K. It is evident that the thermodynamic limit $K << N \to \infty$ which is used in MLPs cannot be applied to RBFN, thus we have to introduce a different approach which allows us to derive from the discrete-time process a continuous-time description for the probability distribution of the net parameters, and, under some conditions, to obtain some ordinary differential equation (ODE) for the mean value and for the fluctuations of these parameters. This derivation is briefly described in the next section.

3 The equations of motion

We start from the discrete-time learning process defined in eqn. (2). Eqn. (2) defines a stochastic process since at each step the vector $\boldsymbol{\xi}^\mu$ is chosen random in the training set. Let introduce the probability density $p_\mu(\boldsymbol{J})$ to find the net parameters in the state $\boldsymbol{J}$ at discrete iteration step μ. In terms of this microscopic probability density the process (2) can be written as:

$$p_{\mu+1}(\boldsymbol{J}) = \int d\boldsymbol{J}' W[\boldsymbol{J}; \boldsymbol{J}'] p_\mu(\boldsymbol{J}') \tag{5}$$

with the transition probability densities: $W[\boldsymbol{J}';\boldsymbol{J}] =< \delta(\boldsymbol{J}'-\boldsymbol{J}-\eta\boldsymbol{F}[\boldsymbol{J},\boldsymbol{\xi}]) >_{\boldsymbol{\xi}}$. A continuous-time description can be obtained from (5) through the assignment of random values Δt to the time interval between two succeeding iteration steps [3]. The probability $P(\boldsymbol{J},t)$ that the net is in the state $\boldsymbol{J}$ at time t can be differentiated with respect to time, yielding the master equation:

$$\frac{\partial P(\boldsymbol{J},t)}{\partial t} = \int d\boldsymbol{J}'\{W[\boldsymbol{J};\boldsymbol{J}'] - \delta(\boldsymbol{J}-\boldsymbol{J}'))\}P(\boldsymbol{J}',t) \tag{6}$$

The cost to pay for this passage from discrete to continuous time is the introduction of an uncertainty in the time axis[6]. We will denote the distribution of the states $\boldsymbol{J}$ at time t by $\Xi(t)$. The symbol $<>_{\boldsymbol{\xi}}$ indicate the average over the input space, while $<>_{\Xi(t)}$ indicate the average over the distribution $P(\boldsymbol{J},t)$. Using the Master equation (6) we can obtain some exact evolution equations for the quantities we are interested in. Since the exact equations are in general insolvable, we use some stochastic approximations. We can imagine that the stochastic process can be viewed as an average, ***deterministic trajectory***, with ***stochastic fluctuations*** around this trajectory. We make a small-fluctuation Ansatz (fluctuations of order $\sqrt{\eta}$ around the deterministic part $\phi(t)$), in order to derive[6] in the limit $\eta \to 0$ a Fokker-Plank equation for the fluctuations, and a deterministic equation for $\phi(t)$. The mean value of $\boldsymbol{J}$ and the expectation value of fluctuations $\Sigma_{ij} =< (J_i- < J_i >_{\Xi(t)})(J_j- < J_j >_{\Xi(t)}) >_{\Xi(t)}$ evolve according to the following equations of motion:

$$\frac{1}{\eta}\frac{d < J_i >_{\Xi(t)}}{dt} = F_i(< \boldsymbol{J} >_{\Xi(t)}) \tag{7}$$

$$\frac{1}{\eta}\frac{d\Sigma_{ij}}{dt} = \sum_k G_{ik}(< \boldsymbol{J} >_{\Xi(t)})\Sigma_{kj} + \Sigma_{ki}G_{jk}(< \boldsymbol{J} >_{\Xi(t)}) + \eta D_{ij}(< \boldsymbol{J} >_{\Xi(t)})$$

where $\boldsymbol{J}$ is the set of student parameters, $\boldsymbol{F}_{\boldsymbol{J}}(\boldsymbol{J},\boldsymbol{\xi})$ is defined in (3), and

$$\boldsymbol{F}(\boldsymbol{J}) =< \boldsymbol{F}_{\boldsymbol{J}}[\boldsymbol{J},\boldsymbol{\xi}] >_{\boldsymbol{\xi}}, \quad D_{ij} =< F_i[\boldsymbol{J},\boldsymbol{\xi}]F_j[\boldsymbol{J},\boldsymbol{\xi}] >_{\boldsymbol{\xi}}, \quad G_{ik}(\boldsymbol{J}) = \frac{\partial F_i(\boldsymbol{J})}{\partial J_k}$$

The small-fluctuation approximation turn out [6] to be valid in the attraction regions $(F'(< J >_{\Xi(t)}) < 0)$, where the energy surface have positive curvature, or on relatively short time scales $(O(1/\eta))$ outside these attraction regions; if there are multiple minima, it provides a description around each of them, but not a global description. The performance of the student RBF network on the given task during evolution is measured by the generalization error, which in our approximation [6] is expressed in terms of the teacher parameters and of the expectation value of the student parameters.

4 Results

The equations (7) can be analytically computed in the general case in which $\boldsymbol{J}$ is the set of student parameters of the RBF net $\boldsymbol{\gamma}, \boldsymbol{u}\,\bar{\boldsymbol{x}}_i$, $i = 1\ldots K$, and the teacher parameters are $\boldsymbol{\beta}, \mathbf{v}, \bar{\boldsymbol{x}}_i^0$, $i = 1 \dots M$. The averages require the evaluation of Gaussian integrals that can be easily carried out. So, they provide a tool for analyzing the learning process for a general RBF network. In the following we simplify the computation by assuming that the centers $\bar{\boldsymbol{x}}_i$ of the student functions are fixed and are settled equals to the values of the centers $\bar{\boldsymbol{x}}_j^0$ of the teacher basis functions.

We focus here on the dynamics of the student net parameters $\boldsymbol{J} = \{\boldsymbol{\gamma}, \boldsymbol{u}\}$, while $\bar{\boldsymbol{x}}_i; \bar{\boldsymbol{x}}_i^0, \boldsymbol{\beta}, \mathbf{v}$ are assumed fixed. The mean tendency of the learning system is described by a set of differential equations [6] which we solve numerically. To demonstrate the validity of the theoretical results, we compare the evolution of the system found by numerically solving the equations, to empirical results found by training real RBF network via on-line gradient descendent. The empirical value of $\{\boldsymbol{\gamma}, \boldsymbol{u}\}$ are calculated during the training. Generalization error was empirically estimated via the average error on a 800 point test set. The components of training input vectors $\boldsymbol{\xi}_\mu$ are chosen as uncorrelated Gaussian random variables of mean 0 and variance $\frac{1}{2\sigma}$. It is reasonable to choose centers of the teacher activation functions $\bar{\boldsymbol{x}}_i^0$ distributed in the input space, as Gaussian random variables. We compare the theoretical prediction for the mean values with the the empirical result of single-run simulations, without averaging over many realizations of the simulated network.

Fig. 1 shows generalization error of four runs with the same initial conditions, in each run we train the RBF network with the same weights initialization, via on-line gradient descendent, on different extractions of examples, generated from the teacher network. Results are compared with the mean evolution of the system found by numerically solving the differential equations. Fig. 2 shows evolutions of the net weights γ_i and u_i during the learning process. The theoretical prediction of the mean value is compared with one single-run simulation result. All the single run simulations stay around the mean value prediction, with only small fluctuations around it. In this example the input space dimension is $N = 2$, and the number of hidden nodes $K = 9$. We see two different relaxation times in the generalization error, the fast decay towards a non-optimal solution (see fig. 1a) is followed by another exponential decay (see inset fig. 1a), with a larger characteristic time, towards optimal solution. The correspondence between theory and simulations seems good, not only in the asymptotic decay, but also in the transient phase. In between one exponential decay and the other, simulations shows slightly larger fluctuations, around the theoretical prediction of the mean value. This is easily understood if we remember that in the small learning-rate limit the small-fluctuations ansatz is assured only in the positive curvature regions, or on relatively short time scales ($O(1/\eta)$) outside of these attraction regions. However in all cases we have analyzed the fluctuations are never such as to invalidate our description, when the learning rate is small enough.

References

[1] T. H. Heskes and B. Kappen , *Mathematical Foundations of Neural Networks* Taylor Ed., Amsterdam 1993. Elsevier, pages 199-233. *Physical Review* **A** 44, 1991

[2] Leen T K, Schottky B and Saad D 1998, *Advanced in Neural Information Systems* Vol. **10**, ed. by Jordan, Kearns, Solla (Cambridge, MA: MIT Press) p 301

[3] D. Bedeaux, K. Lakatos-Lindberg and K. Shuler, *Journal of mathematical Physics* **12** , 2116-2123, 1971.

[4] J. Freeman and D. Saad 1997, *Physical Review* **E** 56 , 907-918.

[5] Saad D (ed) 1998 *Online learning in neural networks* London:Cambridge University Press. Saad D and Solla S A 1995 *Phys. Rev. Lett.* **74** 4337.

[6] Scarpetta S 1998, *Ph. D. thesis*, Salerno University

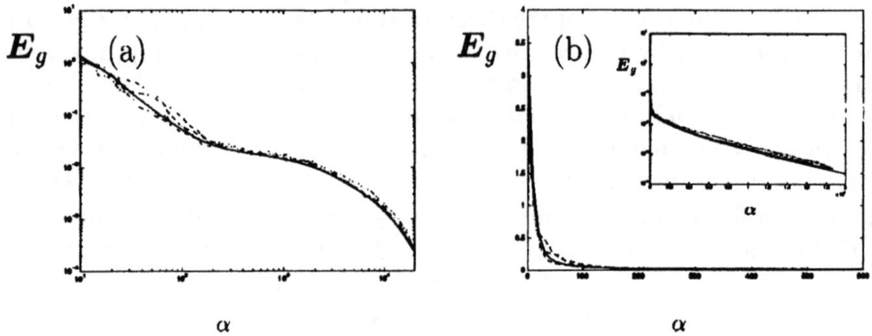

Figure 1: (a,b) The dashed and dot-dashed lines show the empirical generalization error when we train the γ and u parameters with gradient descent ($\eta = 0.1$), in a $K = 9$ RBF network learning a noiseless task, in a bidimensional input space $N = 2$, always with the same initial conditions. We compare simulations results with the theoretical prediction of the expectation value given by the solid line. Task parameters are $\sigma = 2$, $\beta_i = \{2.1, 2.5, 2.7, 3.3, 3.6, 3.7, 4.0, 4.2, 4.3\}$, ($i = 1 \ldots 9$), all $v_i = 1$, and student initial conditions $\gamma_i = 5$,$u_i = 0.0$, for each $i = 1, \ldots, K$. In (a) is shown the evolution in a log-log plot, in (b) the start up of the evolution (transient phase) and in the inset the asymptotic phase in log-lin plot.

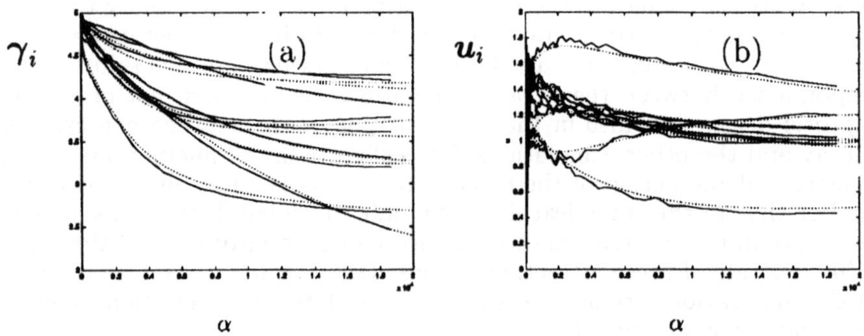

Figure 2: The solid lines show a single run simulation results, dotted line are the theoretical results. (a) shows evolution of student spread parameters γ and (b) the student hidden-output weights u in the same learning scenario of the two previous figures.

Harmony Theory and Binding Problem

Eliano Pessa Maria Pietronilla Penna

ECONA - Centro Interuniversitario di Ricerca sull'Elaborazione
Cognitiva nei Sistemi Naturali e Artificiali
c/o Dipartimento di Psicologia, Università di Roma "La Sapienza"
Via dei Marsi,78, 00185 Roma Italy. email: pessa@axcasp.caspur.it

Abstract

We introduce a neural network architecture based on Smolensky's Harmony Theory in order to solve a particular form of Binding Problem in a visual scene containing 4 subpatterns. The network performance is studied through computer simulations. The results obtained evidence how Harmony Theory can be used to solve Binding Problem, but its performance, even if significatively greater than chance level, is still too low.

1.Introduction

The Binding Problem is one of the most difficult problems to solve for neural network architectures. It arises when multiple patterns are contained within a single scene and this latter can be analyzed only through feature detectors, able to report the presence of a particular feature, but not its spatial location. In this case it is very difficult to design the architecture of the detecting system in such a way as to capture the spatial relationships between the different patterns present within the scene. There are heuristic arguments (see, e.g., [2]) supporting the claim that neural network architectures will never be able to do a structural description of a scene whatsoever, presented as a visual input pattern. Here the term "structural" (cfr. [10]) refers to the network of spatial relationships between the subpatterns (objects) contained within it. According to these arguments (see also [1]), structural descriptions, and whence the solution of Binding Problem and, *a fortiori*, the ability to make abstractions, should be a prerogative only of systems endowed with high-level logical rules, such as the ones expressed through predicate calculus.
On the other hand, the earlier connectionist proposals (see, e.g., [6]) leave open the way to a solution of Binding Problem, as they distinguish between microscopic and macroscopic features of cognition. These latter, and whence their high-level logical structure, should emerge as collective effects from the cooperative interaction between microscopic cognitive units. In particular, this could be the case for the "binding" relationship between the different subpatterns belonging to a given visual scene, that could emerge as a macroscopic collective effect from the interaction of multiple microscopic constraints. There were serious attempts to build connectionist models of cognitive processing, implementing the aforementioned ideas. Between such attempts, one of the most interesting is Smolensky's Harmony Theory [7]. In this paper we will explore the possibility of using Harmony Theory to solve Binding Problem, by measuring the performance of a neural network architecture based on it

in a particular problem of this type. We feel that the theoretical knowledge so far acquired about Harmony Theory be insufficient to let us forecast in advance its ability to solve Binding Problem, so that we are forced, in order to measure such an ability, to resort to numerical simulations.

2. Harmony Theory

Harmony Theory essentially provides us with a network architecture designed to do completion tasks, i.e. cognitive tasks in which the full internal representation of a state of external world has to be reconstructed (or completed) on the basis of a limited number of cues (i.e. of known features) given from outside. In this theory each representation is described through a vector of feature activation values, so that completion task is equivalent to the one of filling the representation vector with the correct activation values, starting from the initial knowledge of some of these values. Such a process is driven by the interaction between feature activations and knowledge atoms activations. Each knowledge atom encodes a particular microcognition or, so to speak, a particular microscopic constraint on cognitive processing. This interaction is designed so as to give rise to a completion dynamics wich is equivalent to a process of relaxation toward a representational state associated to the maximum attainable value of a function, called *Harmony function* (to be viewed, signum apart, as a sort of Ljapunov function).
The implementation of Harmony Theory proposed by Smolensky makes use of a two-layered neural network, containing a *feature layer,* in which each unit is deputed to code, through its activation level (which can be only 1 or -1), the presence of a particular feature, and a *knowledge layer*, in which each unit is associated to a particular knowledge atom, whose activation level can be only 1 or 0. Each feature unit is connected to each knowledge atom by a bidirectional connection, characterized by the same connection weight values in both directions. More precisely, as each knowledge atom is represented by a particular static vector of feature activation values, if we assume that all knowledge atoms have unit strength, the weight of the connection relating the i-th knowledge unit to the j-th feature unit is given by:

$$w_{ij} = k_{ij} / \Sigma_j | k_{ij} | \tag{1}$$

where k_{ij} is the activation value assumed by the j-th feature unit in the vector representing the i-th knowledge atom. The Harmony function chosen by Smolensky has the form:

$$H = \Sigma_i \, a_i \, [\, (\Sigma_j \, r_j w_{ij}) - \kappa \,] \tag{2}$$

where r_i is the vector of actual activation values of feature units, whereas κ is a suitable constant, whose value must be chosen by the experimenter within the interval between 0 and 1. It is possible to show that the state characterized by the maximum allowable value of H can be attained through a dynamical relaxation process driving activation values both of feature units and of knowledge atoms towards an equilibrium state. Such a process, ruled by an asynchronous dynamics, follows a probabilistic updating law, stating that the probability that the i-th unit have an output equal to 1 is given by:

$$P_i = 1/[\,1 + exp\,(-net_i / T)\,]\,, \tag{3}$$

where T is a suitable "temperature" parameter which, during relaxation process, is varied according to a simulated annealing schedule (cfr. [3]). It can be shown that, in order to maximize H, the net input net_i has to be defined, in the case of knowledge units, as:

$$net_i = \Sigma_j\, w_{ij}\, r_j - \kappa \;, \tag{4}$$

whereas, in the case of feature units, as:

$$net_i = 2\, \Sigma_j\, w_{ij}\, a_j \;. \tag{5}$$

We remark that Harmony Theory, after the initial proposal by Smolensky, was not so much studied within neural network literature (cfr. for more recent applications [8], [9]). Whence we know very little about its performance in concrete completion tasks.

3. The Binding Problem to be solved

In order to measure performance of Harmony Theory, we choose a particular Binding Problem, arising in digitalized scene images of dimensions 12x10 pixels, in which each pixel was characterized only by 1 or 0 grey level values. Each scene contained four 6x5 subpatterns, arranged according to the scheme depicted in Figure 1.

b1	*b2*
b3	*b4*

Figure 1

The set of available subpatterns contained only digitized representations of ciphers or alphabetic letters, so that each subpattern could be characterized by a single feature, to be identified with the letter or the cipher it was representing. The scene was scanned through a 6x5 attentional window, starting from a particular subpattern and performing three successive scannings. Only four scanning directions were allowed: up, down, to the right, to the left. Each scanning was constrained in such a way as to let a motion of the attentional window from the starting subpattern only to the immediately adjacent subpattern along the scanning direction chosen. The direction of such a motion was used to individuate the spatial relationship existing between the adjacent subpattern and the starting one. We focussed our attention only on sequences of three successive scannings, constrained in such a way that each scanning, except the first one, should have, as a starting subpattern, the one obtained at the end of the previous scanning, and that every subpattern could not be encountered twice during the three scannings. It is easy to show that, for each scene, there are only 8 different sequences of three scannings satisfying these requirements. At the end of each sequence of three successive scannings we presented to the network, as inputs, the category to which the last subpattern encountered was belonging, and the spatial relationship connecting this subpattern to the subpattern from which the scanning sequence was started. The network must complete this input, by giving as answer the right category of the starting subpattern. As a direct scanning from this latter to the final subpattern was not performed in the previous scanning sequence, such an answer required that the network, on the basis of the knowledge acquired during scanning sequence, be able to perform a "binding" between categories associated to different spatial locations. Thus the problem of giving a correct answer to the previous question is a particular form of Binding Problem.

In order to solve such a problem, we introduced a neural network architecture consisting of the following modules: 1) a scanner controlled by an iconic memory; 2) a categorizer; 3) a knowledge encoder; 4) a knowledge retrieval system (based, of course, on Harmony Theory). We will refer to a previous paper [5] for a detailed description of the scanner and of its operation. Here it is enough to say that it was designed to give, as input to the categorizer, the 2-dimensional image of the subpattern from which the scanning operation started or where the scanning operation ended. As regards the categorizer, it was implemented through a two-layered network with feedforward connections, whose 2-dimensional input layer had the same dimension of the subpatterns contained within the visual scene, whereas the upper layer contained categorizing units. A 2-dimensional pattern x_{jk} was classified as belonging to the category i for which the distance

$$d_i = \Sigma_{jk} (x_{jk} - w_{ijk})^2 \tag{6}$$

was minimal. Here w_{ijk} denotes the connection weights of input lines to the i-th categorizing unit. They were determined by a previous supervised learning procedure, based on LVQ learning rule (cfr. [4]), in which each subpattern should be classified as a function of the letter or the cipher it was representing.

The indices of winning units of the categorization layer were stored into a categorization register. When this latter was containing at least two informational units, the knowledge encoder entered into play, by building two knowledge atoms each one characterized by 10 binary features. For the first atom 4 of these were coding the index of winning unit (i.e the category) of the last subpattern scanned, 2 were coding its spatial relationship to the subpattern from which the scanning started, and 4 were coding the category of this latter. The second atom was built in reverse order (taking care of the new spatial relationship thus implied). In this way, after three successive scannings, 6 different knowledge atoms were built. They constituted the knowledge base for the operation of the network based on Harmony Theory.

4.Simulations

We used a sample of 6 different subpatterns. Starting from random combinations of them we built 100 different scene images. In our simulations we used as independent variables both the parameter κ and the annealing schedule. As regards this latter we introduced two different annealing schedules, lasting, respectively, for 1500 steps (Figure 2a), and for 1700 steps (Figure 2b).

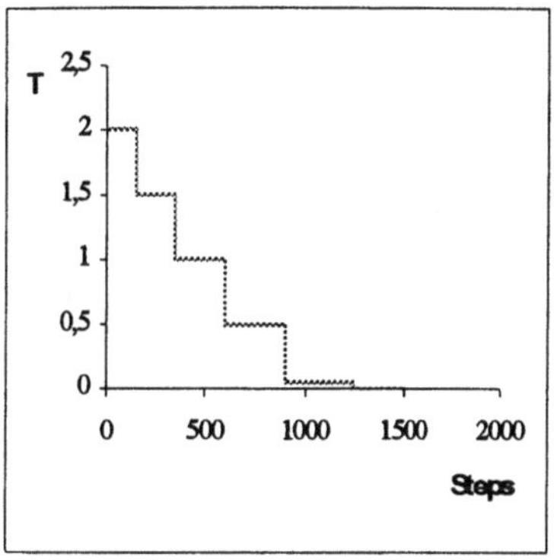

Figure 2a

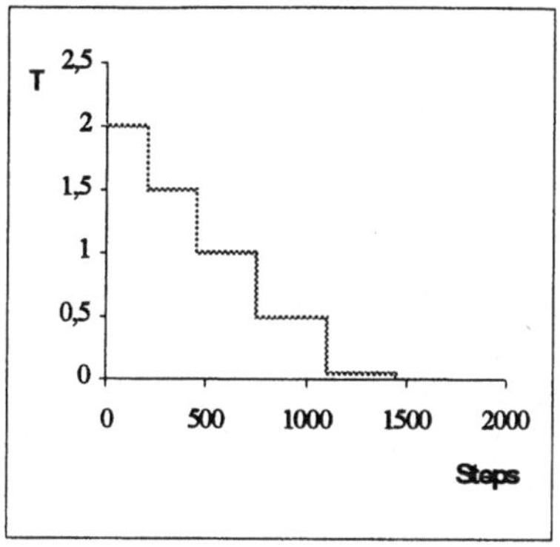

Figure 2b

The chosen values of κ were .1, .2, .3, .4, .5, .65, .7, .8. In the case of the first annealing schedule we added two further values of κ, i.e. .9 and .99. For each annealing schedule and for each value of κ we performed on each scene the 8 different admissible groups of the three successive scannings, for a total of 800 different scannings over all 100 scenes. As dependent variable we used the total number of correct answers over all 800 scannings to the Binding Problem described in the third paragraph. We remark that the probability of obtaining through a random choice the 4 correct binary values of feature activations coding the correct answer to our Binding Problem is 1/16. This implies that, over 800 different scannings, we should expect, on a basis of a purely random choice, (1/16)x800 = 50 correct answers. The results of our simulations (total number of correct answers vs κ) are shown in Figure 3a for the annealing schedule depicted in Figure 2a, and in Figure 3b for the annealing schedule depicted in Figure 2b.

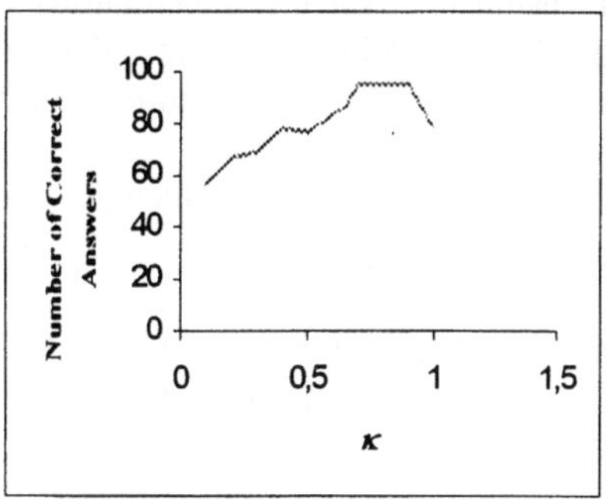

Figure 3a

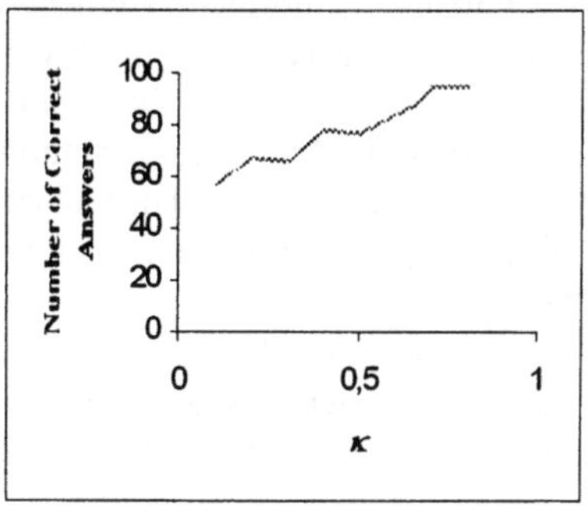

Figure 3b

5. Discussion and conclusions

The outcomes of our numerical experiments, as depicted in the Figures 3a, 3b show, first of all, that the total number of correct answers is always greater than chance level. However, in all cases such a number is still very small, if compared with the total number of scannings performed (800). The form of the graphs contained in Figures 3a, 3b suggests that the situation cannot be improved by choosing a suitable value of κ. Namely, for values of κ greater than or equal to 0.7 we observe a sort of saturation, as the number of correct answers doesn't increase with increasing κ. Even the annealing schedule don't seem to have a strong influence on this phenomenon. As a matter of fact, an analysis of the numerical values attained and attainable by Harmony function H showed us that our Binding Problem is characterized by a number of local maxima of H, very close one to another, so that simulated annealing procedure becomes not so effective in finding the global maximum of H. Such a shortcoming could be avoided by introducing a different definition of H. We are currently exploring this possibility.

References

[1] J.Fodor, E.Lepore. The red herring and the pet fish: why concepts still can't be prototypes. *Cognition*, 58:253-270,1996.

[2] J.E.Hummel, I Biederman. Dynamic binding in a neural network for shape recognition. *Psychological Review*, 99:480-517,1992.

[3] S.Kirpatrick, C.D.Gelatt, M.P.Vecchi. Optimization by simulated annealing. *Science*,220:671-680,1983.

[4] T.Kohonen. *Selforganization and associative memory*. Springer,Berlin-Heidelberg-New York, 1984.

[5] E.Pessa, M.P.Penna. Neural architectures able to give a description of 2-dimensional visual scenes. In *IAPRVA 98-Artificial Intelligence and Pattern Recognition Techniques for Computer Vision*, pages 206-211, Ferrara, Avril 1998.

[6] J.L.Mc Clelland,D.E.Rumelhart, editors, *Parallel Distributed Processing: Explorations in the Microstructure of Cognition*. MIT Press, Cambridge MA, 1986.

[7] P.Smolensky. Information Processing in Dynamical Systems: Foundations of Harmony Theory . In J.L.Mc Clelland, D.E.Rumelhart, editors, *Parallel Distributed Processing: Explorations in the Microstructure of Cognition*, vol. I, pages 154-281, MIT Press, Cambridge MA, 1986.

[8] T.Tambouratzis. A simulated annealing artificial neural network implementation of the n-queens problem. *International Journal of Intelligent Systems*, 12: 739-751, 1997.

[9] P.Tse, D.Wang, D.Atherton. Harmony Theory yields robust machine fault diagnostic system based on Learning Vector Quantization Classifier. *Engineering Applications of Artificial Intelligence*, 9: 487-498, 1996.

[10] P.H.Winston. *The Psychology of Computer Vision*, Mc Graw-Hill, New York, 1975.

Online Learning with Adaptive Local Step Sizes

Nicol N. Schraudolph
nic@idsia.ch

IDSIA, Corso Elvezia 36
6900 Lugano, Switzerland
http://www.idsia.ch/

Abstract

Almeida *et al.* have recently proposed *online* algorithms for local step size adaptation in nonlinear systems trained by gradient descent. Here we develop an alternative to their approach by extending Sutton's work on linear systems to the general, nonlinear case. The resulting algorithms are computationally little more expensive than other acceleration techniques, do not assume statistical independence between successive training patterns, and do not require an arbitrary smoothing parameter. In our benchmark experiments, they consistently outperform other acceleration methods as well as stochastic gradient descent with fixed learning rate and momentum.

1 Introduction

In recent years, many advanced optimization techniques have been developed or adapted for neural network training. In situations where online learning is required or preferable, however, the vast majority of these techniques are not applicable, and stochastic gradient descent remains the algorithm of choice. The central problem here is how to set the local learning rates for rapid convergence. *Normalization* methods [1, 2, 3] *calculate* the optimal local step size under simplifying assumptions — which may or may not model a given situation well. Even such "optimal" algorithms as Kalman filtering can thus be outperformed by *adaptation* methods which *measure* the effects of finite step sizes in order to incrementally improve them [4].

Unfortunately most of the existing step size adaptation algorithms for neural networks adapt only a single, global learning rate [5, 6], can be used only in batch training [7, 8, 9, 10, 11], or both. Given the well-known advantages of stochastic gradient descent, it would be desirable to have online methods for adapting local step sizes. The first such techniques that we are aware of have recently been proposed by Almeida *et al.* [12]. Here we develop an alternative to their approach by extending work by Sutton [4, 13] from linear systems to the general, nonlinear case.

2 Basic Algorithm

Given a sequence $\vec{x}_0, \vec{x}_1, \ldots$ of data points, we minimize the expected value of a twice-differentiable loss function $f_{\vec{w}}(\vec{x})$ with respect to its parameters $\vec{w}$ by stochastic gradient descent:

$$\vec{w}_{t+1} = \vec{w}_t + \vec{p}_t\,\vec{\delta}_t\,, \qquad \text{where} \quad \vec{\delta}_t \equiv -\frac{\partial f_{\vec{w}_t}(\vec{x}_t)}{\partial \vec{w}} \tag{1}$$

The local learning rates $\vec{p}$ are best adapted by exponentiated gradient descent [14], so that they can cover a wide dynamic range while staying strictly positive:

$$\begin{aligned} \ln(\vec{p}_t) &= \ln(\vec{p}_{t-1}) - \mu\,\frac{\partial f_{\vec{w}_t}(\vec{x}_t)}{\partial \ln(\vec{p})} \\ \vec{p}_t &= \vec{p}_{t-1}\exp(\mu\,\vec{\delta}_t\,\vec{v}_t)\,, \qquad \text{where} \quad \vec{v}_t \equiv \frac{\partial \vec{w}_t}{\partial \ln(\vec{p})} \end{aligned} \tag{2}$$

and μ is a global meta-learning rate. The definition of $\vec{v}$ rests on the assumption that each element of $\vec{p}$ affects $f_{\vec{w}}(\vec{x})$ only through the corresponding element of $\vec{w}$. With considerable variation, (2) forms the basis of most local rate adaptation methods found in the literature. In order to avoid an expensive exponentiation [15] for each weight update, we typically use the linearization $e^u \approx 1 + u$, valid for small $|u|$, giving

$$\vec{p}_t = \vec{p}_{t-1}\max(\varrho,\, 1 + \mu\,\vec{\delta}_t\,\vec{v}_t)\,, \tag{3}$$

where we constrain the multiplier to be at least (typically) $\varrho = 0.1$ as a safeguard against unreasonably small — or negative — values.

Definition of $\vec{v}$. It is tempting to look at (1), declare $\vec{w}_t$ and $\vec{\delta}_t$ to be independent of $\vec{p}_t$, and quickly arrive at $\vec{v}_{t+1} = \vec{p}_t\,\vec{\delta}_t$, which (up to normalization) is the approach of Almeida *et al.* [12]. However, this fails to take into account the incremental nature of gradient descent: a change in $\vec{p}$ affects not only the current update of $\vec{w}$, but also future ones. Some authors account for this by setting $\vec{v}$ to an exponential average of past gradients [6, 8]. We prefer to extend Sutton's exact method [4] for iterative calculation of $\vec{v}$ to nonlinear systems:

$$\begin{aligned} \vec{v}_{t+1} &= \frac{\partial \vec{w}_t}{\partial \ln(\vec{p})} + \frac{\partial(\vec{p}_t\,\vec{\delta}_t)}{\partial \ln(\vec{p})} \\ &= \vec{v}_t + \vec{p}_t\,\vec{\delta}_t - \vec{p}_t\left[\frac{\partial^2 f_{\vec{w}_t}(\vec{x}_t)}{\partial \vec{w}\,\partial \vec{w}^T}\,\frac{\partial \vec{w}_t}{\partial \ln(\vec{p})}\right] \\ &= \vec{v}_t + \vec{p}_t\,(\vec{\delta}_t - H_t\vec{v}_t)\,, \end{aligned} \tag{4}$$

where H_t denotes the instantaneous Hessian of $f_{\vec{w}}(\vec{x})$ at time t. Note that there is an efficient $O(n)$ method to calculate the matrix-vector product $H_t\vec{v}_t$ without ever computing or storing H_t itself [16].

Meta-level normalization. The gradient descent in $\vec{p}$ at the meta-level (2) may of course suffer from ill-conditioning as much as the descent in $\vec{w}$ at the main level (1); in a sense we have only pushed the problem of choosing the appropriate step size down one meta-level. In many step size adaptation algorithms, this issue is addressed by incorporating squashing functions such as cosine [7] or sign [8, 9, 10] in the update for $\vec{p}$. Unfortunately such nonlinearities do not preserve the zero-mean property of a skewed density; this makes these methods unsuitable for online learning, whose equilibrium is characterized by zero-mean, skewed noise in the stochastic gradient. Notably, Almeida *et al.* [12] avoid this pitfall by using a running estimate of the gradient's noise variance as their normalizer (*cf.* [3]).

Our iterative definition of $\vec{v}$ has the advantage of serving as a highly effective conditioner for the meta-descent (2): the fixpoint of (4) is given by

$$\vec{v}_t \;=\; H_t^{-1}\vec{\delta}_t\,, \tag{5}$$

a stochastic Newton step which scales with the inverse of the gradient. Consequently, we can expect the product $\vec{\delta}_t\vec{v}_t$ in (2) to be a very well-conditioned quantity. Experiments confirm that our algorithm does not require (and in fact is hampered by) additional meta-level normalization as in [12].

3 Special Cases

The $\vec{v}$ update introduced above forms the basis for a number of existing and novel step size adaptation methods. Simplifying (4) by setting $H_t = \lambda I$ produces algorithms closely related to those of Almeida *et al.* [12, $\lambda = 1$] and Jacobs [8], respectively. Applying (4) to the linear (LMS) system

$$f_{\vec{w}_t}(\vec{x}_t, y_t) \;\equiv\; \frac{1}{2}(y_t - a_t)^2, \qquad \text{where} \qquad a_t \;\equiv\; \vec{w}_t^{\,T}\vec{x}_t\,, \tag{6}$$

diagonalizing the instantaneous Hessian $H_t = \vec{x}_t\vec{x}_t^T$ to $H_t \approx \mathrm{diag}(\vec{x}_t^2)$, and adding a positive-bounding operation, yields Sutton's IDBD algorithm [13]:

$$\vec{v}_{t+1} \;=\; \vec{v}_t\,(1 - \vec{p}_t\,\vec{x}_t^2)^+ + \vec{p}_t\,\vec{\delta}_t \tag{7}$$

For a normalized LMS system, the same approach produces his K1 algorithm [4]. Our ELK1 algorithm [17] extends K1 by adding a sigmoid function σ to the system's output, giving

$$\begin{aligned} f_{\vec{w}_t}(\vec{x}_t, y_t) \;&\equiv\; \frac{1}{2}(y_t - \sigma(a_t))^2, \qquad \vec{w}_{t+1} \;=\; \vec{w}_t \;+\; k_t\,\vec{p}_t\,\vec{\delta}_t\,, \\ \vec{v}_{t+1} \;&=\; (\vec{v}_t \;+\; k_t\,\vec{p}_t\,\vec{\delta}_t)\,[\,1 \;-\; k_t\,\vec{p}_t\,\vec{x}_t^2\sigma'(a_t)^2]\,, \\ &\text{where} \quad k_t \;\equiv\; \frac{1}{1 + \vec{x}_t^T(\vec{p}_t\,\vec{x}_t)\,\sigma'(a_t)^2} \end{aligned} \tag{8}$$

The ELK1 update can be applied independently to each node in a multi-layer perceptron, providing a computationally attractive diagonal approximation to the full Hessian update for $\vec{v}$ given in (4).

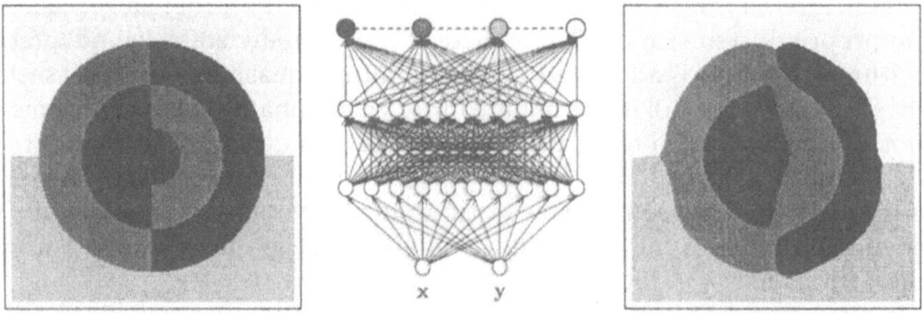

Figure 1: The four regions classification task (left), the neural network trained on it (center), and a typical solution after 50 000 training patterns (right).

4 Benchmark Results

We evaluated our approach on the "four regions" classification task [18]. A fully connected feedforward network with 2 hidden layers of 10 units each (tanh nonlinearity) is to classify two continuous inputs (range [-1,1]) as illustrated in Figure 1. We used "softmax" output units and a negated cross-entropy loss function E. We compared five online algorithms: "SMD" — equations (1), (3), and (4), "ELK1" — equations (3) and (8), "ALAP" — the normalized step size adaptation method of Almeida *et al.* [12], "vario-η" — a learning rate normalization method [3], and "momentum" — stochastic gradient descent with a fixed learning rate and momentum. The free parameters of each method were tuned for best performance; this was easier for SMD and ELK1, which have only the initial and meta-learning rates as free parameters, than for ALAP, which in addition requires a smoothing rate constant.

We trained each algorithm 25 times on a uniformly random sequence of 50 000 patterns, starting from different sets of uniformly random initial weights (range $[-0.3, 0.3]$; biases initialized to zero). Since each pattern was seen only once, the empirical loss E provided an unbiased estimate of expected loss, and hence generalization ability. Figure 2 (left) shows the average value of $\bar{E}$ — an exponential trace (smoothing parameter 0.999) of E — during training. While all accelerated methods convey a similar initial increase in convergence speed over gradient descent with momentum, there are clear differences in their later performance. Our full Hessian update method (SMD) converges slightly but reliably faster than ELK1, the diagonalized, node-decoupled version. Both clearly outperform ALAP, the only other local step size adaptation method for online learning that we know of. Finally, vario-η behaves like a typical normalization method, combining the best initial with the worst asymptotic performance. (We surmise that this shortcoming of vario-η could be addressed by combining it with a global learning rate adaptation mechanism.)

There are other differences between the above algorithms. For instance, ALAP is derived under the assumption that successive training patterns are

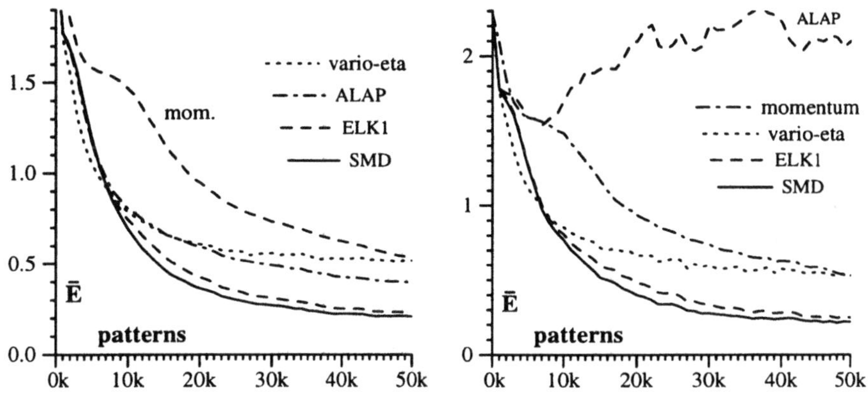

Figure 2: The average smoothed loss $\bar{E}$ during online training on the four regions task, plotted against the number of training patterns, for uniformly random (left) *vs.* super-uniform Sobol sampling (right).

statistically independent — a condition that may be hard to meet *e.g.* in *situated* applications, where the input is provided by a real-time environment over which the learner does not have complete control. We repeated the above experiment while sampling the input space according to the Sobol sequence [19, pp. 311–314], which achieves a super-uniform covering of the input space at the expense of introducing strong short-term correlations between samples. Figure 2 (right) shows that while the other methods remained largely unaffected, ALAP failed to converge in all 25 runs. This casts some doubt on ALAP's reliability in situations where independence between successive training patterns cannot be guaranteed.

References

[1] S. Becker and Y. LeCun, "Improving the convergence of back-propagation learning with second order methods", in *Proceedings of the 1988 Connectionist Models Summer School*, D. Touretzky, G. Hinton, and T. Sejnowski, Eds., Pittsburg 1988, 1989, pp. 29–37, Morgan Kaufmann, San Mateo.

[2] N. N. Schraudolph and T. J. Sejnowski, "Tempering backpropagation networks: Not all weights are created equal", in *Advances in Neural Information Processing Systems*, D. S. Touretzky, M. C. Mozer, and M. E. Hasselmo, Eds. 1996, vol. 8, pp. 563–569, The MIT Press, Cambridge, MA, `ftp://ftp.idsia.ch/pub/nic/nips95.ps.gz`.

[3] R. Neuneier and H. G. Zimmermann, "How to train neural networks", in *Neural Networks: Tricks of the Trade*, vol. 1524 of *Lecture Notes in Computer Science*, pp. 373–423. Springer Verlag, Berlin, 1998.

[4] R. S. Sutton, "Gain adaptation beats least squares?", in *Proc. 7th Yale Workshop on Adaptive and Learning Systems*, 1992, pp. 161–166, `ftp://ftp.cs.`

umass.edu/pub/anw/pub/sutton/sutton-92b.ps.gz.

[5] Y. LeCun, P. Y. Simard, and B. Pearlmutter, "Automatic learning rate maximization in large adaptive machines", in *Advances in Neural Information Processing Systems*, S. J. Hanson, J. D. Cowan, and C. L. Giles, Eds. 1993, vol. 5, pp. 156–163, Morgan Kaufmann, San Mateo, CA.

[6] N. Murata, K.-R. Müller, A. Ziehe, and S.-i. Amari, "Adaptive on-line learning in changing environments", in *Advances in Neural Information Processing Systems*, M. C. Mozer, M. I. Jordan, and T. Petsche, Eds. 1997, vol. 9, pp. 599–605, The MIT Press, Cambridge, MA.

[7] L.-W. Chan and F. Fallside, "An adaptive training algorithm for back propagation networks", *Computer Speech and Language*, **2**:205–218, 1987.

[8] R. Jacobs, "Increased rates of convergence through learning rate adaptation", *Neural Networks*, **1**:295–307, 1988.

[9] T. Tollenaere, "SuperSAB: fast adaptive back propagation with good scaling properties", *Neural Networks*, **3**:561–573, 1990.

[10] F. M. Silva and L. B. Almeida, "Speeding up back-propagation", in *Advanced Neural Computers*, R. Eckmiller, Ed., Amsterdam, 1990, pp. 151–158, Elsevier.

[11] M. Riedmiller and H. Braun, "A direct adaptive method for faster backpropagation learning: The RPROP algorithm", in *Proc. International Conference on Neural Networks*, San Francisco, CA, 1993, pp. 586–591, IEEE, New York.

[12] L. B. Almeida, T. Langlois, J. D. Amaral, and A. Plakhov, "Parameter adaptation in stochastic optimization", in *On-Line Learning in Neural Networks*, D. Saad, Ed., Publications of the Newton Institute, chapter 6. Cambridge University Press, 1999, ftp://146.193.2.131/pub/lba/papers/adsteps.ps.gz.

[13] R. S. Sutton, "Adapting bias by gradient descent: an incremental version of delta-bar-delta", in *Proc. 10th National Conference on Artificial Intelligence*. 1992, pp. 171–176, The MIT Press, Cambridge, MA, ftp://ftp.cs.umass.edu/pub/anw/pub/sutton/sutton-92a.ps.gz.

[14] J. Kivinen and M. K. Warmuth, "Additive versus exponentiated gradient updates for linear prediction", in *Proc. 27th Annual ACM Symposium on Theory of Computing*, New York, NY, May 1995, pp. 209–218, The Association for Computing Machinery.

[15] N. N. Schraudolph, "A fast, compact approximation of the exponential function", *Neural Computation*, **11**(4):853–862, 1999, ftp://ftp.idsia.ch/pub/nic/exp.ps.gz.

[16] B. A. Pearlmutter, "Fast exact multiplication by the Hessian", *Neural Computation*, **6**(1):147–160, 1994.

[17] N. N. Schraudolph, "Online local gain adaptation for multi-layer perceptrons", Tech. Rep. IDSIA-09-98, Istituto Dalle Molle di Studi sull'Intelligenza Artificiale, Corso Elvezia 36, 6900 Lugano, Switzerland, 1998, ftp://ftp.idsia.ch/pub/nic/olga.ps.gz.

[18] S. Singhal and L. Wu, "Training multilayer perceptrons with the extended Kalman filter", in *Advances in Neural Information Processing Systems. Proceedings of the 1988 Conference*, D. S. Touretzky, Ed., San Mateo, CA, 1989, pp. 133–140, Morgan Kaufmann.

[19] W. H. Press, S. A. Teukolsky, W. T. Vetterling, and B. P. Flannery, *Numerical Recipes in C: The Art of Scientific Computing*, Cambridge University Press, second edition, 1992.

Section 5
Pattern Recognition and Signal Processing

A Feed-Forward Neural Network for Robust Segmentation of Color Images

C. Amoroso, A. Chella*, V. Morreale, P. Storniolo
Department of Electrical Engineering, University of Palermo
Palermo, Italy

*Department of Electrical Engineering, University of Palermo and
Centro di Studio sulle Reti di Elaboratori, CNR
Palermo, Italy

Abstract

A novel approach for segmentation of color images is proposed. The approach is based on a feed-forward neural network that learns to recognize the hue range of meaningful objects. Experimental results showed that the proposed method is effective and robust even in presence of changing environmental conditions. The described technique has been tested in the framework of the Robot Soccer World Cup Initiative (RoboCup). The approach is fully general and it may be successfully employed in any intermediate level image-processing task, where the color is a meaningful descriptor.

1 Introduction

Segmentation of color images is one of the most difficult tasks in robot vision. Nevertheless, color is a powerful descriptor that greatly simplifies object identification and extraction [7]. Several approaches based on structural methods and decision-theoretic methods have been proposed in the literature (see Gonzalez and Wintz [2] for a review). These methods are usually based on the assumption of a simple analytic expression for the distribution functions of the pattern classes, such as the Gaussian probability density function. This assumption may be unrealistic, especially in the case of images acquired by a mobile robot vision system, owing to the presence of illumination changes, shadowing and obstacles in the robot working environment.

To accomplish the problem of a robust segmentation of color images oriented to the vision system of a mobile robot, we employ a novel method based on a feed-forward neural network [6] that learns to recognize the hue range of meaningful objects. A neural network does not require the exact knowledge of the statistical distribution of input items, as it generates an internal representation of input data distribution by a suitable learning phase. Furthermore, a neural network deals effectively with noisy and partial data, as it is able to generalize to new and unexpected situations.

The described approach has been developed in the framework of the Robot Soccer World Championship Initiative (RoboCup, see Kitano et al. [3]). In this

Figure 1: Images of the adopted experimental framework. The soccer field is green, the walls and the field lines are near white, the goal area is blue and the ball is red.

framework, a mobile robot acquires the *RGB* color images (width 160 x height 120) by a CCD camera fixed on its top. The images are then processed by a PC on board of the robot. The scene is made up by a green soccer field, near-white walls and field lines, a blue goal area and a red ball (see Fig. 1).

Although the colors of the objects in the scene are clear and well defined, several disturbing conditions arise during the RoboCup competitions, that impose tight constraints to the segmentation step. The light conditions of the field vary during the day: in the morning it is illuminated by sun light and in the afternoon it is illuminated by artificial light. The persons and the other robots generate unexpected shadows effects. Different game fields may be characterized by similar but not equal colors, e.g., the green color of the training field may not be exactly equal to the green color of the game field.

The presented technique is part of the vision system adopted by the soccer player robots of the Italian robot team ART (Azzurra Robot Team) (Nardi, Clemente and Pagello [4]), a robot team in "F2000" category (middle size league).

2 Analysis of the histogram

It is well known that the *HSI* (Hue, Saturation, Intensity) color model of an image is an useful tool for developing image-processing algorithms. In particular, the hue is the angle of a vector with respect to a reference axis lying on the iso-intensity plane; the vector's magnitude is proportional to the saturation value of the pixel. Hence, the hue is a real number in the range $[0° - 360°]$; its histogram is a periodic function in the same range [2].

In our proposal, we segment the color images by analyzing their hue histograms. This choice allows us to work on a 1D space. This simplify the search for significant clusters that identify meaningful regions of pixels. Hence, it avoids the needs of working in a 2D (such as *IQ*, a^*b^*, u^*v^*) or a 3D (such as *RGB*, *CieXYZ*) color space. Furthermore, the analysis of the hue histogram allows us to employ descriptors that exhibit invariant features over a range of lighting conditions [5]. It should be noted that also the needs for time consuming algorithms, as the ones related to the filtering processes in color spaces, is avoided.

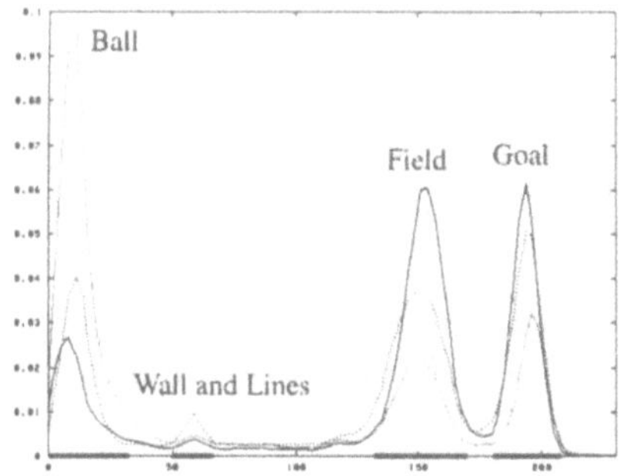

Figure 2: Typical hue histograms related to images of the game field.

Fig. 2 shows typical hue histograms of color images of the chosen framework. For the sake of simplicity, the reported histograms are in the range $[0° - 230°]$; their values outside are near zero. It should be noted that the color modes (the peaks in the histograms) extend themselves within similar hue ranges. Each color mode corresponds to an object in the scene: the first one is related to the ball, the second one indicates the wall and the lines, the third one is related to the game field and the last one indicates the game goal. In the Fig., the hue ranges corresponding to the color modes are highlighted.

3 The neural network approach

As previously explained, the hue range of a meaningful color mode of the histogram is not clearly defined because it greatly depends on illumination conditions, shadows effect and so on.

In this section, we show how to associate the hue ranges of the color modes extracted from the image with the correct class of objects. Briefly, we extract the hue ranges by using an efficient technique based on the convolution of the hue histogram with a suitably defined *valley detection kernel* (see [1] for details). Each extracted hue range is presented as input to a multilayer perceptron [6] that associates it with the corresponding object.

To code the hue ranges of the histogram to feed them as input to the neural network, we compared three different approaches: the *bit range*, the *Gaussian* and the *value* coding. Let L be the number of the considered hue sampled levels. Each level corresponds to one of the input units of the neural network. In the *bit range* coding, we set to 1 all the input units corresponding to the hue levels belonging to the considered range, and we set the other units to 0. In the *Gaussian* coding, we consider a Gaussian distribution of probability over the considered hue range. Then, we set to 1 the input units corresponding to the hue levels belonging to the range $[m - \sigma^2, m + \sigma^2]$, where m is the mean value and σ^2 is the variance of the hue of the considered color mode. The units outside the range are set to 0. In the *value* coding, we set the input units to the effective values of the color mode of the considered hue range; the units

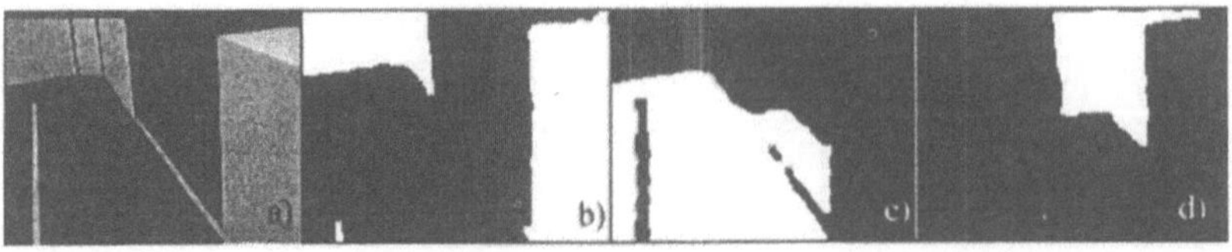

Figure 3: An example of the operation of the network: a) is the starting image; b) is the *wall* area; c) is the *field* area, and d) is the *goal* area.

that do not belong to the considered hue range are set to 0.

The choice of the number L of sampled hue levels corresponding to the input nodes of the network is an important issue. We have chosen $L = 32$ according to the width and the resolution of the hue segments, corresponding to the classes of objects of interest. Experiments showed that for $L > 32$ the network does not improve their performances, while for $L < 32$ the hue segments overlap and the performances of classification degrade.

In the considered application, we recognize the following four classes of object: the *field*, the *wall and lines*, the *goal* and the *ball*. Hence, we chosen the number O of the units in the output layer so that each node represents one of the above mentioned classes; i.e., $O = 4$.

We chosen the number H of hidden nodes by considering the variation of performances of the network by varying H. The number of hidden nodes was assumed to be the lowest one that achieves the minimum error on the test set. Experiments showed that the best performances were reached with $H = 16$.

In conclusions, we experimented the best performances with a feed-forward neural network with 32 input nodes, 16 hidden nodes and 4 output nodes. We chose the sigmoid activation function for every node. The learning rate used was 0.5.

After the classification of the hue ranges performed by the neural network, the obtained classified ranges are used to label each pixel in the image. The last step of our technique is the extraction of the connected components sharing the same label.

A visual example of the obtained results is reported in Fig. 3. Fig. 3a shows the acquired image. Fig. 3b shows the pixels classified as belonging to the *wall* area; Fig. 3c shows the pixels classified as belonging to the *field* area, and Fig. 3d shows the pixels classified as belonging to the *goal* area.

4 Experimental results

To analyze in details the characteristics of the network, we performed three different tests: *Test A*, *Test B* and *Test C*, described below. In all of the tests, the training and the test sets are non-overlapping sets of patterns. We repeated the tests for all the three input coding described in the previous Sect.

In *Test A*, we analyzed the performances of the network by using training and testing images drawn by the same soccer field. To generate the training

	bit range	*Gaussian*	*value*
Test A (%)	0.33	0.67	0.67
Test B (%)	10.0	19.0	17.0
Test C (%)	6.17	7.83	9.33

Table 1: Misclassifications of the neural network (see text).

an the testing sets, we used 300 images acquired by the robot moving in the soccer field built in our laboratory. Different lighting conditions have been considered during acquisition in order to provide the network with a high degree of robustness.

We divided the training set into 5 mutually exclusive groups of 60 patterns. In each group, there were 15 patterns of each of the four classes. The network has been trained using 4 of these groups and tested using the fifth group. This train-and-test process has been repeated 5 times, using all groups of 60 test patterns. At the end, a mean misclassification error has been evaluated: the results are shown in the first row of Tab. 1. Results show the high reliability of the network when it works with pattern generated from images acquired in the same environment of the training phase.

To verify the capabilities of the network to work in different environmental conditions, during *Test B* we used the same training set of *Test A* but the test set was made up by images acquired by the robot during the previous RoboCup competition held in Paris in 1998. Therefore, environmental conditions, tonality of colors and illuminations in test set were very different with respect to the training set. The corresponding misclassification errors are shown in second row of Tab. 1. It should be noted that the error is maintained within acceptable limits. These results confirm that our method is satisfactory with respect to severe environmental changes.

To evaluate the performances of the network working on noisy data, during *Test C* we used the same training set of the previous tests, and a test set made up by noisy images. We produced the test set as follows: each pattern of the training set has been modified by operating a right shift and a left shift of the input values. Hence, for any training pattern, we produced two new test patterns. Altogether, 600 new patterns have been generated, on which the network has produced the results shown in third row of Tab. 1. Once again, experimental results highlight that the network is robust with respect to the noise.

By analyzing Tab. 1 with respect to the input coding, it should be noted that the *bit range* always performed better in terms of misclassification error with respect to the other coding. This is due to the fact that the *bit range* coding do not make any assumption on the form of the hue histogram. It therefore provides a low sensibility with respect to the form of the color modes. The result is an improvements in terms of robustness and adaptation to the environmental situations.

5 Conclusions

The problem of automatic object segmentation in color images acquired by a camera placed on a mobile robot platform has been discussed. A feed-forward neural network has been employed to classify hue ranges corresponding to color modes extracted from histograms. Experimental results showed that the proposed approach is effective and robust with respect to environmental conditions, as shadows and illumination variations. Although this method has been applied in the RoboCup experimental framework, it is an effective intermediate-level step in generic image interpretation tasks where the color is a meaningful descriptor.

Acknowledgements

Authors would like to thank Edoardo Ardizzone and Salvatore Gaglio for stimulating discussions on the topics of the papers. Authors would also like to thank all the persons involved in the Azzurra Robot Team and especially the coordinator Daniele Nardi for their encouragement. This work has been partially supported by MURST "Progetto Cofinanziato CERTAMEN".

References

[1] C. Amoroso and V. Morreale. Artificial vision for mobile robot. Technical report, CSAI Lab, Dept. of Electrical Engineering, Università di Palermo, 1999.

[2] R.C. Gonzalez and P. Wintz. *Digital Image Processing, II ed.* Addison-Wesley, 1987.

[3] H. Kitano, M. Asada, Y. Kuniyoshi, I. Noda, E. Osawa, and H. Matsubara. RoboCup: a challenge problem for AI. *AI Magazine*, 18(1):73–85, 1997.

[4] D. Nardi, G. Clemente, and E. Pagello. ART: azzurra robot team. Technical report, DIS Università di Roma "La Sapienza", 1998.

[5] F. Perez and C. Koch. Toward color image segmentation in analog VLSI algorithm and hardware. *International Journal of Computer Vision*, 12(1):17–42, 1994.

[6] D.E. Rumelhart, G.E. Hinton, and R. Williams. Learning internal representation by error propagation. In D.E. Rumelhart and J.L. McClelland, editors, *Parallel Distributed Processing*, volume 1. MIT Press, Cambridge, MA, 1986.

[7] G. Sharma and H.J. Trussell. Digital color imaging. *IEEE Transactions on Image Processing*, 6(7):901–932, 1997.

Parameter Identification Using Aspects
Application to the Human Cardiovascular System

Alexander Asteroth, Jens Frings-Naberschulte
Dept. of Computer Science, Bonn University
Bonn, Germany

Knut Möller
Dept. of Medical Engineering, Univ. of Applied Sciences Furtwangen
Villingen Schwenningen, Germany

Abstract

The human cardiovascular system (CVS) is a complex dynamical system. The determination of models for individual subjects/astronauts is of particular interest in (space) medicine. CVS models complexity and the large number of free parameters cause standard system identification methods to fail. A concept named *aspect* is introduced. This concept represents independent components of dynamical systems. Aspects are closely related to projection pursuit, independent component analysis and model selection. Using this concept it was possible to identify individualized CVS models that reproduce the shift in cardiovascular reaction due to exposure to weightlessness. The application of MLPs to approximate single *aspects* of the CVS (model) allows to perform identification in real time.

1 Introduction

Alterations in the cardiovascular system (CVS) of astronauts are a major topic in current space medical research [5]. After exposure to weightlessness (μ-G), orthostatic intolerance or even circulatory collapse may occur. Particularly the reaction-patterns to orthostatic stress – such as standing in an upright position – differ significantly pre- and post-flight. Similar changes are also noticed in clinical patients with long bed rest periods.

The identification of exact causes is difficult because potential sources for the observed reactions are not directly accessible to measurements. Moreover data is particularly scarce due to the limited number of subjects. Modeling shall be employed to determine hypotheses on cardiovascular alterations explaining the observed reaction-patterns [1]. If consistent models can be determined, prediction, online monitoring and evaluation/optimization of countermeasures [10] are further areas of application. Online monitoring poses time constraints on the identification process and demands for real time identification.

The cardiovascular response of a test subject to a certain stimulus is influenced by a huge number of variables. Therefore special provocative tests were designed to keep this number as small as possible. In particular tilt-table-test

(TT) and lower-body-negative-pressure-experiments (LBNP) are performed before, after and, in case of LBNP, during space flight.

In TT the subject is fixed on a table in supine position. Data, like blood-pressure, heart-rate, etc. is recorded continously. After 15 minutes in supine position, the subject is tilted upright, yielding a typical response in physiological parameters. It should be mentioned that this response, while typical for a particular subject, differs significantly inter-individually. This also applies to the μ-G induced alterations.

During LBNP-experiments the subject is wearing a special device, that applies negative pressure to the lower body. The cardiovascular response pattern is comparable to TTs but is applicable in μ-G too. Though, some differences exist and demand for the inclusion of further parameters.

The before mentioned inter-individual differences require individualized models. While a large number of average-case models exist, e.g. [4, 9], individualized models have not been thoroughly investigated so far. Parameter identification techniques have to be applied and the models' complexity provokes considerable computational costs.

The functional space has to be restricted to allow the determination of probable solutions in the absence of large amounts of data. Physiological knowledge on the underlying mechanisms is exploited to construct a system of parameterized differential equations. Search can thereupon be restricted to the parameter space of this complex dynamical system.

The dynamics of the human CVS are nonlinear and, in particular, cardiovascular control is even non-continuous [9]. Most algorithms from system identification [7] are therefore not applicable and unconstrained methods [6] have to be used. these methods rely on model's derivatives whose efficient determination is not possible due to non-continuity. The complexity of their estimation by model evaluation hindered successful identification with these straightforward approaches.

In current work we are investigating the usage of structural information, gathered in advance, to decompose and simplify the learning task. The structures used therein were called *aspects* and represent a formalization of the very same term in informal language. It is closely related to projection pursuit methods, independent component analysis techniques and Model selection criteria such as AIC or NIC (see [11]) but focuses especially on *local* independent factors because of their importance in gradient descent.

The following sections will briefly describe the methods used, and results achieved for tilt-table-experiments.

2 Methods

Independent factors of the dynamical system help to decompose the learning task into several independent subtasks. Those factors might only exist in a local environment of some point in parameter space but because gradient descend is a local method this suffices to reduce the dimensionality of the learning task.

Factors of a dynamical system correspond to the term aspect in informal

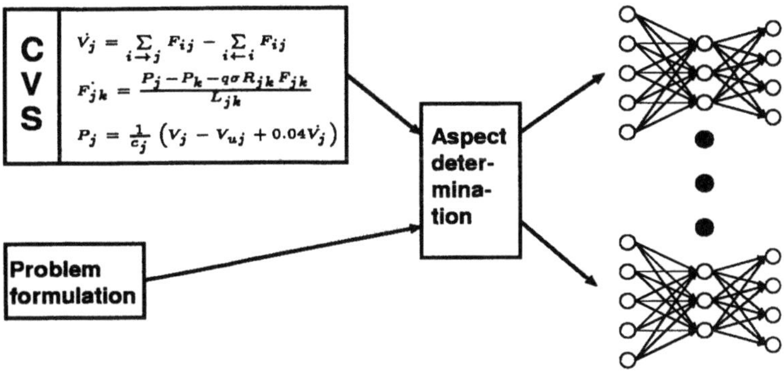

Figure 1: Suitable aspects of the dynamical system may be approximated by MLPs.

language which we may formalize by the following

Definition 1: *Let $\Sigma = (T, X, \varphi, P)$ be a parameterized system (see [8, 3]), where P is the set of parameter values, X is the set of system's states, T is the time axis and $\varphi : P \to (T \times X)$ is a functional that maps parameters to state sequences.*

If $\hat{X}$, $\hat{P}$ and $\hat{T}$ are sets, and $\xi : X \to \hat{X}$, $\pi : P \to \hat{P}$ and $\tau : T \to \hat{T}$ are surjective maps, then the graph $A(\Sigma, \pi, \tau, \xi)$:

$$A = \{ (\hat{p}, \hat{t}, \hat{x}) \in (\hat{P} \times T \times \hat{X}) \,|\, \exists p \in P, t \in T : \hat{p} = \pi(p), \hat{t} = \tau(t), \hat{x} = \xi(\varphi(p,t)) \}$$

is called **aspect** *of Σ with regard to (π, τ, ξ).*

Thus an *aspect* is a projection of a system's graph into a lower dimensional space. Independence of *aspects* may then be defined by the following commutative diagram:

$$\begin{array}{ccc} P & \xrightarrow{\varphi} & T \times X \\ \pi \downarrow & & \downarrow \tau\times\xi \\ \hat{P} & \xrightarrow[\hat{\varphi}]{} & \hat{T} \times \hat{X} \end{array}$$

If a function $\hat{\varphi}$ exists, such that the diagram commutates, $A(\Sigma, \pi, \tau, \xi)$ is a *commutative aspect* and corresponds to an independent factor of the system (see [3]).

The decomposition into relevant factors often yields non-dynamical aspects. While dynamical systems are not representable by feed-forward perceptrons the latter usually are, thus allowing their approximation by MLPs (comp. figure 1) and providing a large amount of fast and robust identification algorithms.

If the approximated *aspect* is *commutative*, parameter identification can be based on MLP yielding an *aspect* of the true parameter vector. Aspect's commutativity ensures that re-projection produces a valid set of parameters. This set in turn fulfills the corresponding *aspect* of the parameter identification problem. By intersecting the mentioned parameter sets of suitable *aspects* the solution is obtained.

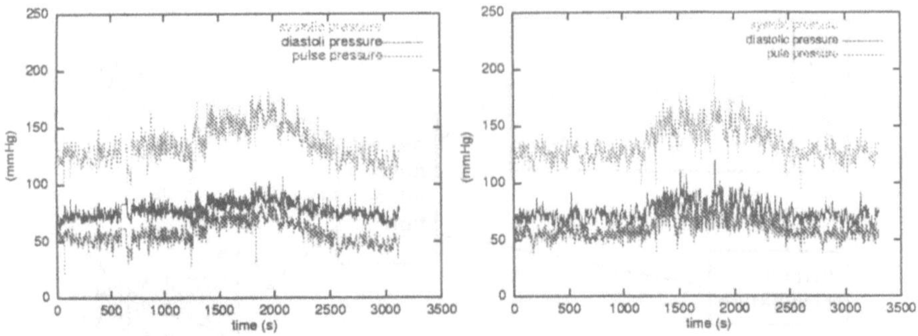

Figure 2: Left (a): Subject's typical tilt response in systolic diastolic and mean arterial pressure. Right (b): simulated response (with added colored noise)

Besides robustness and fastness MLPs (with appropriate nonlinearities) possess a further advantage. They are infinitely often differentiable. By approximating an aspect of the system they encode the corresponding structural information in a convenient fashion. Higher order derivatives of the MLP [2] may be used to express local structural properties of the system.

Thus MLPs may provide a tool for decomposing the learning task even if parameter identification may not be performed on the MLP directly (due to insufficient accuracy).

3 Results

In this section we will describe the successful identification of an individualized CVS-model for TT. In this particular case, problem relevant aspects were convergent and could be approximated by MLPs. While the introduction of aspects made the determination of a solution possible their approximation by MLPs lead to the fulfillment of the real time property.

3.1 Scenario

In the german MIR'97-mission TT-experiments were carried out pre (2×) and post (3×) space flight. Preflight experiments yielded typical cardiovascular responses — for this particular subject (comp. figure 2 (a)). However a significant change in this reaction pattern was observed in postflight experiments.

Space-medical experts from the DLR proposed that a shift in cardiovascular regulation may account for these alterations. Therefore search was restricted to the corresponding aspect of the model, to prove the possibility of this hypothesis. From recorded data average systolic, diastolic and mean arterial pressure values as well as heart rate were selected for reproduction by the model, resulting in a yet smaller aspect of the CVS.

The usual sum-squared-distance between aspect-model and the system's (subject's) aspect was used to guide gradient search. Figure 3 (a) shows this two-dimensional projection of the error surface. Only the peripheral resistance (PR) and myocardial contractility (MC) controller parameters are varied. The remaining parameters are fixed to their optimal values.

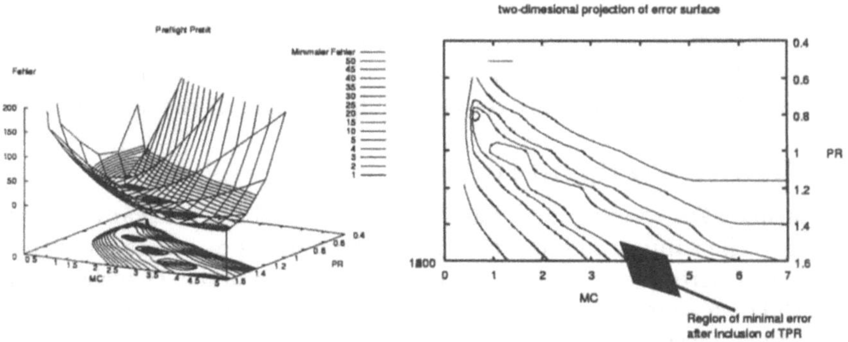

Figure 3: Left (a): sum squared error as a function of PR and MC. Right (b): two-dimensional projection of (a). The rhombus marks the area of error < 1 after inclusion of TPR.

A narrow valley with a flat nonlinear bottom and a wide optimal region is observed. Taking into account the signal to noise ratio, the possible solution space consists of the entire bottom of the valley . Obviously the ξ-projection of the aspect was too restricted to lead to a unique solution. We therefore (re-) included an estimate of the total peripheral resistance (TPR), thus enhancing the aspect. The resulting region of tolerable error is depicted in figure 3 (b).

The narrow valley still exists, after the inclusion of TPR, which demands for carefully chosen learning parameters. Thus the identification problem, not using MLPs, was not robust. However with five different manually chosen learning-parameter settings learning succeeded (comp. figure 2 (b))within 500 steps – 5 days – for each experiment.

But MLPs could be applied for approximation as the underlying functional dependences are sufficiently simple which could be expected by the smoothness of the error surface. With MLPs we were able to achieve the same results (qualitatively) as with direct model evaluation. The solution however was determined in two minutes approximately (20.000 steps). No adjustments of learning parameters were necessary and convergence proved to be robust.

4 Conclusions and Future Work

Robust real-time identification of individualized CVS-models became feasible by two steps. First, solvability was achieved by restricting parameter identification to problem relevant features. The procedure however was neither fast nor robust. The concept ***aspect*** proved valuable in formalization. Second robustness was obtained by approximating the aspect-model with a differentiable function approximator such as MLPs. A speedup factor of 3600 approx. and real-time characteristics were advantages of this approach.

In current research we are investigating LBNP-experiments because the number of LBNP subjects is sufficiently large to allow statistically relevant statements. In contrary to the tilt-table experiments aspect-approximations by

MLPs were not sufficiently accurate to allow parameter identification in the approximations directly. The possibility to determine derivatives of arbitrary order in MLPs [2], allows a further subdivision in local commutative aspects. Those shall in turn be employed to speed up learning in the true system.

5 Acknowledgments

The author likes to thank Dr. habil Joachim Hertzberg for many fruitful discussions. The support in important medical questions by members of the DLR institute of aerospace medicine, in particular Dr. Jürgen Drescher and Luis Beck PhD was particularly useful. The authors work is supported by a scholarship from Wernher von Braun-foundation. This study was partly supported by grant 50 WB 96286 from DARA GmbH Germany to K. Möller.

References

[1] A. Asteroth, K. Möller, J. Frings, L. Beck, and J. Drescher. Model based characterization of μg induced alterations of cvs-regulation. *J. Grav. Phys.*, 1998.

[2] A. Asteroth. A calculus for higher order derivatives in multi layer perceptrons. Technical report, Institut für Informatik I, Universität Bonn, 1998.

[3] A. Asteroth. System decomposition by aspect modelling. Technical report, Institut für Informatik I, Universität Bonn, 1998.

[4] J. E. W. Beneken and B. DeWit. A physical approach to haemodynamic aspects of the human cardiovascular system. In *Physical Bases of circ. trans.* W. B. Saunders, 1967.

[5] C. G. Blomqvist, J. C. Buckey, F. A. Gaffney, L. D. Lane, B. D. Levine, and D. E. Watenpaugh. Mechanisms of post-flight orthostatic intolerance. *Journal of Gravitational Physiol.*, pages 122–124, 1994.

[6] J. E. Dennis, Jr. and Robert B. Schnabel. *Numerical Methods for Unconstrained Optimization and Nonlinear Equations.* Prentice-Hall, Englewood Cliffs, NJ 07632, USA, 1983.

[7] R. Isermann. *Identifikation dynamischer Systeme*, volume I + II. Springer Verlag Berlin, Heidelberg, 1988.

[8] R. E. Kalman, P. L. Falb, and M. A. Arbib. *Topics in Mathematical System Theory.* McGraw-Hill Books Company, 1969.

[9] M. S. Leaning, H. E. Pullent, E. R. Carson, and L. Finkelstein. Modelling a complex biological system: the human cardiovascular system. *Trans. Inst. M. C.*, 5, 1983.

[10] K. Möller, L. Beck, F. Baisch, A. Diedrich, and J. Drescher. Mobocol: Model based optimization of countermeasures using lbnp. In *Research Prg. of the German-Russian MIR '97.* WPF, Köln, 1997.

[11] B. D. Ripley. *Pattern Recognition and Neural Networks.* Cambridge University Press, Cambridge, 1996.

A Neural Network-based ARX Model of Virgo Noise.

F. Barone, R. De Rosa, A. Eleuteri, F. Garufi, L. Milano
Dipartimento di Scienze Fisiche, Università di Napoli "Federico II"
Istituto Nazionale di Fisica Nucleare, sez. Napoli,
Complesso Universitario di Monte Sant'Angelo,
via Cintia, I-80126 Napoli Italia

R. Tagliaferri
Dipartimento di Matematica ed Informatica,
Università di Salerno, and INFM unità di Salerno,
via S. Allende, 84081 Baronissi (SA) Italia
and IIASS "E. R. Caianiello", Vietri s/m Italia

June 2, 1999

Abstract

In this paper a Neural Networks based approach is presented to identify the noise in the VIRGO context. VIRGO is an experiment to detect Gravitational Waves by means of a Laser Interferometer. Preliminary results appear to be very promising for data analysis of realistic Interferometer outputs.

1 Introduction

Neural Networks (NN's) have become in the last years a very effective instrument for solving many difficult problems in the field of Signal Processing due to their properties like non-linear dynamics, adaptability, self-organization and high speed computational capability (see for example [4] and the papers therein quoted).

Aim of this paper is to show the feasibility of the use of NN's to solve difficult problems of signal processing regarding the so called VIRGO project. Gravitational Waves (GW's) are travelling perturbations of the space-time predicted by the theory of General Relativity, emitted when massive systems are accelerated. Up to now, there is only an indirect evidence of their existence, obtained by the observations of the binary pulsar system PSR 1913+16. Moreover, the direct detection of GW's is not only a relevant test of General Relativity, but the start of a new picture of the Universe. In fact, GW's carry complementary information with respect to electromagnetic and optical waves, since the GW's are practically not absorbed by the matter.

The aim of the Virgo experiment is the direct detection of gravitational waves and, in joint operation with other similar detectors, to perform gravitational waves astronomical observations. In particular, the VIRGO project is designed for broadband detection from $10Hz$ to $10kHz$. The principle of the detector is shown in figure 1.

A 3 km arm-length Michelson interferometer with suspended mirrors (test masses) is used. The phase difference $\Delta\phi$ between the two arms is amplified using Fabry-Perot cavities of Finesse 50 in each arm. Aiming for detection sensitivity of $3.10E-23\frac{1}{Hz}$ at $100Hz$, VIRGO is a very delicate experimental challenge because of the competition between various sources of noise and the very small expected signal. In fact, the interferometer will be tuned on the dark fringe, and then the signal to noise ratio will be mainly limited, in the above defined range of sensitivity, by residual seismic noise, thermal noise of the suspensions photon counting noise (shot noise). In figure 2 the overall sensitivity of the apparatus is shown. In this figure it is easy to see the contribution of the different noise sources to the global noise.

In this context we use a Multi-Layer Perceptron (MLP) NN with the back-propagation learning algorithm to model and identify the noise in the system, because we experimentally found that FIR NN's and Elman NN's did not work in a satisfying manner.

Both the FIR [11] and Elman [12] models proved to be very sensible to overfitting and were not stable. Furthermore the Elman network required a great number of hidden units, while the FIR network required a great number of delay terms. Instead, the MLP proved succesfull and easy to train because we used the Bayesian learning paradigm.

2 Neural networks for time-domain system identification

NN's are massively parallel, distributed processing systems. They are composed of a large number of processing elements (called nodes, units or neurons) which operate in parallel. Scalars (called weights) are associated to the connections between units and determine the strength of the connections themselves. Computational capability is due to the connections between the units and to their collective behaviour. Furthermore, information representation is distributed, i.e. no single unit is charged with specific information. NN's are well-known for their universal approximation capability [1].

System identification consists in finding the input-output mapping of a dynamic system. A discrete-time Multi-Input Multi-Output (MIMO) system (or a continuous-time sampled-data system) can be described with a set of non-linear difference equations of the form (***input-state-output*** representation):

$$\begin{cases} \mathbf{z}(n+1) = F(\mathbf{z}(n), \mathbf{u}(n)) \\ \mathbf{y}(n) = G(\mathbf{z}(n)) \end{cases} \tag{1}$$

where $\mathbf{z} \in R^L$ is the state vector of the system, $\mathbf{u} \in R^M$ is the input vector and $\mathbf{y} \in R^N$ is the output vector. Since we can not always access the state vector of the system, therefore we can use an input-output representation of the form:

$$\mathbf{y}(n) \quad = \quad \mathcal{F}[\mathbf{y}(n-1), \mathbf{y}(n-2), \ldots, \mathbf{y}(n-n_y),$$

$$\mathbf{u}(n-n_d),\ldots,\mathbf{u}(n-n_u-n_d+1)] \tag{2}$$

where n_u and n_y are the maximum lags of the input and output vectors, respectively, and n_d is the pure time-delay (or dead-time) in the system. This is called an ARX model (autoregressive with exogenous inputs,) and it can be shown that a wide class of discrete-time systems can be put in this form [2]. To build a model of the form (2), we must therefore obtain an estimation of the functional $\mathcal{F}[\cdot]$, which generally is nonlinear.

Given a set of input-output pairs, a neural network can be built [4] which approximates the desired functional $\mathcal{F}[\cdot]$. Such a network has $n_y N+(n_u-n_d)M$ inputs and N outputs (see figure 3).

A difficulty in this approach arises from the fact that generally we do not have information about the model order (i.e. the maximum lags to take into account) unless we have some insight into the system physics. Furthermore, the system is non-linear. Recently [3] a method has been proposed for determining the so-called *embedding dimension* of nonlinear dynamical systems, when the input-output pairs are affected by very low noise. Furthermore, the lags can be determined by evaluating the *average mutual information* (AMI)[8]. Such methodologies, although not always successful, can be nevertheless used as a starting point in model design.

3 Virgo

In the VIRGO data analysis, the most difficult problem is the gravitational signal extraction from the noise due to the intrinsic weakness of the gravitational waves, to the very poor signal-to-noise ratio and to their not well known expected templates. Furthermore, the Virgo detector is not yet operational, and the noise sources analyzed are purely theoretical models (often stationary noises), not based on experimental data. Therefore, we expect a great difference between the theoretical noise models and the experimental ones. As a consequence, it is very important to study and to test algorithms for signal extraction that are not only very good in signal extraction from the theoretical noise, but also very adaptable to the real operational conditions of Virgo.

For this reason, we decided the following strategy for the study, the definition and the tests of algorithms for gravitational data analysis. The strategy consists of the following independent research lines.

The first line starts from the definition of the expected theoretical noise models. Then a signal is added to the Virgo noise generated and the algorithm is used for the extraction of the signal of known and unknown shape from this noise at different levels of signal-to-noise ratio. This will allow us to make a number of data analysis controlled experiments to characterize the algorithms.

The second line starts from the real measured environmental noise (acoustic, electromagnetic, ...) and tries to identify the noise added to a theoretical signal. In this way we can test the same algorithms in a real case when the noise is not under control.

Using this strategy, at the end, when in a couple of years Virgo will be ready for the first test of data analysis, the procedure will be moved to the real system, being sure to find small differences from theory and reality after having acquired a large experience in the field.

4 A neural network-based model of the Virgo system

As we have seen in the introduction, the Virgo interferometer can be characterized by a sensitivity curve, which expresses the capability of the system to filter undesired influences from the environment, and which could spoil the detection of gravitational waves (such a noise is generally called seismic noise). The sensitivity curve has the following expression:

$$S(f) = \begin{cases} \frac{S_1}{f^5} + \frac{S_2}{f} + S_3\left[1 + \left(\frac{f}{f_k}\right)^2\right] + S_\nu & f \geq f_{\text{min}} \\ S(f_{\text{min}}) & f < f_{\text{min}} \end{cases} \tag{3}$$

where:

- $f_{\text{min}} = 4Hz$
- $f_k = 500Hz$ is the shot noise cut-off frequency
- $S_1 = 1.8 \cdot 10^{-36}$ is the pendulum mode
- $S_2 = 0.33 \cdot 10^{-42}$ is the mirror mode
- $S_3 = 3.24 \cdot 10^{-46}$ is the shot noise

The contribution $S_\nu(f)$ of violin resonances is given by:

$$S_\nu(f) = \sum_i^n \frac{1}{i^4}\frac{f_i^{(c)}}{f}\frac{C_c\phi_i^2}{\left[\left(\frac{f}{if_i^{(c)}}\right)^2 - 1\right]^2 + \phi_i^2} + \frac{1}{i^4}\frac{f_i^{(f)}}{f}\frac{C_f\phi_i^2}{\left[\left(\frac{f}{if_i^{(f)}}\right)^2 - 1\right]^2 + \phi_i^2} \tag{4}$$

where the different masses of close and far mirrors are taken into account:

- $f_i^{(c)} = i \cdot 327Hz$
- $f_i^{(f)} = i \cdot 308.6Hz$
- $C_c = 3.22 \cdot 10^{-40}$
- $C_f = 2.82 \cdot 10^{-40}$
- $\phi_i^2 = 10^{-7}$

Note that we used a simplified curve for our simulations, in which we neglected the resonances (see figure 4).

Samples of the sensitivity curve $\{S_i\}_i$ can be obtained by evaluating the expression (3) at a set of frequencies $\{f_i\}_i$, $f_i \in [10, 10000]$. The samples of the sensitivity curve allow us to obtain the system transfer function (in the frequency domain), $N(j2\pi f)$, such that:

$$|N(j\omega)|^2 = S(j\omega) \tag{5}$$

¿From this, by means of an inverse Discrete Fourier Transform, samples of the system transfer function (in the time domain) can be obtained. Our aim is to build a model of the system transfer function (5).

Assuming that the interferometer input noise is a zero mean Gaussian process, by feeding it to the system (i.e. filtering it through the system transfer function) we obtain a coloured noise. The so obtained white noise-colored noise pairs can then be used to train an MLP, as shown in figure 3.

4.1 Experimental Results

The first step in building an ARX model is the model order determination. To determine suitable lags which describe the system dynamics, we used the AMI criterion [8]. This can be seen as a generalization of the autocorrelation function, used to determine lags in linear systems. A strong property of the AMI statistic is that it takes into account the non-linearities in the system. Usually, the lag is chosen as the first minimum of the AMI function. The result is reported in figure 5, in which the first minimum is at 1. To find how many samples are necessary to unfold the (unknown) state-space of the model (the so called *embedding dimension* [8]) we used the method of [3], the Lipschitz decomposition. The result of the search is reported in figure 6.

From the figure we can see that, starting from lag three, the order index decreases very slowly, and so we can derive that the width of the input window is at least three. In order to test the NN's capability in solving the problem, we chose a width of 5, both for input and output (i.e. $n_y = n_u = 5$). In this way, we obtained a NN with a simple structure. Furthermore, some preliminary experiments showed that the system dead-time is $n_d = 0$; this gives the best description of the system dynamics.

Another fundamental issue is the NN complexity, i.e. the number of units in the hidden layers of the NN. Usually the determination of the network complexity is critical because of the risks of overfitting. Since the NN was trained following a Bayesian framework, then overfitting was of no concern; so we directed our search for a model with the minimum possible complexity. In our case, we found a hidden layer with 6 *tanh* units is optimal.

The Bayesian learning framework (see [10] and [9]) allows the use of a *distribution* of NN's, that is, the model is a realization of a random vector whose components are the NN weights. The so obtained NN is the *most probable* given the data used to train it. This approach avoids the bias of the cross-validatory techniques commonly used in practice to reduce model overfitting [6]. To allow for a smooth mapping which does not produce overfitting, several regularization parameters (also called *hyperparameters*) embedded in the error criterion have been used:

- one for each set of connections out of each input node,
- one for the set of connections from hidden to output units,
- one for each of the bias connections,
- one for the data-contribution to the error criterion.

Usually, the hyperparameters of the first three kinds are called *alphas*, while the last is called a *beta*.

The approach followed in the application of the Bayesian framework is the "exact integration" scheme, where we sample from the analytical form of the distribution of the network weights. This can be done if we assume an analytic form for the prior hyperparameters distribution, in particular a distribution uniform on a logarithmic scale:

$$p(\ln \alpha_i) = \frac{1}{\alpha_i}.$$

This distribution is *non-informative*, i.e. it embodies our complete lack of knowledge about the values the hyperparameters should take.

The chosen model was trained using a sequence of little less than a million of patterns (we sampled the system at 4096Hz for 240s) normalized to zero mean and unity variance with a *whitening* process. Note that the input-output pairs were processed through discrete integrators to obtain pattern-target pairs, as shown in figure 3. The NN was then tested on a 120s long sequence.

The NN was trained for 370 epochs, with the hyperparameters being updated every 15 epochs. A close look at the α hyperparameters shows that all the inputs are relevant for the model (they are of the same magnitude; note that this further confirms the pre-processing analysis). The β hyperparameter shows that the data contribution to the error is very small (as we would expect, since the data are synthetic).

The simulations were made using the MATLAB© language, the Netlab Toolbox [7] and other software designed by us.

In figure 7, the PSDs of the target and the predicted time series are shown; in the lower part of the figure is reported the PSDs of the prediction residuals.

In figure 8, the PSD of a 100Hz sine wine added to the noise is shown, with the signal extracted by the network. As can be seen, the network recognizes the frequency of the sine wave with the maximum precision allowed by the residuals.

5 Conclusion

In this paper we have shown some preliminary tests on the use of NN's for signal processing in the VIRGO Project. Some observations can be elicited from the experimental results:

- In evaluating the Power Spectral Densities (PSDs), we made the hypothesis that the system is sampled at $4096Hz$. It is only a work hypothesis, but it shows how the network reproduces the system dynamics up to $2048Hz$. Note that the PSDs are nearly the same also if we were near the Nyquist frequency.

- The PSD of the residuals shows a nearly-white spectrum, which is index of the model goodness (see [5]).

The next steps in the research are:

- to increase the system model order and to test if there are significant differences in prediction;

- to test the models with a greater number of samples to obtain a better estimate of the system dynamics;

- to model the noise inside the system model to improve the system performance and to allow a multi-step ahead prediction (i.e. an output-error model)

References

[1] K. Hornik, M. Stinchcombe, H. White, "Multilayer Feedforward Networks are Universal Approximators", Neural Networks, Vol. 2, pp. 359–366, 1989.

[2] S. Chen, S. A. Billings, "Modelling and Analysis of Nonlinear Time Series", Int. J. Control, Vol. 50, No. 6, pp. 2151–2171, 1989.

[3] X. He, H. Asada, "A New Method for Identifying Orders of Input-Output Models for Nonlinear Dynamic Systems", Proc. of the American Control Conf., San Francisco, California, 1993.

[4] F. Luo, R. Unbehauen, "Applied Neural Networks for Signal Processing", Cambridge University Press, Cambridge, 1997.

[5] L. Ljung, "System Identification Toolbox User's Guide", The MathWorks Inc, 1995.

[6] C. M. Bishop, "Neural Networks for Pattern Recognition", Clarendon Press, Oxford, 1996.

[7] C. M. Bishop, I. T. Nabney, Netlab Toolbox, Neural Computing Research Group, Aston University, Birmingham, 1996.

[8] H. D. I. Abarbanel, "Analysis of Observed Chaotic Data", Springer, 1996.

[9] R. M. Neal, "Bayesian Learning for Neural Networks", Ph.D. thesis, University of Toronto, Canada, 1994.

[10] D. J. C. MacKay, "Hyperparameters: optimise or integrate out?", Maximum Entropy and Bayesian Methods, Dordrecht, 1993

[11] A. C. Tsoi, A. D. Back, "FIR and IIR synapses, a new neural network architecture for time series modelling", Neural Computation Vol.3, No. 3, pp. 375–385, 1991

[12] J. Elman, "Finding structure in time", Cognitive Science Vol. 14, pp. 179–211, 1990

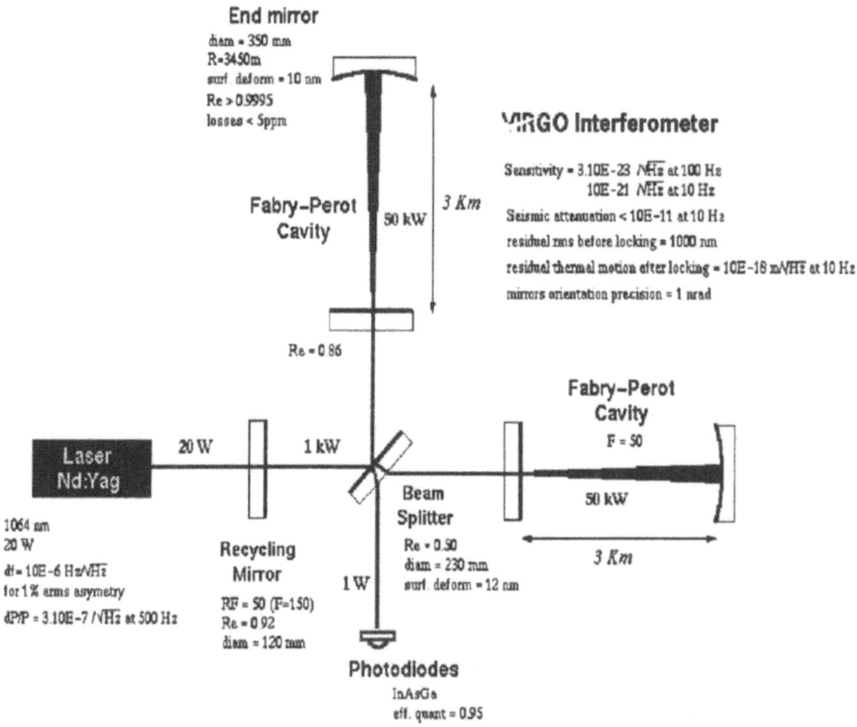

Figure 1: Scheme of the Virgo interferometer.

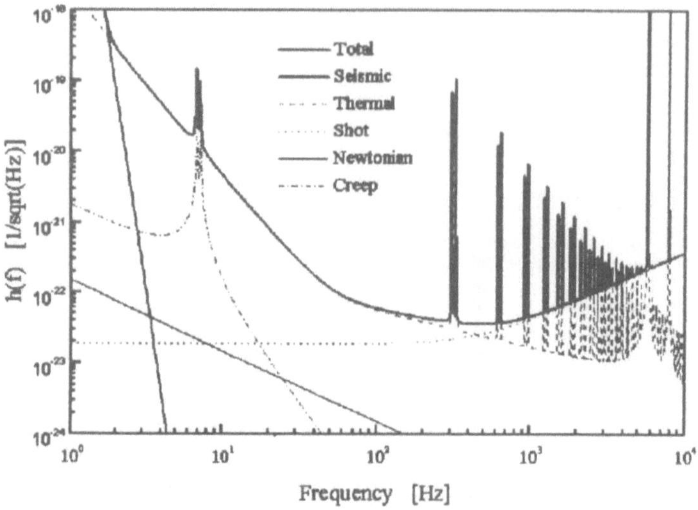

Figure 2: Sensitivity curve of the Virgo interferometer.

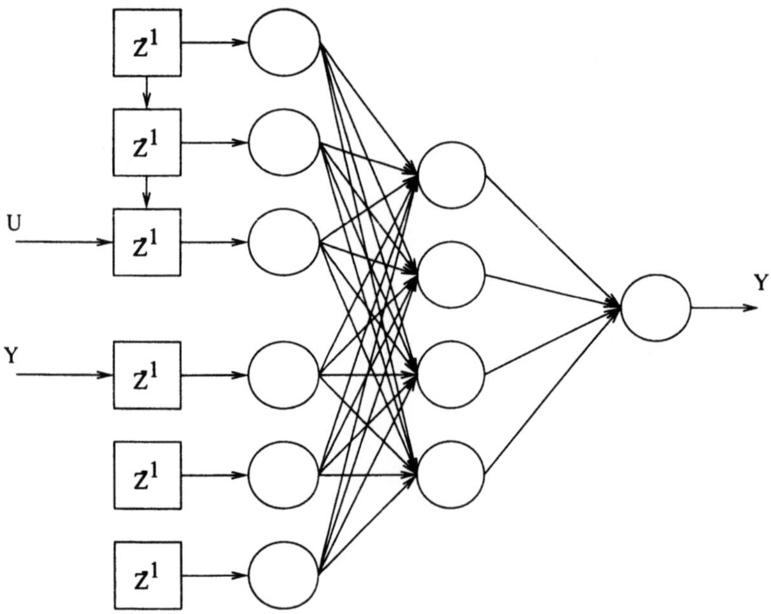

Figure 3: A neural network ARX model.

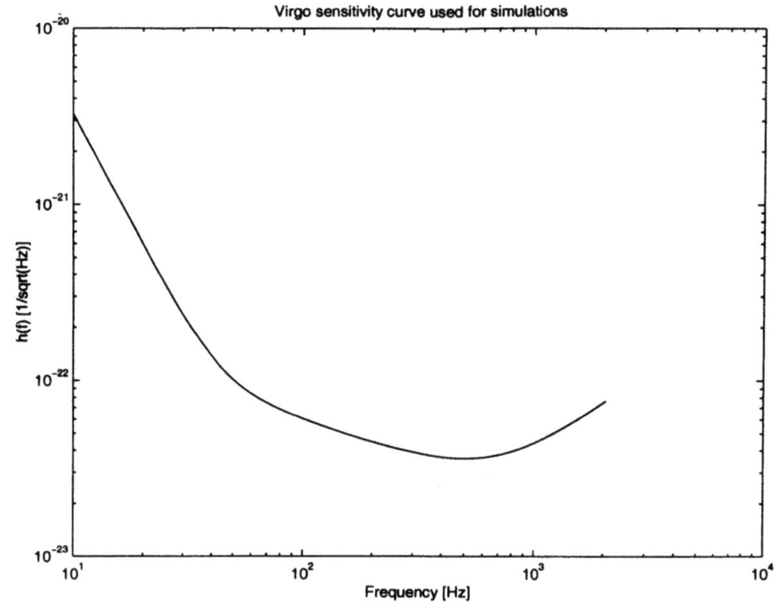

Figure 4: The simplified sensitivity curve used for the experiments.

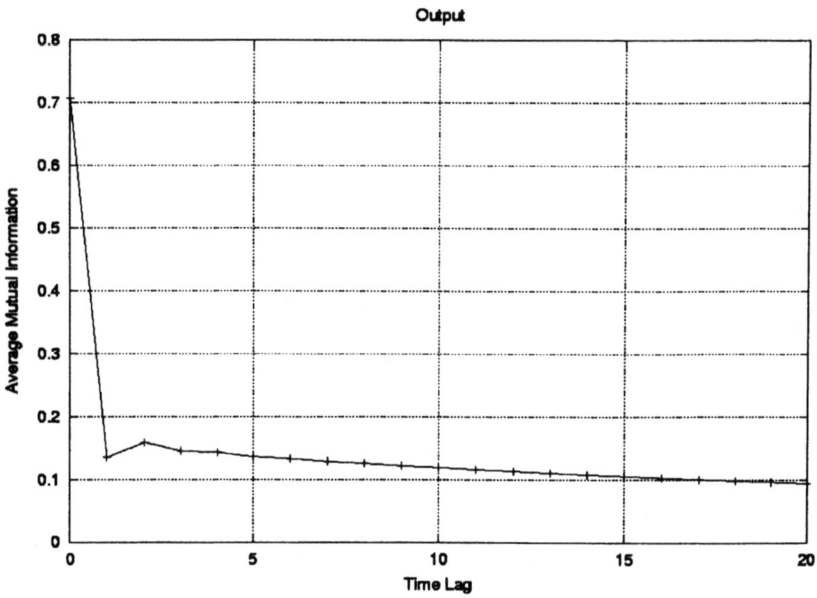

Figure 5: Average mutual information.

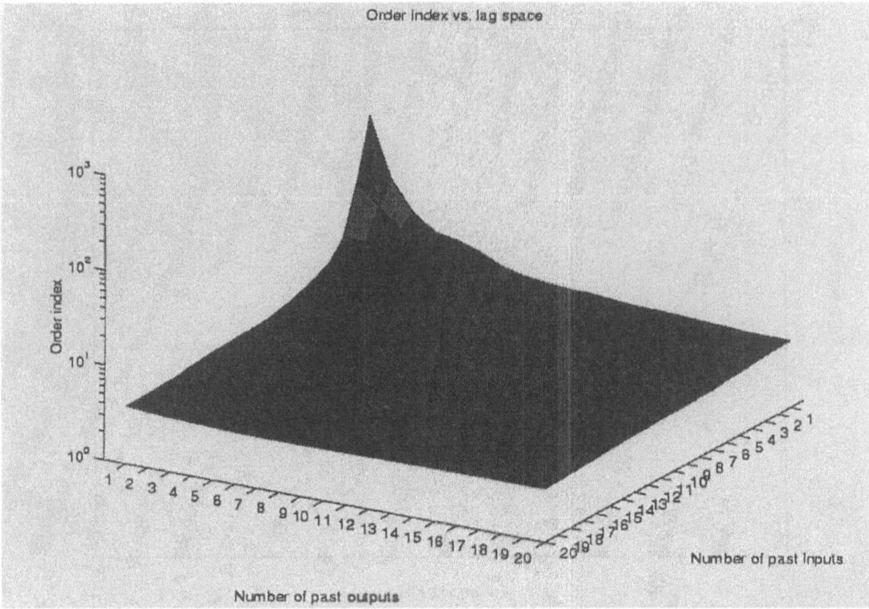

Figure 6: Determination of the embedding dimension.

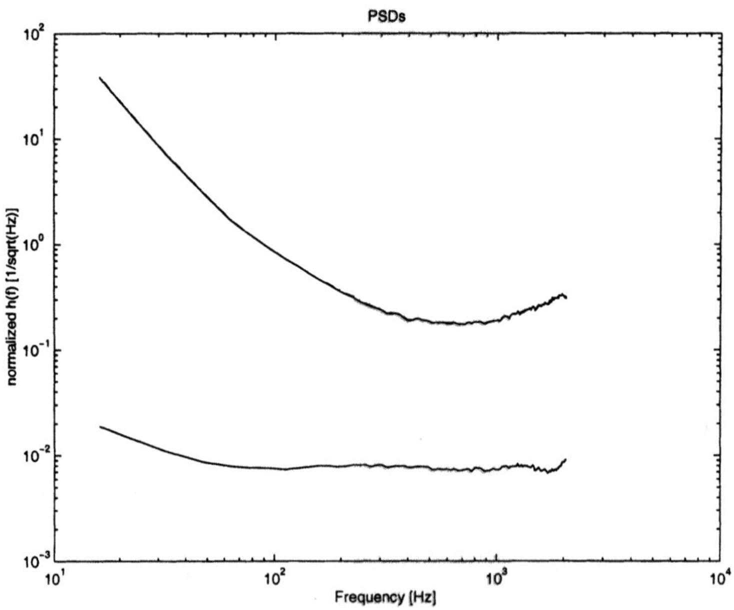

Figure 7: Results of the training.

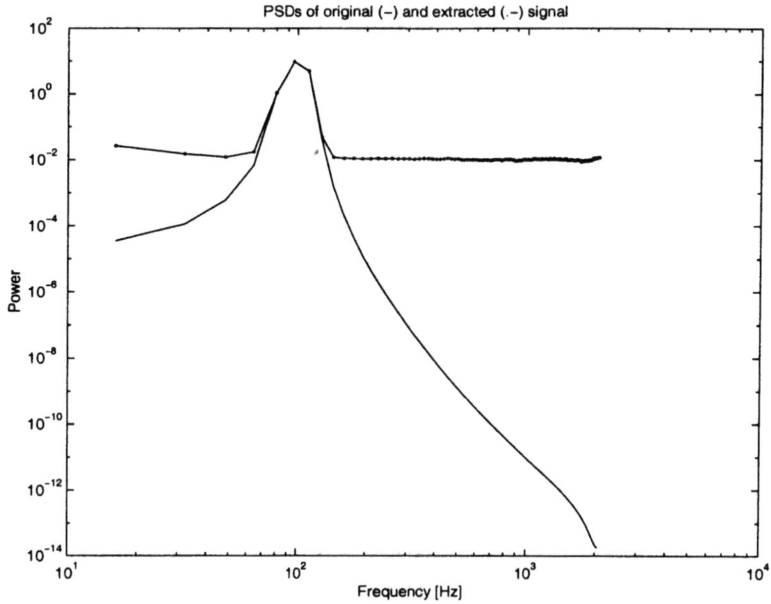

Figure 8: Example of signal extraction.

Local Wavelet Decomposition and its Application to face Reconstruction

N.A. Borghese

Laboratory of Human Motion Study and Virtual Reality, INB - CNR
Via f.lli Cervi 83, 20090 Segrate (Milano), Italy, borghese@inb.mi.cnr.it.

S. Ferrari and V. Piuri

Department of Electrical Engineering, Politecnico of Milano,
p.zza Leonardo da Vinci, 32 – 20133 Milano, [ferrari, piuri]@elet.polimi.it

Abstract

Wavelets are a powerful tool for multi-resolution analysis as they combine spatial and frequency locality. In this paper an efficient procedure to compute the Wavelet coefficients, called lifting schema, is illustrated. It can be applied efficiently to construct Wavelet networks. Results on face reconstruction operated at different resolution are reported and discussed.

1 Introduction

Wavelet theory provides a unifying framework for different techniques, developed independently. In particular it provides an efficient tool for a Multi-Resolution reconstruction of surfaces whose frequency content is non-stationary [1, 2]. The Wavelet Transform (WT) decomposes the surface onto a set of basis functions, called *Wavelets,* which are obtained from a single prototype, the *mother Wavelet*, through dilation, contraction and shift. This decomposition is achieved by projecting the surface onto the base set. An efficient implementation requires that the surface is projected iteratively on a base of decreasing scale. The analytical key operation is the internal product of the surface with the Wavelet base, which is theoretically extended over the entire R^2 plane. To cope with real data, which have a finite support, some approximations have to be made: the contribution of a Wavelet basis is supposed to vanish over a certain distance from its centre, or windowing of the input surface is performed [3]. A simpler analytical solution has been recently proposed under the name of lifting schema [4]. It implements only local operations and implicitly performs Wavelet windowing. Results on the application of this schema to surface reconstruction are reported and discussed.

2 Multi-Resolution Analysis and Wavelet Transform

The Wavelet Transform is based on the Multi-Resolution Analysis (MRA). This is obtained projecting $f(x) \in L^2(R)$ over a suitable set of bases which form a Riesz base for $L^2(R)$. These projections give an approximation of $f(x)$: $P_j[f(x)]$, at a resolution j. The bases are obtained by scaling a given function, called *Scaling Function*, $\varphi(x)$, such that each base is orthogonal to the previous one. The MRA produces a sequence of approximations of $f(x)$ such that its limit coincides with $f(x)$ itself:

$$\lim_{j \to \infty} P_j[f(x)] = f(x) \tag{1}$$

If we move into the discrete (sampled functions and discrete resolutions), the concept of filter bank tree substitutes the concept of Multi-Resolution. In particular, if we define:

$$R_j = P_j[f(x)] - P_{j-1}[f(x)] \tag{2}$$

it can be shown that it is possible to define a base, the *Wavelet*, such that projecting $f(x)$ over this basis, R_j is obtained. If we call $Q_j[f(x)]$ this projection, the following relationship holds:

$$Q_{j-1}[f(x)] = P_j[f(x)] - P_{j-1}[f(x)] \tag{3}$$

From Eq. (2), it follows that the series of the $Q_j(.)$ gives the function itself:

$$\sum_j Q_j[f(x)] = f(x) \tag{4}$$

and that $f(x)$ can be approximated ad libitum, by a finite number of the $Q_j[f(x)]$. The *Wavelet Transform* is defined as the coefficients, $\{\gamma_{j,k}\}$, which produce $Q_j[f(x)]$:

$$f(x) = \sum_j Q_j[f(x)] = \sum_{j,k} \gamma_{j,k} \psi_{j,k}(x) \tag{5}$$

The bases $\psi_{j,k}(.)$ are orthogonal and are constructed by dilation, from a single Wavelet function, called *mother Wavelet*. The $\{\psi_{j,k}(.)\}$ constitute a Riesz base for $L^2(R)$. The lifting schema, see Section 4, gives a simple and local method to set the values of the $\{\gamma_{j,k}\}$. Similarly, the approximation at the resolution j, can be written as:

$$P_j[f(x)] = \sum_{j,k} \lambda_{j,k} \varphi_{j,k}(x) \tag{6}$$

It can be shown [5] that the coefficients $\{\gamma_{j,k}\}$ and $\{\lambda_{j,k}\}$ can be computed recursively from the set of coefficients at the higher resolution, $j+1$. This decomposition is exemplified in Fig. 1 and takes the name of pyramid algorithm or Mallat algorithm.

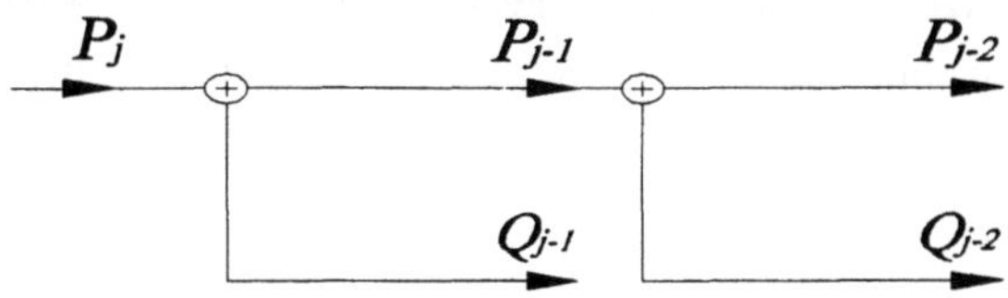

Fig. 1. The decomposition of a function $f(x)$: at each resolution or scale j, a pair of Approximation $P_j(.)$ and Residual $Q_j(.)$ is obtained. $Q_j(.)$ can be obtained by projecting the function, $f(x)$, over the Wavelet of scale j, $\psi_j(.)$. $P_0[f(x)]$ is usually assumed equal to $f(x)$.

3 Computing the Wavelet Transform

Haar is the simplest orthogonal Wavelet and it will be used to exemplify the construction machinery. If we consider a finite support of length 1, the basis-function, $\psi_{j,0}(.)$, is $2^{j/2}$ on a subinterval of length 0.5 * 2^{-j} and $-2^{j/2}$ on the next subinterval. There are therefore 2^j wavelets at level j. Its graphical representation is a square wave. The coefficients of the Haar Wavelet Transform are computed projecting the function $f(x)$ on this square wave as:

$$\gamma_{j,k} = \langle f, \psi_{j,k} \rangle = \int_{-\infty}^{+\infty} f(x) 2^{j/2} \psi(2^j t - k) dt \tag{7a}$$

The Scaling function, which has to be orthogonal to the Wavelet, is a box function which is constant and equal to $2^{j/2}$ in the interval of length 2^{-j}, starting at $t = k2^{-j}$. The coefficients of the scaling function at the resolution j, can be computed by projection as:

$$\lambda_{j,k} = \langle f, \varphi_{j,k} \rangle = \int_{-\infty}^{+\infty} f(x) 2^{j/2} \varphi(2^j t - k) dt = \sum_{k=0}^{k=2j} 2^{j/2} f(x_k) \tag{7b}$$

As can be seen, the Haar Scaling function, $\varphi_{j,k}$ computes the local mean value of $f(x)$ at the resolution j. This can be recast into a recursive schema where $\lambda_{j,k}$ is computed as:

$$\lambda_{j,k} = \frac{\sqrt{2}}{2} \left(\lambda_{j+1,2k} + \lambda_{j+1,2k+1} \right) \tag{8a}$$

which is the mean value between two consecutive samples at the higher resolution. It is usually assumed, as for the other Scaling functions: $\varphi_{0,k} = f(x_k)$. From the definition of Haar Wavelet, the following equation for the coefficients $\{\gamma_{j,k}\}$ holds:

$$\gamma_{j,k} = \frac{\sqrt{2}}{2} \left(\lambda_{j+1,2k} - \lambda_{+1j,2k+1} \right) \tag{8b}$$

Eq. (8b) is used to compute the coefficients of the Haar Wavelet Transform. Given the coefficients $\{\lambda_{j,k}\}$ and $\{\gamma_{j,k}\}$, the higher resolution coefficients $\{\lambda_{j+1,k}\}$ can be recovered by:

$$\lambda_{j+1,2k} = \frac{\sqrt{2}}{2} \left(\lambda_{j,k} + \gamma_{j,k} \right) \quad \text{and} \quad \lambda_{j+1,2k+1} = \frac{\sqrt{2}}{2} \left(\lambda_{j,k} - \gamma_{j,k} \right) \tag{8c}$$

The (8c) formalize the Inverse Wavelet Transform (IWT). The algorithm described above can be generalized to perform a general WT. Every wavelet can be characterized by a set of four finite filters, $\{h, g, \tilde{h}, \tilde{g}\}$, which can be used to move between different resolutions. The general form of Eqs. (8a-c) is:

$$\lambda_{j,l} = \sqrt{2} \sum_k \tilde{h}_{k-2l} \lambda_{j+1,k} \tag{9a}$$

$$\gamma_{j,l} = \sqrt{2} \sum_k \tilde{g}_{k-2l} \lambda_{j+1,k} \,. \tag{9b}$$

$$\lambda_{j+1,l} = \sqrt{2}\sum_l h_{k-2l}\lambda_{j,k} + \sqrt{2}\sum_l g_{k-2l}\gamma_{j,k} \tag{9c}$$

In the following, simple filters used for some of the most common Wavelets are reported [3]: Haar Wavelet: $h = \tilde{h} = [1/2, 1/2]$, $g = \tilde{g} = [1/2, -1/2]$; Hat (linear) Wavelet: $h = [1/4\ 1/2\ 1/4]$, $\tilde{h} = [-1/8, 1/4, 3/4, 1/4, -1/8]$, $g = [-1/8, -1/4, 3/4, -1/4, -1/8]$, $\tilde{g} = [-1/4, 1/2, -1/4]$; Cubic Wavelet: $h = [-1/32\ 0\ 9/32\ 1/2\ 9/32\ 0\ -1/32]$, $\tilde{h} = [1/64\ 0\ -1/8\ 1/4\ 23/32\ 1/4\ -1/8\ 0\ 1/64]$, $g = [1/64\ 0\ -1/8\ -1/4\ 23/32\ -1/4\ -1/8\ 0\ 1/64]$, $\tilde{g} = [1/32\ 0\ -9/32\ 1/2\ -9/32\ 0\ 1/32]$.

4 The Lifting schema

The lifting schema, recently proposed by [4], is a very efficient way to compute the Wavelet Transform. We will illustrate here how does it work when the Hat (linear) Wavelet is used. Let us start with a sampled function: $\{f(x_k)\}$, $k \in Z$. Subsampling by a factor two we obtain the even and the odd sequence. These can be seen as the approximation of $f(x_k)$, the even samples, and the residual, $r(x_k)$, the odd samples. This schema produces a Lazy Wavelet where:

$$\lambda_{-1,k} = \lambda_{0,2k} = f(x_{0,2k}) \tag{10a}$$

are the coefficients of the Scaling functions $\varphi_{-1,k}(.)$, and:

$$\gamma_{-1,k} = \lambda_{0,2k+1} = f(x_{0,2k+1}) \tag{10b}$$

are the coefficients of the Wavelet, $\psi_{-1,k}(.)$. A more efficient coding of $f(x_k)$ is achieved by considering that the correlation between neighbour samples is usually high, at least for smooth functions. The odd samples can therefore be computed as the deviation from the average of the value of pairs of neighbour even samples. The Wavelet coefficients are thus:

$$\gamma_{1,k} = \lambda_{0,2k+1} - 1/2(\lambda_{1,k} + \lambda_{1,k+1}) \tag{11a}$$

As these coefficients encode the deviation from linearity of the function $f(x_k)$, they will usually be small. A good reconstruction of $f(x_k)$ can be obtained even if the small coefficients are discarded. We can now improve the coding of the coefficients of the Scaling function, $\{\lambda_{j,k}\}$. In fact, if the given function samples are composed by 1's interleaved with 0's (i.e. $\{\lambda_{0,k}\} = [1\ 0\ 1\ 0\ 1\ 0\ 1\ 0\ 1\ 0\ 1\ 0\ 1]$), the coding reported above will suffer from aliasing and poor approximation, as the approximation at the first level (j=1) will result in a sequence of 1's. A better schema to compute the $\{\lambda_{j,k}\}$ is the following:

$$\lambda_{j,k} = \lambda_{j+1,2k} + 1/4(\gamma_{j,k-1} + \gamma_{j,k}) \tag{11b}$$

which guarantees that the average of the $\lambda_{j,k}$ is at the same at every level of resolution j. Eqs. (10) and (11) describe the lifting schema for the Hat Wavelet. This schema can be generalised to other WTs and it allows to cut the computational time to half with respect to the classical computation which performs a convolution. It does not require additional memory, as the calculation can be carried out in place. Moreover, the algorithm is well suited for SIMD parallelisation into SIMD machines. In literature, the WT was heavily linked to Fourier transform, and its use supposes regularly spaced samples. Since the lifting schema does not depend on frequency analysis, it allows to generalise the WT to deal directly with range data.

5 Reconstruction of a human face

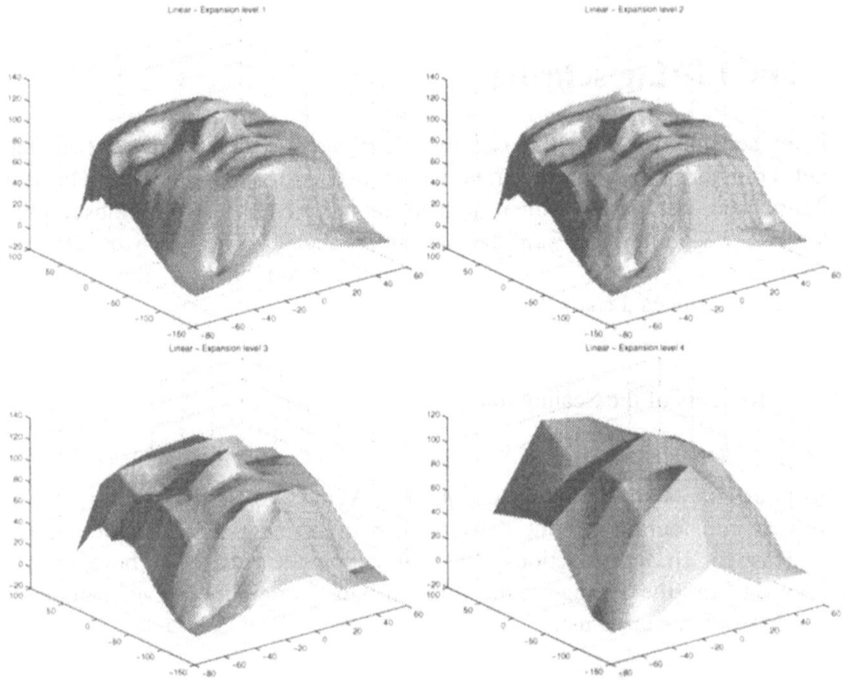

Fig. 2. The reconstruction of a human face obtained at four different scales, using the Hat wavelet.

The WT can be applied to MRA and reconstruction of 3D surfaces ($z = f(P)$, with $P \in R^2$ and $z \in R^1$). Provided that the sampled data are organised into a matrix, it could be shown that the WT can be realised transforming first all the rows and, successively, all the columns of the matrix, i.e. considering alternatively the rows and the columns as a battery of 1D functions. This schema has been applied to the reconstruction of human faces at different resolutions. The results are obtained applying the Hat wavelet, but the results can be generalised easily to quadratic and cubic Bases. The face in Figure 2 has been reconstructed starting from 12,641 data points sampled over the face using the Autoscan system [6]. A Hierarchical RBF network [7][8] has been used to resample the data points with a spacing of 1,5mm. The Hat Wavelet has been applied to this set of data.

6 Conclusion

Wavelets perform a local analysis in the combined frequency/space domain. In this respect they are similar to the Short-Time-Fourier-Transform with the difference that WT uses a short window at high frequency and a large window at low frequency while the STFT uses a fixed window. It is similar also to other MRA schemas like HRBF with the difference that in HRBF the number of coefficients increases one level after the other while in the Wavelet transform the number of coefficients decreses. On the other side, the WT does not have the redundant representation in the frequency domain which the HRBF has. Wavelets are made even more powerful by the lifting schema which gives to the MRA more freedom than the classical approach as different metrices can be used in computing the coefficients, windowing can be implemented locally, saving in computational time is achieved and only local operations are carried out.

References

1. Kovacevic J. and Daubechies I., Proceedings of the IEEE, Special Issue on Wavelets, Vol. 84(4), April, 1996.
2. Sweldens W. and Schroder P., Course Notes on Wavelets, Siggraph'95, 1995.
3. Strang G. and Nguyen T., Wavelets and Filter Banks, Wellesley, Cambridge Press, 1996.
4. Sweldens W., The Lifting Scheme: a Custom-Design Construction of Biorthogonal Wavelets, Siam J. Math. Anal., Vol. 29(2), pp. 511-546, (1997).
5. Mallat S.G., A Theory fro Multiresolution Signal Decomposition: The Wavelet Representation, IEEE Trans. on PAMI, Vol. 11(7), pp. 674-693, 1989.
6. N.A. Borghese, G. Ferrigno, G. Baroni, R. Savarè, S. Ferrari and A. Pedotti "AUTOSCAN: A flexible and portable scanner of 3D surfaces", *IEEE Computer Graphics & Applications*, May/June, pp. 38-41 (1998). Borghese and Ferrari.
7. N.A. Borghese, S. Ferrari "Hierarchical RBF networks in function approximation" *Neurocomputing* Vol. 19(1-3), pp. 259-283 (1998).
8. Borghese N.A. and Ferrari S. 3D (1998) Scanning and Reconstruction of human body parts. Proc. Eurographics98, Section Short Presentations, pp. 3.5.1-3.5.2.

A Multilayer Perceptron for Fast Interpolation of JPEG/MPEG Coded Images

Sergio Carrato
D.E.E.I., University of Trieste
v. Valerio, 10, 34100, Trieste, Italy

Abstract

We present a novel approach for the reconstruction of coded images and video, which permits fast recovery of the data with good quality in case of limited channel bandwidth. It is based on the use of a neural network which is able to interpolate the DC components of the coded data with higher accuracy and lower artifacts with respect to standard interpolators.

Introduction

A situation which often occurs in multimedia environments is the need for fast retrieval of large sets of images or video data, for example when browsing from a database; this applies to both commercial (e.g. product catalogues for shops which sell via Internet) and professional (e.g. video indexing in a broadcasting company) applications. If the channel bandwidth is limited, this operation may be slow, due to the huge amount of data involved. During a preview phase, however, the user can often accept an image of low quality; consequently, it may be useful to be able to quickly reconstruct an approximate version of the required images, while deferring the exact decoding to a later stage, possibly on demand.

Images and video are often stored using compressed formats, as JPEG [1] and MPEG2 [2], where each image is subdivided into 8×8 blocks, and each block is DCT transformed. Consequently, the DC components of the blocks are readily available[1] and can be used to roughly reconstruct the image using an $8\times$ interpolation both horizontally and vertically. The simple visualisation of the DC components yields to very small images, while pixel repetition obviously leads to heavy blocking effects. Linear interpolators as [4] generally give lower artifacts, mainly along oblique edges, at the cost however of increased blurring. In [5], a nonlinear interpolator has been proposed, where a rational filter which interpolates by a $2\times$ factor is applied recursively for three times.

In this paper, we propose a novel algorithm based on a multilayer perceptron (MLP). It is well known, in fact, that MLPs are excellent universal

[1] or, for P and B frames in a MPEG bistream, can be easily computed [3].

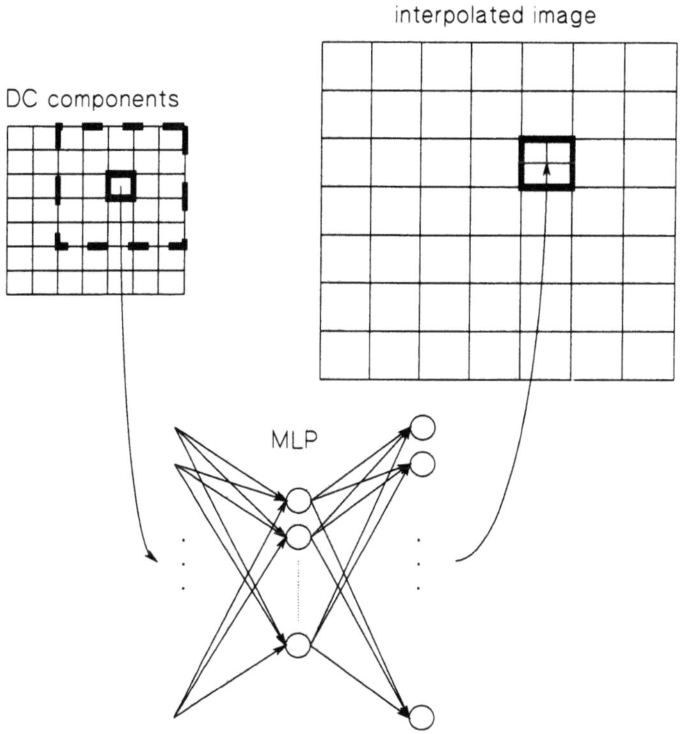

Figure 1: Interpolation scheme of the DC components of a coded image (in this drawing, 2× interpolation is considered for visualization purpose). Data belonging to a neighborhood (dashed line) of the block to be reconstructed (solid line) are fed into an MLP. Interpolated data are obtained at the output layer of the neural network.

approximators, so that a suitably trained MLP is expected to be able to accurately interpolate the DC components of DCT-coded images yielding good quality results.

Image reconstruction from the DC components using a multilayer perceptron

As already explained, the task is the interpolation of an image to its full size from its DC components only. The approach we propose is very simple. We start from the (small) image formed by all the DC components. For each 8×8 block to be reconstructed, a small window centered on the corresponding DC component is considered, and these data are fed into the input layer of an MLP, which is suitably trained in order to provide at its output an approximation of the original 8×8 block (Fig. 1).

The MLP can be dimensioned as follows. The input window must be large

algorithm	*Lenna*	*Pepper*	*Boats*
MLP, $h = 16$	267.0	273.8	362.4
MLP, $h = 32$	253.8	249.1	353.3
MLP, $h = 64$	251.9	247.4	353.6
linear interp. [4]	379.8	501.2	445.9
nonlinear interp. [5]	260.4	267.6	371.6

Table 1: Mse of several interpolated images using the proposed interpolator with different number h of hidden nodes, and two non MLP-based interpolators.

enough to give the MLP sufficient information on the surrounding data, but it is apparent that an excessively large window includes DC components of pixels which are too far from those of the central block; a reasonable choice can be to use a 5×5 window. The number of hidden layers can be restricted to one, without reduction of the approximation capabilities of the network [6]. The number of hidden nodes, h, may be optimized experimentally taking into account both the resulting image quality and the computational complexity. The number of output nodes has to be 8×8; consequently, the MLP dimensions are $25 \times h \times 64$.

An observation has to be done, finally, about the kind of activation functions to be used. The sigmoid never reaches the extreme values of its range, and this fact can create problems for the reconstruction of very dark or very light pixels. For this reason, we use a linear activation function (with saturation) instead of the sigmoid for the output layer, and the backpropagation rule is consequently modified.

Experimental results

The MLP has been trained using four 512×512 pixel real-world images, using backpropagation [7] with exponentially decaying learning factor and fixed momentum; training duration has been 50 epochs. The system has been tested using three 512×512 images, *Lenna*, *Pepper*, and *Boats*, which do not belong to the training set.

In Tab. 1, the mse of the reconstructed images is reported for the MLP, for a standard linear interpolator [4], and for the nonlinear interpolator based on the rational filter [5]. It may be seen that the MLP significantly outperforms the standard linear interpolator. Its performances, for $h = 32$ and $h = 64$, are also better than those of the nonlinear interpolator; moreover, an advantage of our approach is given by the possibility of a very fast hardware implementation. In fact, the MLP structure permits a fully parallel realization, and only one interpolation step is needed instead of three.

As an example, in Fig. 2b the reconstructed image *Lenna* is shown for $h = 32$. The absence of annoying artifacts can be appreciated; in particular, the edges, e.g. of the shoulder, are sharp and with reduced blocking effect. As a comparison, the corresponding images obtained by simple pixel repetition and using standard linear interpolation [4] are shown in Fig. 3.

(a) (b)

Figure 2: (a) DC components of *Lenna* (b) *Lenna* reconstructed from its DC components using the proposed MLP-based approach.

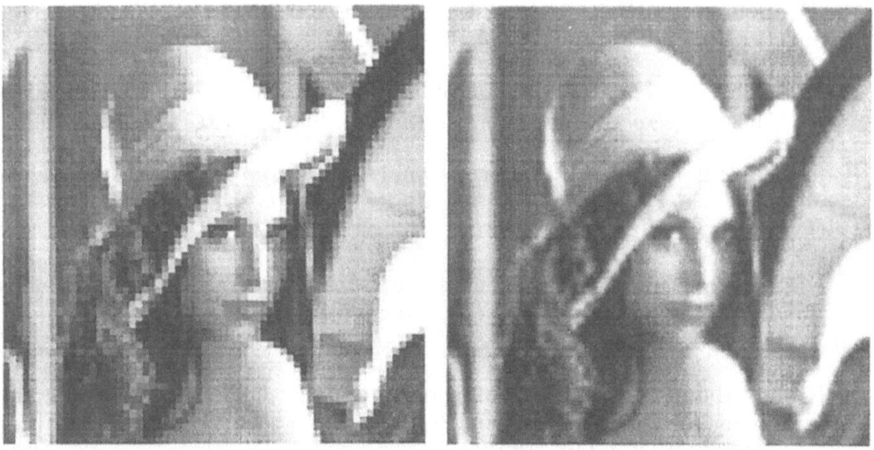

Figure 3: *Lenna* reconstructed from its DC components using simple pixel repetition and a standard linear interpolator [4].

It may be concluded that the MLP, thanks to its good interpolation properties, is effectively capable of reconstructing the coded data yielding an image which has an acceptable quality for applications as browsing or previewing.

Finally, it may be worthwhile to note that the proposed algorithm can also be considered as an interesting approach to the problem of image coding at very low bit rates for transmission systems where the computational complexity is of concern at the coder side. A possible example in the consumer electronics area can be a set-top box which receives images from a video camera and transmits coded video using a narrow bandwidth upstream channel. In this cases, it may be convenient to leave to the coder only the evaluation of the DC components and their coding, while the reconstruction of the image at the decoder is obtained with acceptable quality using the proposed MLP-based interpolator. For example, arithmetic coding of the DC components of the image in Fig. 2a, linearly quantised with 128 levels, requires 0.1 bpp; the PSNR of the resulting image after interpolation (Fig. 2b) is 24.1 dB. It may be added, moreover, that for very low bit rates JPEG provides images with a slightly superior PSNR(25.5 dB at the same bit rate) but with evident blotching in the smooth areas; in fact, very coarse quantization tables are used for the DC levels, in order to have some bits available for the detailed areas. In turn, if no AC components are considered, the DC components can be more finely quantised, and our interpolator can yield images which have fewer details but look more natural due to the absence of artifacts.

Acknowledgements

Work partially supported by MURST 60% grant.

References

[1] W. Pennebaker, "JPEG technical specification, revision 8." Working Document No. JTC1/SC2/WG10/JPEG-8-R8, Aug. 1990.

[2] L. Chiariglione, "The development of an integrated audiovisual coding standard: MPEG," *Proceedings of the IEEE*, vol. 83, pp. 151–157, Feb. 1995.

[3] B.-L. Yeo and B. Liu, "Rapid scene analysis on compressed video," *IEEE Trans. on Circuits and Systems for Video Technology*, vol. 5, pp. 533–544, Dec. 1995.

[4] R. G. Keys, "Cubic convolution interpolation for digital image processing," *IEEE Trans. Acoust., Speech, Signal Processing*, vol. ASSP-29, no. 6, pp. 1153–1160, 1981.

[5] G. Ramponi and S. Carrato, "Interpolation of the DC component of coded images using a rational filter," in *Proc. ICIP'97*, (Santa Barbara, California), pp. I–389–I–392, Oct. 1997.

[6] S. Tamura and M. Tateishi, "Capabilities of a four-layered feedforward neural network: four layers versus three," *IEEE Trans. on Neural Networks*, vol. 8, pp. 251–255, Mar. 1997.

[7] R. P. Lippmann, "An introduction to computing with neural nets," *IEEE ASSP Magazine*, pp. 4–21, Apr. 1987.

The automatic detection of microcalcification clusters in the CALMA project: status and perspectives

Pasquale Delogu
Dipartimento di Fisica, Università di Pisa and INFN
Piazza Torricelli 2
I-56100 Pisa, Italy
Tel: (+39)050911275 Fax: (+39)05048277 E-mail: delogu@pi.infn.it

Abstract

A status report is presented on microcalcification clusters search in the frame of the CALMA project. CALMA's purpose is to build a mammographic database and to develop automatic tools for the detection of breast cancers. A microcalcification is a very brilliant object, but rather small (0.1 to 1.0 mm in diameter). Some of them, either grouped in cluster or diffused may indicate the presence of a tumour. In the following 250 images with microcalcifications from our database have been analyzed, and a CAD tool has been designed to detect clusters, reaching a correct classification of 88%.

1 CALMA Project

CALMA is a collaboration among the INFN sections of Bologna, Pisa, and Udine , and the Hospitals of Bari, Bologna, Livorno, Massa and Udine .

The purpose of this project is a CAD (Computer Assisted Diagnosis) system based on neural networks, in screening mammography.

Traditional film and screen mammography is widely recognized as the only imaging modality for the early detection of breast cancer in asymptomatic women. Typically, sensitivity ranges from 73% to 88%, while specificity are of the order of 83% to 96% [1]. These figures are shown to increase of 7%-10% and 2%-5% respectively if the image is analyzed independently by two radiologists. The aim of CALMA is to improve these merit figures of diagnosis behaving as "the second radiologist" in the screening. Other important advantages are the speed of the method and the reproducibility of the diagnosis.

The first step to be accomplished is to collect a set of useful images to train the system, organized in an easy-to-access database. Mammographic images are digitized with 50-100 microns per pixel and with 12-bit grey levels, and each one is described by a homogeneous and coherent diagnosis. Our aim is to collect 20000 images with these requirements in a reasonable amount of time. At this moment we have digitized about 2200 mammographic images, corresponding to about 600 patients. Images come from the collaborating hospitals where mammograms are scanned and transferred to CD-ROM. Given the average dimensions of the images, without any compression, about 60 images can be written on a single CD-ROM. Presently, our database is the biggest european database, public and free.

2 Microcalcifications

In many cases (30% to 50%) an early breast cancer diagnosis follows from the detection of a cluster of microcalcifications, i.e. groups of small (as much as 100-200 μm) but very brilliant objects. These clusters infact often appear in the presence of a tumoural lesion.

A single calcification is not necessarily indicating the presence of cancer, but some morphological characteristics of groups of those are normally used in the medical routine as a hint for deeper exams.

3 Available images

CALMA's database at the moment already consists of 2200 images, but of course not all of them contain the pathological structures we are interested in for this search.

Up to now 250 images (85 μm granularity and 12-bit resolution) with microcalcification clusters are available from the hospitals of Bari and Udine.

The images have been produced digitizing film mammograms with a CCD linear scanner. In the next months, thanks to contributions from the hospital of Bari and Udine we expect that the number of images with microcalcifications hiding a patology should be as large as 600 images.

4 Strategy

The raw images as coming after digitization need to be "cleaned" because of the presence of unwanted regions (e.g. edges, lead labels, and so on).

Later on the image is enhanced in order to filter out low contrast details and regions with low spatial frequency.

After the filter small regions are selected and for each region a set of relevant features is extracted by statistical methods.

Finally these features are classified by means of a Feed Forward Neural Network.

4.1 Preliminar treatment

The "cleaning" operation has to be performed in the same way for all the images, even if these unwanted regions may be in different part of the image and of different nature, depending on the devices that were used at the time the image was produced.

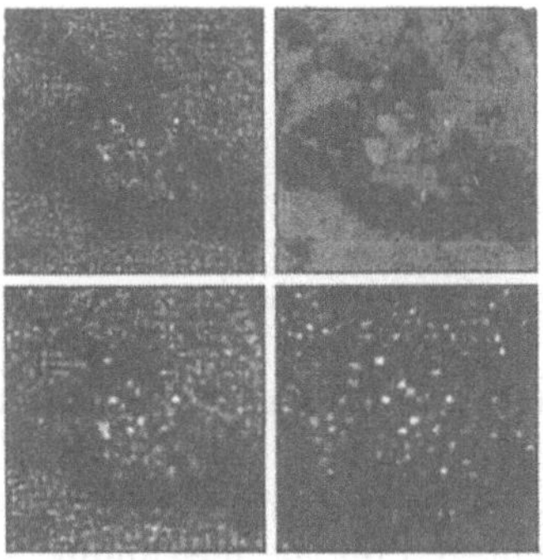

Figure 1: The four images show the same region after four different filtering procedures. From left to right, high to low, the filters are: quadratic gradient, laplacian edge-enhancement, 4×4 generalized edge enhancement, 5×5 generalized edge-enhancement

The cleaning algorithm has been designed in such a way to eliminate large continuous regions of uniform intensity.

4.2 Image enhancement

Classical filters have been used to enhance small high-contrast objects, both gradient-based filters and morphological ones.

Figure 1 shows the effect of the application of four of these filters on the same region. Among those that we tested, there are some standard high-pass filters, such as the quadratic gradient filter or the laplacian edge-enhancement. We also developed dedicated filters that could better match our problem, namely a generalized edge-enhancement filter with kernels of various dimensions (3×3 to 7×7) to enhance calcifications of various sizes.

The kernel of a 5×5 filter is like the following:

$$\begin{matrix} -1 & -1 & -1 & -1 & -1 \\ -1 & \frac{16}{9} & \frac{16}{9} & \frac{16}{9} & -1 \\ -1 & \frac{16}{9} & \frac{16}{9} & \frac{16}{9} & -1 \\ -1 & \frac{16}{9} & \frac{16}{9} & \frac{16}{9} & -1 \\ -1 & -1 & -1 & -1 & -1 \end{matrix}$$

4.3 Feature extraction

At this point square windows of 100×100 pixels are extracted. Dimensions are chosen so to contain the largest clusters. We extract then the window features that we believe can describe the window structures.

These features are:

- absolute maximum intensity
- average

- variance
- number of local maxima
- maximum distance between two local maxima
- minimum distance between two local maxima
- average distance between two local maxima
- average area of local maxima (the area of a maximum is defined as the area of a rectangle centered on the maximum, whose pixels have an intensity larger than a fixed threshold)
- average intensity of local maxima

The 9 features were computed both for the original window image and for the enhanced one.

We think that these parameters represent optimally the characteristics of the image in our windows.

4.4 Classification

Our data set at this point consists of 862 windows. We divided it into a learning set with 724 patterns (276 with clusters) and a test set with 138 patterns (50 with clusters).

We used a Feed Forward Neural Network with 18 input and 1 output neurons trained by backpropagation with momentum. After the optimization the chosen architecture had 2 hidden neurons.

The Net was trained for 40000 cycles.

5 Results

The performance of the Net was evaluated in terms of the following parameters

$$sensitivity = \frac{\text{number of windows with MC detected}}{\text{tot. number of windows with MC}} x100$$

$$specificity = \frac{\text{number of windows without MC detected}}{\text{tot. number of windows without MC}} x100$$

$$correct\ \ classification = \frac{\text{number of windows correctly class.}}{\text{tot. number of windows}} x100$$

We obtained a sensitivity of 86%, a specificity of 90%, with an average correct classification of 88%.

References

[1] A.G.Haus and M.J.Yaffe, Editors, *A categorical course in physics. Technical aspects of breast imaging.* Radiological Society of North America, 1993, presented at the 79th scientific assembly and annual meeting of RSNA.

[2] R.A.Smith, "Epidemiology of breast cancer" in *A categorical course in physics. Imaging considerations and medical physics responsibilities*, Madison, Wisconsin, 1991. Medical Physics Publishing.

[3] V.M.Marzulli, "The CALMA project: Computer Assisted Library for MAmmography", WIRN98 Proceedings, Springer-Verlag (1998).

[4] A.Zell et al., "SNNS - Stuttgart Neural Network Simulator - V4.1", University of Stuttgart Report No. 6/95 (1995).

Signal classification by subspace neural networks

M.Di Giacomo, G.Martinelli

INFO-COM Dpt, University I, Rome Via Eudossiana 18, 00184 Rome, Italy
E-mail:martin@infocom.ing.uniroma1.it

Abstract

Classification problems involving signals can benefit from the application of subspace neural networks. In order to fully exploit them, a constructive approach based on learning theory is mandatory. A possible method having these characteristics is proposed in the present contribution. It yields satisfactory performances, as illustrated by the results obtained with the well-known sonar benchmark.

1 Introduction

The expansion in a linear combination of basis functions is a common and efficient technique for analyzing signals. The basis is often directly related to the physical mechanism which originates the signal, while the coefficients of the combination depend on the exciting conditions. Since they vary more or less stochastically, little information can be drawn from them. What must be perceived from a classifier is the basis which is the representation of the mechanism. There are several mechanisms which operate simultaneously. The problem to be solved by the classifier is the determination of the corresponding bases. Assuming to digitize the signals under consideration to produce feature vectors (patterns), the basis defines a subspace of the space where the above vectors are represented as points. An unknown pattern is classified in terms of its distance from the subspace associated with the mechanism which originates the corresponding signal. Subspace methods of classification are motivated by the previous remarks. The determination of the set of subspaces characterizing an actual application could be directly pursued when physical nature of the mechanism originating the signals were closely understood. This is however not the usual case. What is available is a set of examples . The subspaces must be determined from a training set, as in the case of the neural approach. Despite its evident convenience, the latter has been rarely followed in the technical literature. The most relevant contribution is the ASSOM proposed in [1].

The basis of subspace neural networks (SNN) is a particular partition of the input space in a suitable number of linear subspaces of same dimension. Each pattern of the training set is associated with the subspace closest to it, i.e. to the subspace where its orthogonal projection is the largest. The classes are usually represented by several subspaces in order to accommodate their different features. After training, the neural net operates by projecting the input pattern onto all its subspaces . The largest projection yields the response of the classifier as the class of the corresponding subspace.

The architecture of the net is shown in fig.1. It is constituted by a layer of neurons. These neurons represent the subspaces. The connection of the input to a neuron represents the orthogonal projection operator onto the corresponding subspace. The output neuron implements a winner-takes-all operator.
The determination of the neural net from the training set regards both the architecture (number of neurons = number of subspaces) and the weights of the connections (projection operators onto the subspaces). The learning algorithm suggested in [1] is limited to the latter. A procedure which also considers the former is suggested in [2]. However it presents several inconveniences as:

- The determination of the complexity of the neural network is not obtained by the robust methods of learning theory.
- There are no controls on the closeness of subspaces
- The successive inclusion or exclusion of the patterns into the clusters formed for constructing the subspaces may leave unused an excessive number of patterns of the training set.

The present contribution suggests a new learning algorithm which is inspired by the techniques proposed in [1] and [2]. The construction of the net is guided by a cross-validation technique, which allows the control on the level of generalization acquired by the net under construction. The generation of subspaces is pursued by using rules and techniques in part similar to those introduced in [1] and [2].
The proposed algorithm will be described in sect.2 and its efficiency is illustrated in sect.3 in relation to the well-known sonar benchmark.

2 Proposed learning algorithm

Preliminaries : The training set is partitioned in clusters. To each cluster is associated a linear subspace of a fixed dimension D. The orthogonal basis of the subspace is obtained by considering the first D principal components of the correlation matrix of the patterns of the associated cluster. During training some patterns can be misclassified. The following notations help to deal with them. The generic pattern $\mathbf{x}$ of the training set belongs to a class $\mathbf{T}_i$. This class will be denoted as its "correct class". If the largest orthogonal projection of $\mathbf{x}$ is in correspondence to subspace $\mathbf{C}_{k,j}$ of class $\mathbf{T}_j$, then the classifier assigns $\mathbf{x}$ to class $\mathbf{T}_j$, which is its "wrong class". $\mathbf{C}_{k,j}$ is the "wrong subspace". The subspace $\mathbf{C}_{h,i}$, where the projection of $\mathbf{x}$ is the largest with respect to all the subspaces of $\mathbf{T}_i$, is denoted as the "correct subspace".

Outline of the algorithm : The proposed method is constructive, based on the objective function derived by a validation set as follows. The training set TS is divided in H parts constituted by a same number of examples of the classes of interest. Each part in turn plays the role of validation set, while the rest of TS is used as the true training set. If MSC_k is the number of misclassifications obtained by the neural net obtained with the k-th partition, then the objective function is

(1) $$F = \frac{1}{H} \sum_{k=1}^{H} MSC_k$$

The optimal choice of H is possible in the asymptotic case [3]. In actual situations its value is determined by trials.

The construction of the neural net is undertaken iteratively by increasing the dimension D of the subspaces by a unitary step and controlling the reduction of the objective function F defined in (1) achieved by this process. In correspondence to each step, the initial configuration of the neural net is characterized by one subspace per class. Then, one new subspace at a time is inserted under suitable controls for avoiding both the introduction of insignificant subspaces and overfitting.

Generic step of the algorithm : The strategy followed by the learning algorithm in its generic step aims to reduce misclassifications by increasing the number of subspaces. For this purpose, the misclassified patterns are grouped in homogeneous sets on the basis of their correct and wrong classes. We denote the number of patterns present in a set as its cardinality. The set Ω with the largest cardinality is the more appropriate candidate for yielding the new subspace to be introduced. The new subspace is associated to the correct class of Ω.

The success of the previous strategy is hampered by two inconveniences: the excessive proliferation of subspaces and the introduction of insignificant subspaces. The first of them is easily removed by imposing a limit on the maximum number of subspaces per class. The latter requires to control the validity of the set Ω for yielding the new subspace. A first control at this regard concerns with the minimum number MP of patterns present in Ω . The value of MP should exceed or be equal to D in order to avoid the introduction of arbitrary vectors in the base of the new subspace. Without this constraint insignificant subspaces could be created. However, the necessity of using a high threshold determines another inconvenience. The sets of misclassified patterns which do not meet the previous condition are not corrected. The number of misclassifications therefore could be excessive. For this reason, it is convenient to rely on Ω only at the beginning of the learning step and then to group the misclassified patterns in a less stringent way. Namely, we will consider in succession, in place of Ω , the sets Ω_1 and Ω_2 defined as follows. The set Ω_1 coincides with the set $\mathbf{S}_{ab}$ such that

(2) $$\mathrm{card}\,(\mathbf{S}_{ab}) = \mathrm{Max}_{r,i}\,(\mathrm{card}\,(\mathbf{S}_{ri}))$$

where $\mathbf{S}_{ri}$ is the set of misclassified patterns having $\mathbf{C}_{ri}$ as the correct cluster of the correct class i without paying attention to their wrong classes. The set Ω_2 coincides with the set $\mathbf{S}_c$ such that

(3) $$\mathrm{card}\,(\mathbf{S}_c) = \mathrm{Max}_i\,(\mathrm{card}\,(\mathbf{S}_i))$$

where S_i is the set of misclassified patterns having the correct class i.

The consideration of the previous very general sets Ω_1 and Ω_2 for introducing a new subspace must be strictly controlled in order to avoid the formation of subspaces very similar to others already introduced. This control ascertains the independence between the new subspace and the others of the same class.
In conclusion the procedure for introducing a new subspace is as follows.

Procedure for introducing a new subspace: Use the set Ω for introducing the new subspace of its correct class if its cardinality is larger than or equal to MP. On the contrary, use the set Ω_1. If its cardinality is larger than or equal to MP, determine the first D principal components of the correlation matrix of its patterns. Introduce a new subspace having them as orthogonal basis, if at least one of its basis vectors is linearly independent of the basis vectors of all the other subspaces of its correct class. On the contrary, use the set Ω_2 with the same conditions required for Ω_1.

Improved determination of the sets Ω_1 and Ω_2 : The determination of the sets Ω_1 and Ω_2 on the basis of the previous definition does not take into account the disturbing action of misclassified patterns very far from the bulk of them. A simple way for reducing this effect is to disregard them preliminarly. The only inconvenience due to this procedure is the reduction of the misclassifications corrected by the new introduced subspace. However, subsequent subspaces can take care of the disregarded misclassified patterns.
The determination of the disturbing patterns is easily accomplished by considering their projections onto the set under formation and by comparing them with their lengths. A pattern is disregarded if the ratio between its projection and its length is less than a prefixed threshold L. This procedure is applied to both Ω_1 and Ω_2. It is interesting to interpret geometrically the previous procedure. The elimination of the disturbing patterns from the set under formation causes a rotation of the resulting subspace from them toward the remaining patterns. The rotation of subspaces for correcting misclassifications is currently applied in [1] and it is aimed at exploiting as much as possible the fixed number of subspaces there used. In the present case the net does not possess a fixed number of clusters and consequently it is better to give the subspaces as specialized nature as possible. Moreover, the suggested procedure is a very economical way for implementing a rotation.

Summary of the proposed algorithm: The algorithm is iterative. Each step regards a value of the dimension D of the subspaces. The starting value of D is 1 and it is increased by 1 at each step. In correspondence to the generic step, with a current dimension D, the following substeps are carried out by applying them to the H partitions of the training sets.

1) Determine the first D principal components of the correlation matrix of the patterns of each class of the particular partition of the training set under consideration. They define the basis of the subspaces to be initially associated to the classes, one per class.

2) Introduce a new subspace at a time following the above "procedure for introducing a new subspace" and controlling that the number of subspaces per class does not exceed a prefixed threshold. The said procedure requires the determination of the sets Ω_1 and Ω_2 by the rotation-like technique described previously.
3) Determine the number of misclassifications of the neural net obtained in the previous steps when applied to the validation set of the partition of the training set under consideration.
4) Repeat steps from 1 to 3 for all the partitions of the training set and determine the objective function F , defined in (1) .

The objective function F is a function of D. The optimal value of D corresponds to its minimum. With this optimal value, the final version of the net is determined applying substeps 1 and 2 using the entire training set.

3 Performance of the proposed method

The efficiency of the proposed method was tested in connection with the well-known sonar benchmark [4]. This problem addresses the identification of the sonar return from two different obstacles. The data set, available from [5], includes 208 sonar returns (one class of 111 metal cylinder returns and another class of 97 rock returns). They are split into a training and a test data set of size 104 each. Using H=5, L=0.9, MP=D the proposed learning algorithm yields the values of F shown in tab.I, where we consider the range $1 \leq D \leq 6$. The minimum is obtained in correspondence to D=3. The final neural net is characterized by a recognition accuracy on the test set equal to 91.4 % and a number of neurons equal to 30. The algorithm yields the neural net automatically without the necessity of successive trials. It is also simple to apply and it consequently requires a reasonable time for training.
The performances of the resulting net are compared in tab.II with those obtained with the methods of same nature suggested in [2]. We see from the table the gain obtained by our method with respect to them. The dependence of the method on its parameters MP,H,L was investigated in detail in [6] . The sensitivity of the recognition accuracy on them is not critical. The values used in the present example can be reasonably used as default in other cases.

tab.I

D	1	2	3	4	5	6
F	0,548	0,558	0,462	0,491	0,490	0,548

tab.II

Method	Proposed	GSM [2]	GLSM [2]
Accuracy	91.4 %	87.5 %	89.4 %
number of neurons	30	33	32
Normalized training time	1	1.3	26

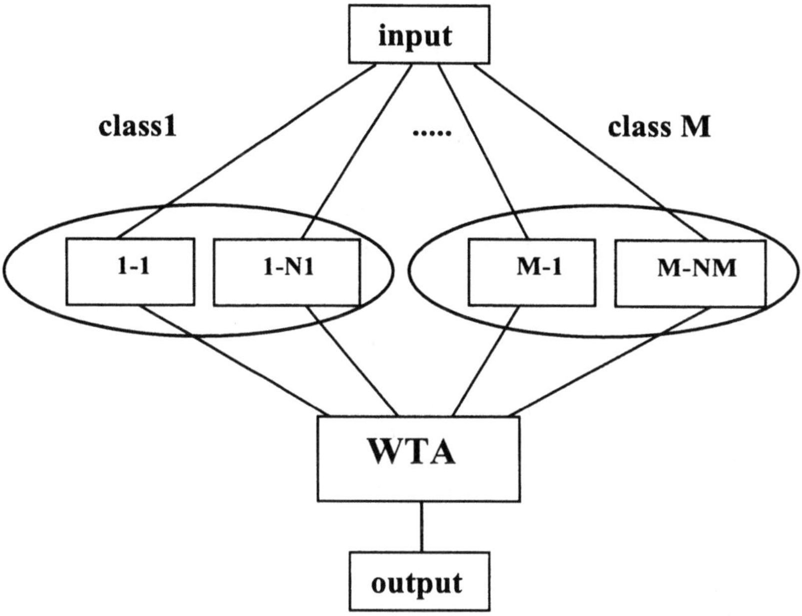

fig.1 The architecture of the neural net. Neurons correspond to subspaces. The output of a neuron coincides with the projection of the input into the corresponding subspace. The output neuron selects the largest projection by a winner-takes-all operator (wta). The output is the label of the corresponding class.

References

[1] T. Kohonen, Self Organizing Maps, (Springer Series in Information Sciences, New York, 1997).

[2] M.Prakash and M.Narishma Murty, Growing Subspace Pattern Recognition Methods and Their Neural Network Models, IEEE Transactions on Neural Networks, 8 (1997) 161-168.

[3] S.Amari, N.Murata, K.R.Müller, M.Finke, H.H.Yang, Asymptotic Statistical Theory of overtraining and Cross-Validation, IEEE Transactions Neural Networks 8 (1997) 985-996.

[4] R.P.Gorman and T.J.Sejnowski, Learned Classification Of Sonar Targets Using a Massively Parallel Network, IEEE Transactions on Acoustics, Speech, and Signal Processing, 36 (1988) 1135-1140.

[5] S.E.Fahlman, CMU Benchmark collection for neural net learning algorithms, School Computing Sciences, Carnegie-Mellon University, Pittsburg, PA, 1993.

[6] M.Di Giacomo, Approccio costruttivo all'ottimizzazione di una rete neurale ASSOM, Engineering Thesis, INFOCOM Department, University "La Sapienza" of Rome, 1997.

Weightless Neural Networks for face recognition

Stanislao Lauria, Dr. Richard Mitchell
Dept. of Cybernetics, The University of Reading, READING, U.K.
s.lauria@reading.ac.uk
http://www.cyber.rdg.ac.uk/research/CIRG/

Abstract

Face recognition is an important application of image processing and a recent paper has compared various methods of processing images of faces. The best of these methods is a complicated linear projection based system, which successfully copes with images having variable expression, pose and facial lighting. This paper considers how a simple weightless neural network, with simple preprocessing, compares with the more complicated method.

1 Introduction

The interest in reliable personal identification using computerized access control has increased in the past few years. Face recognition has the advantage of being a passive, non intrusive system for verifying personal identity. The tests are focused on recognition with varying facial detail, expression, pose. Invariance to high degrees of rotation and scaling is not considered. Moreover, as rapid classification is required, time for extensive pre-processing and normalization has to be small or ideally null.

A general statement of the problem can be formulated as follows: given still or video face images f and using a stored database of faces D, identify f. A *duplicate* is any data representing a face independently of the condition of light or expression or position. The *gallery* set is given to the system as the set of known duplicates during the learning set, whereas a *probe* set a set of duplicates to be identified. The intersection between the gallery and the probe set is empty.

A recent paper [1] has compared various methods of face recognition. These methods include correlation [2], Eigenfaces [3] and Fisherface [1]. In [1] it is shown that the best model is the Fisherface system, which is derived from Fisher's Linear Discriminant (FLD), first developed [4],[5] for taxonomic classification, based on the reduction of the dimensionality of the feature space. The Fisherface method processes a set of N duplicates $x_1,...x_N$ taking values in an n-dimensional image space, assuming that each duplicate belongs to one of c classes $X_1,...X_c$, let consider a linear transformation mapping the original n-dimensional space to an m-dimensional feature space, with $m \leq n$. The new m-dimensional vectors $y_k \in R^m$ are defined by the following linear transformation: $y_k = W^T x_k$ with $k = 1, 2, .., N$. Where $W \in R^{n \times m}$ is a matrix with orthonormal columns. Fisherface seeks the transformation matrix W in such a way that the ratio of the between-class scatter and the within-class scatter is maximized. The m-dimensional space is analysed to determine which of the c classes to which the original image belongs. It is shown [1] that the method can simultaneously handle variation in lighting and expression.

However, all these methods, in order to achieve satisfactory results, need extensive pre-processing of the image in order to normalize or quantize it at the expenses of high computational complexity. As little has been done with Weightless Neural Network (WNNs) in the same field [6], [7], [8], this paper presents an approach using WNNs. Moreover, some of the WNNs typical aspects, such as speed and simple pre-processing, should be very advantageous for such tasks.

2 Weightless Neural Networks

The n-tuple classifier or Weightless Neural Network (WNN) is a memory based method first introduced by Beldsoe and Browing [9] and implemented in hardware by Aleksander [10]. It is one of the oldest practical pattern recognition methods based on memorisation of randomly selected features. The method offers several advantages over other forms of pattern recognition. The training algorithm as a one shot memorisation task, the internal parameters can be set after running a few tests, the computational triviality, are some of the features that make it faster and simpler compared to the other techniques.

These networks consist of a set of neurons, implemented using standard RAM memories, which form a *discriminator*. The basic operation is to sample n Boolean values from the input data; these form a n-tuple which is used to address a location in one RAM neuron. When an input is being taught to the system, a series of tuples are formed from the input, and each tuple is passed to its own RAM neuron, and a '1' is written into each addressed memory location. When an input is being tested to see if the system recognises that input, tuples are formed in the same manner, but the data are read from each addressed location in the various RAM neurons. The number of '1's read indicates whether the input is recognised. In practise, several versions of a given input are taught in this manner, so the system can recognise an input similar to, but not identical to, an input it has been taught.

The generalisation properties of the system arise by teaching many similar images to the network. Thus the system will respond with 100% of the RAMs outputting a '1' when any of these images is presented to the network. In addition when an image is presented to the network which is similar but not identical to any of those it has been taught, then 100% of the RAMs may still respond. Indeed, for example, the first tuple from the image may be from the fifth image taught, the second tuple may be from the ninth image taught and so on. It is also possible to introduce a threshold value p%. Then the image is recognised if the number of RAMs outputting a '1' exceeds the value p%. Therefore it is necessary to set tuple size n and the threshold value p correctly and ensure that an appropriate amount of learning is done to obtain a proper response from the network. For instance, if the tuple size is too small, too many locations in the RAM have a '1' stored so the network 'recognise' images that are too different from those it has taught. On the other hand if the tuple size is too large the system cannot generalize. Although better theoretical work on the n-tuple recognition algorithm would be highly desirable, it must be underlined that the high speed of the algorithm makes it entirely practical to just run few tests with a few plausible parameter setting to find suitable ones.

Different classes of input can be taught into the same discriminator, and the system can report whether the discriminator 'recognises' any input. However, it will not be able to discriminate between the classes. If recognition and discrimination are required, then a network with many discriminators is used and each class of input is taught into its own discriminator. Then, when an input is being analysed, tuples are formed and passed to the respective RAMs of all the discriminators, and the input is 'recognised' as being from the class whose discriminator was found to have the most number of '1's.

When the input set is derived from an analogue source, then some processing is needed to convert the analogue values into the Boolean states required to define the address input to the RAM. The simplest process is to threshold the input, sampling a true or false condition depending on the magnitude of the analogue input relative to a reference level. However, a large amount of information is lost using this method, while the logical address patterns input to the RAM will vary greatly according to the changes of overall amplitude of the source. Indeed, for the same image, but with different conditions of light, the tuple thus formed can be very different from the value previously stored. As a consequence, the change of light causes the network not to recognise the image previously learnt. Another method of processing level data is to use 'thermometer coding' [10]. This involves converting each sampled point of the input into an array of Boolean values, where the greater the amplitude of the source, the more Boolean values are true. This is equivalent to maintaining a number of threshold systems, each with a different threshold. However, it requires more memory and it is slower than the conventional method.

A different technique has been developed at the University of Reading, called Minchinton cells [11]. These are new processing blocks connected between the input and the RAMs. Each cell samples data from the input data, and computes a local function on its sample. The samples are randomly distributed over the input data. The cell functions used here are the types:

$$V(x_i) > V(x_r) \text{ [Type 0 cell]}$$

$$|(|V(x_r) - V(x_r - 1)| - |V(x_i) - V(x_i - 1)|)| < K \text{ [Type A cell]}$$

where K is an arbitrary constant and V is the input data and with $V(x_r)$ and $V(x_r)$ being respectively the value to be processed and a value chosen randomly.

The type 0 cell reports whether one random sample from the image exceeds another random sample. This can be used to make the system tolerant of overall changes in the lighting of an image. If the lighting changes in this manner, the value of each sample is likley to change by a similar amount, but the difference between the samples should be small. Any such difference should be handled by the generalisation properties of a standard WNN. The type A cell looks at the differences between adjacent samples in the image. The idea behind the cell is to try to make the WNN insensitive to zonal changes in the input, such as shadows. This is appropriate in the experiments performed, as the faces being processed are illuminated from different angles. In the experiments described here, WNNs are used with the thermometer and Minchinton cell pre-processing methods and the performance compared against other methods.

3 Database

Using the database constructed at the Yale Center for Computational Vision and Control [1], tests have been designed to compare the different methods under a different range of conditions. For sixteen subjects, ten images (320 × 243 pixels, 256 grey level) were acquired during one session in front of a simple background. Subjects included females and males (some with facial hair) and some wore glasses. The first duplicate was taken under frontal lighting in a neutral facial expression. In the second image, the glasses were removed. If the person normally wore glasses, those were used; if not, a random pair was borrowed. Duplicates 3-5 were acquired by illuminating the face in a neutral expression with the lighting source in three different positions. The last five images were acquired under frontal lighting with different expressions (happy, sad, winking, sleepy and surprised). As the change of light conditions introduces a considerably noise and shadowing in the background (see fig. 1), two different databases have been created: *Face Crop* and *Full Image.* The original images with no changes formed the Full Image database, while the Face Crop ones included images automatically centered and cropped selecting a 150 × 200 window of approximatively the dimension of the face (i.e. excluding the background) in the central area of the image. It is important to underline that no further pre-processing has been introduced. That is, for all the duplicates the same position for the selected window has been used, indipendently from where the particular face is really located in the picture.

Figure 1: **Yale Database.** *Example images for a given subject. It is possible to note the drastic background change under different condition of light.*

4 Experiments

In order to compare the performance of WNNs with the systems presented in [1], the same testing method used in [1] has been adopted here to obtain the error rate. That is, error rates were determined by the "leaving-one-out" strategy [5]. To classify an image, that image was removed from the data set forming the gallery set. As a consequence, the whole Yale database, except that given image, formed the gallery set, whereas the probe set was made by that given image.

Various network configurations have been tested. All the configurations have been based on the same principles. That is, a discriminator for each person has been configurated, where each discriminator has been taught with all the images from the probe set belonging to the same subject. The system thus obtained, with 15 different discriminators, has been then tested with the image from the gallery set (i.e. with the only image from the Yale database not included in the probe test). Recognition was then performed associating the tested image with the discriminator with the highest response. All the images from the database have been tested in this way. It is important to note that the various configurations used in each test may not give the best possible performance as the current best performing system was only obtained as a result of these experiments. Table 1 shows results for the Minchinton cells (type 0 and type A) and 4 bit thermometer coding. Finally, the various responses from 1 have been processed so that an image has been assigned to the class with the highest number of method giving the same answer. Table 2 is a comparison between the results thus obtained and other methods presented in [1].

Coding Methods	Face Crop	Full Image
Minch. Cells (type 0)	12.7	24.8
Minch. Cells (type A)	18.2	23.6
4 Bit Thermometer	23.0	23.0

Table 1: *Yale test for different WNN configurations. Error rate in percentage.*

System	Face Crop	Full Image
Eigenfaces	24.4	19.4
Correlation	23.9	20.0
Linear Subspace	21.6	15.6
WNN(type A and 0)	11.5	15.1
Fisher face	7.3	0.6

Table 2: *Yale test comparisons with different methods. Error rate in percentage.*

Overall, there is a better response when the background has been eliminated as shown in Table 1. From Table 2, it is possible to conclude that only Fisherface methods performed better than WNN. WNN is the only method improving on the images where the background has been excluded and, as also recognised in [1], the recognition rate for full images is less reliable because highly background dependent.

Finally, as shown in Table 3, WNN performed better in terms of computational complexity. Where $t =$ tuple size, $N_i =$ number of training duplicates for a single class, $n =$ image size, $N =$ total number of training duplicates, $c =$ n. of classes.

Methods	Classification complexity	Training complexity
Correlation	$O(nN)$	$O(1)$
Eigenfaces	$O(nN)$	$O(n^2)$
Linear Subspace	/	/
Fisher face	$O(nc)$	$O(n^2)$
WNN	$argmax\{O(n), O(tN_i\}$	$O(nN_i)$

Table 3: *Complexity comparisons.*

It is possible to note that WNN classification complexity is independent from both the total number of classes c and the total number of training images N, being dependent at most on the number of images for each class ($N_i \ll N$). Then, when c increases (and consequently N increases) classification complexity is supposed to increase for all the methods presented in the Introduction, whereas WNN complexity remained invariate. Although correlation training complexity is better than WNN, nevetheless all the other methods perform worse than WNN to train ($N_i \ll n$). Moreover, with a WNN approach there is a further advantage, as the system can be extended to cover new classes or images without retraining the system again. Indeed, the Eigenfaces and Fisher face methods need a completely new training session every time a new class or even a new single image needs to be learnt.

5 Conclusion

The use of Weightless Neural Network for face recognition has been investigated. Various preprocessing methods have been used and their performance compared with the best of a number of other methods on a particular set of data images. The use of the Type A Minchinton cells introduced a better invariance to the change of light as well as a very simple form of control on the sampling process. Overall, the tests described here show that only the Fisherface method can perform better than WNNs. Hence, given the relative simplicity in configuring and using weightless networks, they should seriously be considered in the design of a face recognition system.

References

[1] Belhummeur *et al.* Eigenfaces vs. Fisherfaces: Recognition Using Class Specific Linear Projection **IEEE Trans. on PAMI** vol. 19, No 7, July pp711-720

[2] R. Brunelli, T Poggio. Face recognition: Features vs. Templates. **IEEE Trans. on PAMI** . Vol 15, no. 10, pp. 1042-1053. 1993.

[3] M. Turk and A. Pentland, Eigenfaces for recognition. **J. cognitive neuroscience**, vol3, no.1, 1991.

[4] R.A. Fisher, The use of multiple measures in taxonomic problems. **Ann. Eugenics**, vol 7, pp179-188, 1936.

[5] R. Duda, P. Hart, **Pattern Classification and Scene Analysis.** New York: Wiley, 1973.

[6] Wilkie, B. **A Stand-Alone High Resolution Adaptive Pattern Recognition System**, PhD Thesis, Brunel University, 1983.

[7] Stonham, J. Practical Face Recognition and Verification with WISARD. In Ellis et al. **Aspects of face processing**, Dordrecht: Martinus Nijhoff, 1986.

[8] Lauria S., Mitchell R. J., WNNs for face recognition: a comparison. 1997, in **Neural Networks for Signal Processing VIII.** M. Niranjan, E. Wilson, T. Constantinides, S.-Y. Kung (eds.). IEEE, New York. pp. 539-546.

[9] Bledsoe W. W.,Browning I. Pattern recognition and reading by machine. **Proc. Eastern Jnnt. Computer Conference**. pp225-232. 1959.

[10] Aleksander I., Stonham T.J. A guide to pattern recognition using RAM. **IEE Proc. E, Comput Dig. Tech** 2(1).pp. 29-40. 1979

[11] Bishop, J.M., Minchinton, P.R. and Mitchell, R.J. "Real time invariant grey level image processing using digital neural networks", **Proc IMechE Conf. EuroTech Direct '91 - Computers in the Engineering Industry**, pp 187-8, Birmingham, July 1991

The search for spiculated lesions in the CALMA project: status and perspectives

Rosa Palmiero
Dipartimento di Fisica, Università di Pisa and INFN
Piazza Torricelli 2
56100 Pisa, Italy
tel. +39-050-911311 fax +39-050-48277 e-mail: Rosa.Palmiero@cerd1.difi.unipi.it

Abstract

A status report is presented on the massive lesions search in the frame of the CALMA project. CALMA's main purpose is to collect a database of mammographic images, developing CAD tools to be used as a second radiologist in the classification of breast tumoural disease. Massive lesions are rather large objects to be detected, but they show up with a faint contrast slowly increasing with time. The need for tools able to recognize such a lesion at an early stage is therefore apparent. The work performed on images collected from italian hospitals in the last year, as well as on images available from public databases is here presented, indicating, at this stage, a number of false positives of the order of 3 per image keeping a large sensitivity on our sample. This achievement although improvable, is comparable with results obtained by other groups working on the same subject.

1 Introduction

A massive lesion is a rather "large" object characterized in some cases by peculiar shapes but a faint contrast (few percent).

The early detection of such lesions would enormously improve the success of removing breast cancer, reported to be one of the first causes of women mortality ([1],[2]).

In the following we refer specifically to the so called "spiculated" lesions, whose shape vaguely reminds of a star.

The radiologist purpose is to figure out such objects out of a variable texture, ranging from a fatty breast where background is rather uniform, to nodular ones, where nodules themselves are misidentified as cancers.

In the attempt to perform an early diagnosis therefore is important to limit as much as possible the number of these false positives keeping sensitivity to cancers close to 100%.

This work has been performed in the more general frame of the CALMA project [3, 4], a collaboration among INFN sections of Bologna, Pisa and Udine and some italian hospitals (Bari, Bologna, Livorno, Massa, Udine).

1.1 Available images

Our database at the moment already contains 2200 images, but of course not all of them contain the patological structures we are interested in for this search.

Spiculated lesions available are 10 from the hospital of Livorno (digitized with a 85 μm granularity and a 12-bit resolution), 14 from the MIAS public database (200 μm granularity and 8-bit resolution) and 305 images from the hospital of Bari and Udine (85 μm, 12-bit).

In the following, where cuts or thresholds values are reported they should be understood as applied to the Bari/Udine images, and properly scaled when applied to the rest of our sample.

In the next month, thanks to contributions from the hospital of Bari and Udine we expect that the number of spiculated lesions should increase.

2 Method

The chosen method is not entirely neural.

A Region Of Interest (ROI) hunter has been prepared to select interesting portions of the mammogram, with efficiency of 100% for all the spiculated regions.

The ROI hunter's characterizing parameters can eventually vary as a function of the mammograph texture, i.e. as a function of the tissue patterns. Figure 1 shows two spiculated lesions on two different tissue patterns.

Here we have two possible choices:

- Each portion has been preprocessed to get a set of features that in the final stage enters a Feed Forward Neural Network.
- The ROI hunter itself is adjusted so to cut heavier on false positive regions, accepting a sensitivity lower than 1.

2.1 Preliminar treatment

Images are conventional radiographies, containing together with the breast image a series of additional information on the emulsion, such as a lead label typically located on an image corner. These parts are therefore "cleaned" in such a way to leave only the relevant part to be processed later on.

2.2 Texture Analysis

An early classification based on the tissue structure of the mammograph has turned out to be of utmost relevance in the subsequent classification of the tumoural lesion. Figure 2 shows the six main classes with which radiologists label tissues. They are dense, irregularly nodular, microglandular, fibro-adipose, fibro-glandular, and parvinodular. We wish at least to discriminate among fibro-adipose (i.e. fat tissues), dense, and glandular ones (comprehensive of the other four).

We have tried three different methods to perform automatically the classification:

- Spatial frequencies approach: a 2D-Fourier transform is calculated on image samples of 512 × 512 pixels. We choose then 11 frequencies as an input to an FFNN with 6 or 3 outputs (classification in 6 or 3 classes). These frequencies have been chosen so that their distributions for the 6 classes are the most separated.

  With a net with 5 hidden neurons educated for 40000 cycles we obtain the following classification (in percentage):

	G(true)	A(true)	D(true)
G(rec)	72	28	0
A(rec)	26	74	0
D(rec)	58	22	20

- Statistical approach: image samples of 200 × 200 pixels are extracted from the database and an average pixel-by-pixel is calculated for each class. For each image is later calculated the pixel-by-pixel distance from the class averages. The average of these distances and the standard deviations are later calculated.

  In this way we have 12 inputs to an FFNN, to which we add the age of the patient which turns out to be a relevant feature. The chosen architecture is a 13-10-6 net, trained on 50 images per each class with 200000 cycles and tested on 1431 independent ones. Grouping together all the glandular ones we get (percentage values):

	G(true)	A(true)	D(true)
G(rec)	83.1	16.4	0.5
A(rec)	49.1	50.7	0.2
D(rec)	0.0	4.8	95.2

- Unsupervised Neural Network: we used a Sanger's generalized Hebbian algorithm for principal components analysis, trained whith 90 images (30 for each of 3 principal classes: A, D, G) samples of 512 × 512 pixels, extracted from database and then reduced to 256 × 256. The architecture has 256 × 256 input neurons, no hidden layer and 8 to 15 output neurons.

  The final choice of relevant parameters (output neurons number, training epochs and learning-rate parameter) is done on the basis of the matrix weigth evolution and the reconstructed images.

  Our goal consists in achieving classification just using these components.

  We think that a particularly good result may be obtained in the discrimination between dense tissues and adipose ones. We get for these two classes:

	A(true)	D(true)
A(rec)	100	0
D(rec)	7	93

We are evaluating at the moment the possibility to merge the previous methods into a single hybrid approach.

2.3 ROI hunter

For each pixel in the image, out of ten (to reduce the amount of calculus), whose counts are larger than zero, a spiral is built with a fixed number of turns and a fixed pitch. It has been found that the best result is obtainable with a 30 pixels starting radius, and 100 pixels maximum radius. A sampling of 100 pixels per turn is taken so that a 600 pixels array is built in 6 turns.

In presence of a spiculated lesion in the region centered around the pixel we espect the content of the array being oscillating periodically, while it should be pseudo-random in all the other cases.

A Fourier transform is later applied to the array content, producing the spectrum of pixel intensities. Here, supposing again to be centered on a spiculated lesion, we should find a "peak" corrisponding to the number of arms of the star. We decide to be interested in any peak between 2 and 7 at this stage of the analysis, so a definite integral is calculated between the corresponding values in the spatial frequencies domain. The integral, normalized to the maximum of the current image, is assigned to the pixel around which the spiral was unrolled, and shows to be strictly correlated to the probability of being in the center of a spiculated-like structure.

We then cut out all the pixel, whose probability value shows up to be lower than 80% of the maximum. Two relevant regions are merged if they overlap at least for a fixed percentage of their pixels (here we chose 40%). In this way, having chosen properly all the thresholds and parameters in the calculation, all the spiculated regions are found and up to 30 non-spiculated regions are kept for the next step.

2.4 Preprocessing

Once 30 regions per image are selected by the ROI hunter, we analyze each of them with an algorithm based essentially on the study of iso-intensities curves.

These curves, being fixed their area, should show a larger perimeter the less "circular" they are. So, for each region, we consider the ratio perimeter/area of the iso-intensity curve as a function of the intensity.

We sample 10 to 20 values of this ratio (at different fixed intensities) and use them as the input for our classificator.

2.5 Classification with FFNN

A simple Feed Forward Neural Network with momentum back-propagation is used to perform classification.

We used the SNNS software [5] to simulate the networks.

The number of input neurons has been varied between 10 and 20, with 4 to 10 neurons in a single hidden layer and 2 output neurons.

Images are divided equally into a learning and a test sample. Final classification is performed setting a threshold on the output neurons value. If a 90% sensitivity has to be reached, we find that up to 3 false positives per image are still found.

2.6 Classification with ROI hunter

We found that chosing heavier cuts to reject false positives, eventually as a function of the tissue texture, the ROI hunter itself may be used for the classification.

On adipose textures (A class), we found that it is possible to get a sensitivity [1] per image of 73% (that increases to 90% as specificity per patient) on spiculated lesions, while on glandular tissues (G class) 80% can be reached as a sensitivity per image (85% is the corresponding "per patient" figure), keeping an average of 1.2 false positive per image.

3 Results

Our results should improve, both with the increasing number of images available in the database, and with a better evaluation of the tissue texture to be included possibly in the last step.

In any case these results are somehow encouraging, considering the present status of achievements in this field.

References

[1] A.G.Haus and M.J.Yaffe, Editors, *A categorical course in physics. Technical aspects of breast imaging*. Radiological Society of North America, 1993, presented at the 79th scientific assembly and annual meeting of RSNA.

[2] R.A.Smith, "Epidemiology of breast cancer" in *A categorical course in physics. Imaging considerations and medical physics responsibilities*, Madison, Wisconsin, 1991. Medical Physics Publishing.

[3] V.M.Marzulli, "The CALMA project: Computer Assisted Library for MAmmography", WIRN98 Proceedings, Springer-Verlag (1998).

[4] P.Delogu, "The automatic detection of microcalcification clusters in the CALMA project: status and perspectives".

[5] A.Zell et al., "SNNS - Stuttgart Neural Network Simulator - V4.1", University of Stuttgart Report No. 6/95 (1995).

[1] sensitivity per image is defined as the percentage of images with at least one lesion recognized as such, sensitivity per patient is defined as the percentage of patients with at least a lesion recognized as such

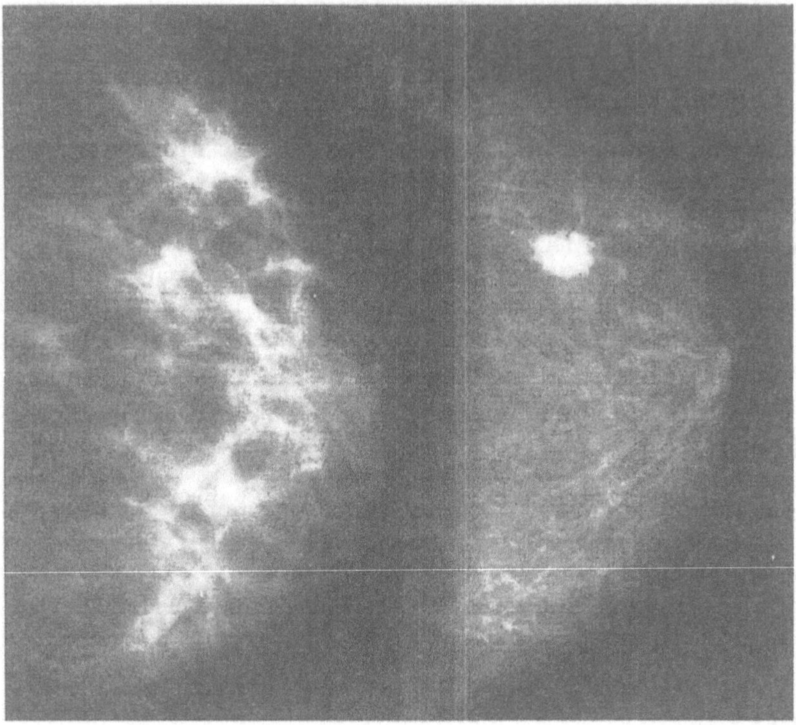

Figure 1: Spiculated lesions on two tissue patterns: glandular and fibro-adipose

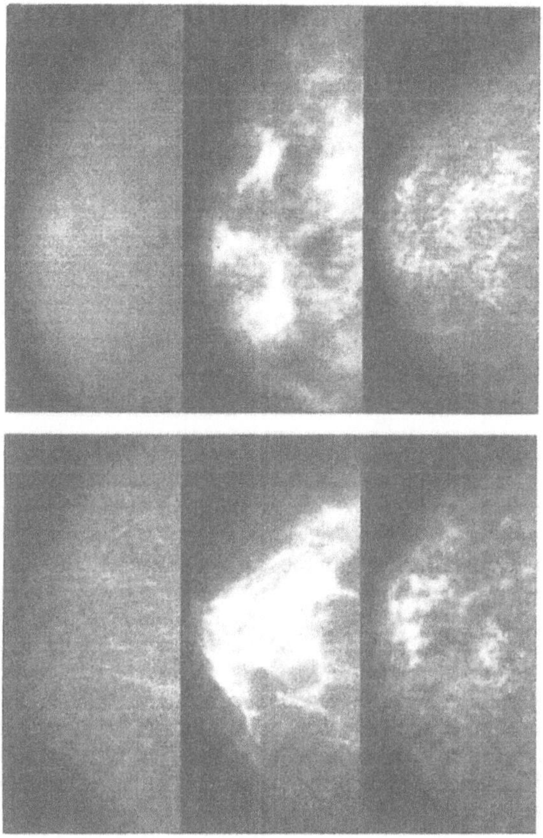

Figure 2: textures classes, from left to right, up to down: dense, irregularly nodular, microglandular, fibro-adipose, fibro-glandular, parvinodular

A Novel Wavelet Filtering Method in SAR Image Classification by Neural Networks

Giovanni Simone[1]

Francesco Carlo Morabito[2]

Università di Reggio Calabria - Facoltà di Ingegneria - DIMET
Via Graziella, Loc. Feo di Vito - I-89100 Reggio Calabria , Italy
e-mail: [1]simoneg@ing.unirc.it, [2]morabito@unirc.it

Abstract

This paper presents a method for classification of Synthetic Aperture Radar (SAR) images [1], based on the joint exploitation of a novel Wavelet Filtering Algorithm [2] and Neural Networks [3]. An illustrative example is presented that shows how the use of the proposed technique of Wavelet Filtering allows both to mitigate negative effects of multiplicative noise (speckle) and to classify the considered image without resolution decrease. The proposed Wavelet Filtering, applied on the image, provides four images (an *approximation* image and three *detail* images); *detail* images have the same resolution as original image. The use of Wavelet Filtering (WF) on the *detail* images jointly to a Multilayer Feedforward Network allows us to reduce some of typical drawbacks of the SAR classification problems, and thus to have a more efficient tool with respect to the traditional approach. The proposed method is tested on a real SAR image, and it is compared with traditional approaches based on MLP use, and on FFT-2D-filtering-MLP technique.

1 Introduction

Information on land-use yields important clues for planning and exploitation of natural resources [1,4]. Classification is the basic block of automated image analysis, since it is in step where different entries of interest are extracted from an image for further processing: the classification aims to classify a considered image, labeling each pixel with one of the classes that form the input image. The SAR image in Fig.1, obtained by the German E-SAR flying over Oberpfaffenhofen (Munich, Germany), shows an urban area in the upper left corner, some forests in the lower regions and in the upper right corner of the image, some cultivated areas and an airport; the imaged scene was illuminated from the top. The basic properties of pixels, on which classification techniques are based, are similarity and discontinuity. SAR images are typically characterized by a texture depending on the structure of the illuminated surface and on the illumination angle, and by a speckle related to the coherency of the image formation process [1,4,5]. In this paper an original Wavelet Filtering Algorithm (WFA) is presented and the structure of details is discussed analyzing the *detail* image construction process; the proposed WFA analyses outputs details, in which we have a non-coherent smoothing of the negative effects of the speckle, that causes a misinterpretation of the image content. We combined the WFA with a Multilayer Feedforward Network (the "Multilayer Perceptron" (MLP)): the interest in using artificial NN to classify remote-sensing data has recently been confirmed by various works [7,8], since this approach overcomes the difficulty with defining a single statistical model for different classes. The aim of this work is to show that, using a suitable filtering algorithm, and classifying the filtered images with a MLP, trained by means of the back-propagation algorithm, we increase considerably the classification performance with respect to the direct application of MLP to the input image, without filtering. A further advantage of our WFA is that the classified image has the same resolution as the input image: in the standard wavelet decomposition scheme [6] the output images have a resolution four times lower than the input image. Our technique is tested on the SAR image of the airport of Oberpfaffenhofen (Fig.1): the outcomes of the proposed approach are compared to the standard direct application of the MLP to the original image.

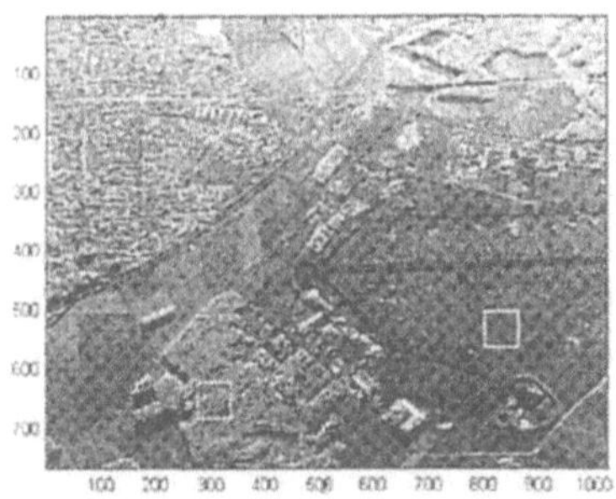

Fig.1: SAR image of Oberpfaffenhofen obtained by the German E-SAR (Courtesy of DLR, Germany). The insets represent the training samples.

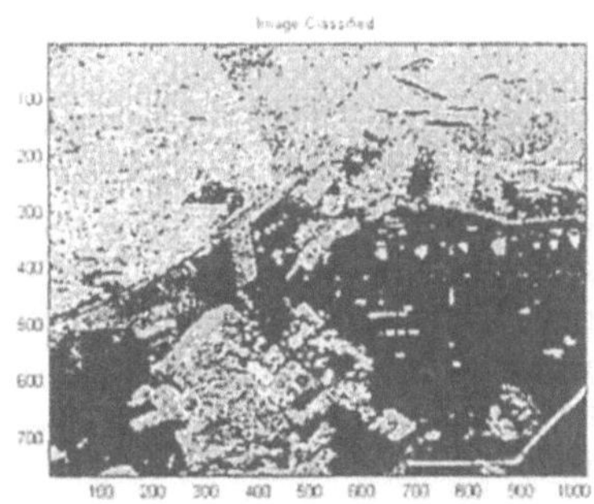

Fig.2: SAR image of Oberpfaffenhofen classified with MLP.

2 SAR Image Classification using MLP and FFT-MLP Techniques

In the context of remote-sensing data classification, the neural-network based approach is particularly attractive, as it needs not for *a-priori* knowledge on statistical distribution of data. In the standard applications [3] of artificial neural networks, the aim is to optimize the network to increase technique performances related to the considered problem. Typically, in the case of MLP, trained by the back-propagation algorithm, the image classification consists of two basic stages:

1. ***learning stage***: from a 64*64 pixels sample for each area, as indicated in Fig.1, we extract a feature vector $\underline{p} = [p_1, p_2, ..., p_n]^T$ (e.g. mean, standard deviation, etc.), with whom the neural network is trained; our MLP architecture has three outputs, that we set to $[0,0,1]^T$ for the forest area, $[0,1,0]^T$ for the

agriculture area, $[1,0,0]^T$ for the urban area; as a convergence criterion, we required a mean square error smaller than 0.005;

2. ***classification stage***: we extract from each pixel the contextual information, with a 5*5 pixels window centered on the pixel; the feature vector $\underline{p}$ is obtained from the window and the neural network produces its outputs; using Bayes decision rule for minimum error, we assign each pixel to be classified, to the most probable estimated class; to this end, we select the maximum output by using a "winner-takes-all" (WTA) decision block.

Fig.2 represents the result of this standard technique, applied to the SAR image of Oberpfaffenhofen (Tab.1).

Urban Area [%]	Forest Area [%]	Agricultural Area [%]	Global [%]
30	90	70	64

Tab.1: The classification rates using the MLP technique.

The main problem of this technique is that the illumination direction produces a misclassification of the considered image; therefore, to reduce the described negative effects, we can apply a high pass filtering in the Fourier domain. Fig.3 represents the high pass component of Oberpfaffenhofen SAR image, obtained transforming the considered image through FFT-2D, applying high pass filtering in Fourier domain, and using IFFT-2D algorithm; the filtered image is classified (Tab.2) with MLP (Fig.4), as in previous case. We would show that, combining the MLP basic scheme with the WFA, the classification performances could be significantly increased.

Urban Area [%]	Forest Area [%]	Agricultural Area [%]	Global [%]
70	75	70	72

Tab.2: The classification rates using the FFT-MLP technique.

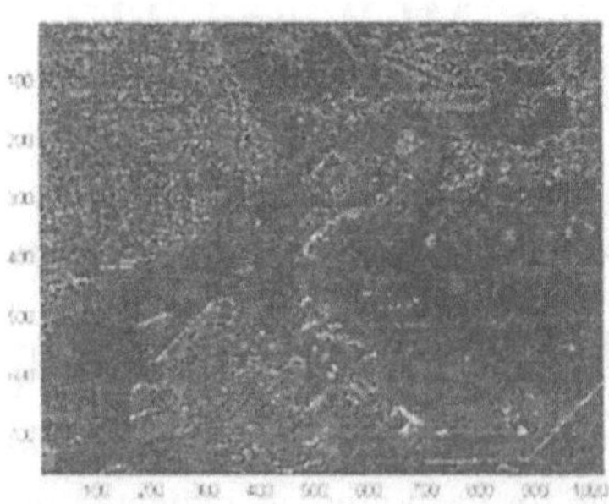

Fig.3: High pass filtered image through FFT algorithm of SAR image of Oberpfaffenhofen.

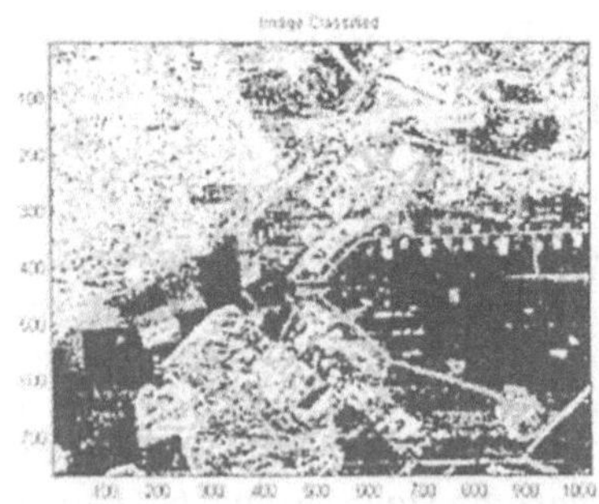

Fig.4: SAR image of Oberpfaffenhofen classified through the FFT-high pass filtered image.

3 SAR Image Classification using WFA-MLP Technique

The WFA is based on the Wavelet Transform (WT) analysis: usually, the WT [2,5] is employed in signals compression, but it offers several advantages in signals and image processing. The WT consists of a decomposition of a n-dimensional signal in n-dimensional sub-signals, that represent the frequency contents of the original signal at different levels of resolution. In the two-dimensional case the WT produces four sub-images from the original image on which it is applied: the rows of the input image are convolved with a one-dimensional wavelet decomposition low pass filter and are sub-sampled taking one pixel out of two adjacent pixels; the result is saved in the left half of the row filtered image; the rows of the input image are convolved with a one-dimensional wavelet decomposition high pass filter, and the result is sub-sampled; the result is saved in the right half of the row filtered image. The same operation is applied on the row-filtered image, but convolving the columns of the row filtered image with the wavelet decomposition filters. The results of the low pass filtering and of the high pass filtering are saved respectively in the up half and in the bottom half of the final image. The final image can be considered as formed by four sub-images, according to the following scheme:

1. the sub-image of approximation is obtained by low-pass filtering in rows and low pass in columns: it represents a sub-sampled approximation of the original image;
2. the sub-image of horizontal details is obtained by high-pass filtering in rows and low-pass in columns: in this sense, it contains horizontal details of the original image;
3. the sub-image of vertical details is obtained by low-pass filtering in rows and high-pass in columns: thus, it contains vertical details of the original image;
4. the sub-image of diagonal details is obtained by high-pass filtering in rows and high-pass in columns: thus, it contains diagonal details of the original image.

The drawback of this approach is that the *detail* sub-image resolution is four times lower than the input image, and this fact is a handicap when we want a direct relation between the filtered images and the original image.

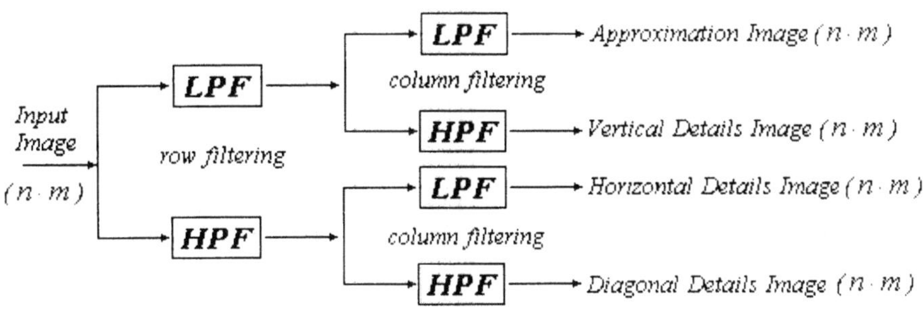

Fig.5: Wavelet Filtering Algorithm Block Scheme.

[6] shows that using a basic mean square error classifier, applied to *detail* sub-images, classification performances increase with respect to the case in which the same classifier is applied to the original image; but it is not possible to identify each

pixel of the classified image in the input image, since the input image has n rows and m columns, and the output image has $n/2$ rows and $m/2$ columns. Our WFA is based on the use of the low pass, and high pass decomposition wavelet filters, as indicated in Fig.5: the filters joint action preserves the non-coherent filtering effect that reduces the speckle, and the high pass filtering removes the negative effects that the illumination direction produces in the classification stage. The *detail* images have the same resolution than the input image, and Fig.6,7,8 shows the SAR image of Oberpfaffenhofen filtered with WFA; if we apply the MLP, we can combine benefits offered by the WFA with those offered by the MLP. In our scheme, we extract a sample from each *detail* image for each area, for the training stage; in the classification stage, the contextual information is obtained extracting features from a pixel-centered window, in each *detail* image. Fig.9 represents the SAR image of Oberpfaffenhofen classified with the proposed technique.

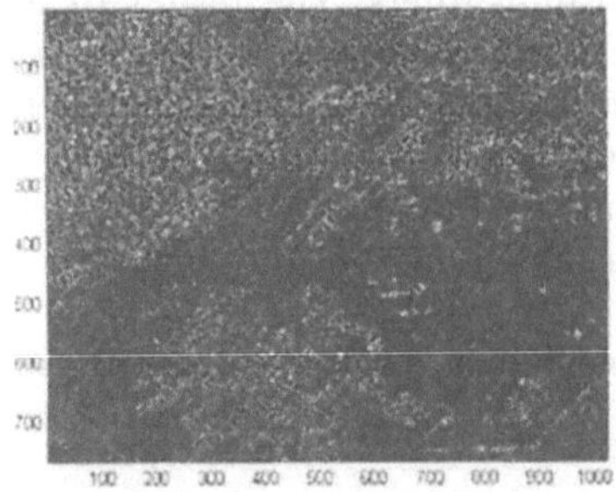

Fig.6: Horizontal detail image of SAR image of Oberpfaffenhofen, filtered through a Daubechies-8 wavelet mother.

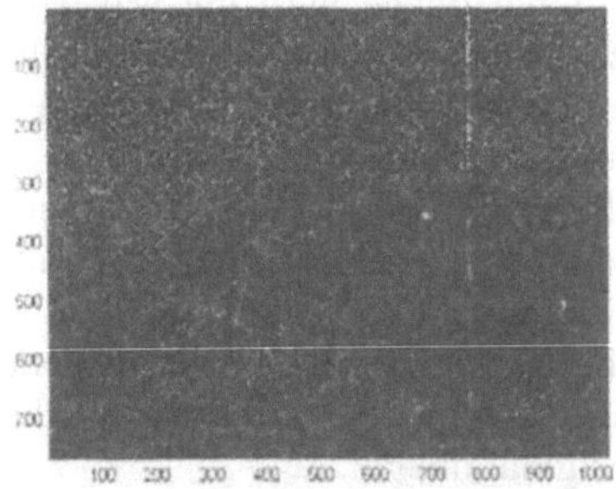

Fig.7: Vertical detail image of SAR image of Oberpfaffenhofen, filtered through a Daubechies-8 wavelet mother.

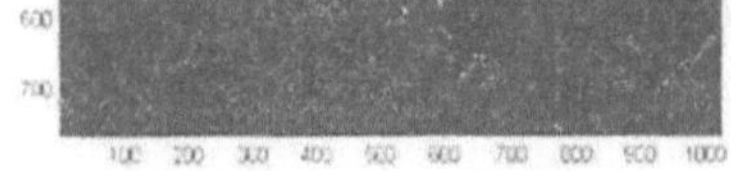

Fig.8: Diagonal detail image of SAR image of Oberpfaffenhofen, filtered through a Daubechies-8 wavelet mother.

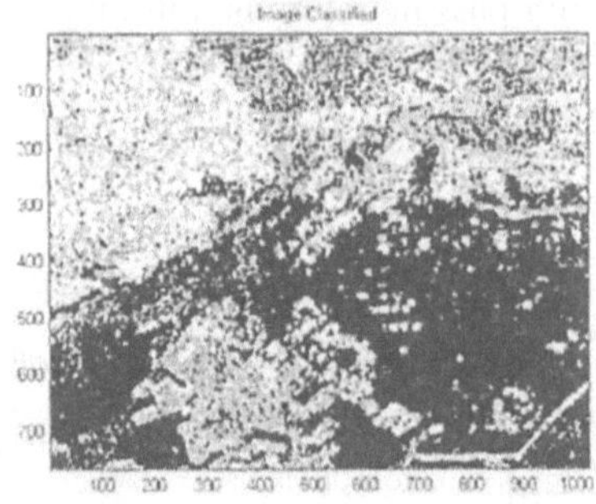

Fig.9: SAR image of Oberpfaffenhofen classified with the WFA-MLP technique.

Urban Area [%]	Forest Area [%]	Agricultural Area [%]	Global [%]
80	80	90	84

Tab.3: The classification rates using the WFA-MLP technique.

4 Conclusions

This paper proposes a system of SAR image classification that combines a novel Wavelet Filtering Algorithm and an artificial neural network MLP within a unique signal processor. The filtering provided by the proposed wavelet-filtering scheme represented in Fig.5, has been used to reduce the speckle and to remove the negative effects of the illumination direction, without reducing the resolution as it happens in traditional approaches. In our method the wavelet filtering is used without sub-sampling the filtering outputs, in order to produce three output images with the same size of the input image. The proposed technique has been compared with two traditional techniques, neural network based: a real SAR image has been used in order to test the MLP method and the MLP method applied on the FT of the input image. The experimental results show that our method improves the classification performances the MLP standard application, and than the MLP application to high-frequency component of considered image, without resolution decrease with respect to the input image.

References

[1] A. Farina, F. Vinelli, "A Center of Excellence for SAR Signal Processing", *Alenia Technologies,* Vol.1, No.1, 1994;

[2] N. Hess Nielsen, M.V. Wickerhauser, "Wavelets and Time-Frequency Analysis", *Proceedings of the IEEE,* Vol.84, No. 4, pp.523-529;

[3] *Neural Networks*, Special Issue on Automatic Target Recognition, Pergamon, Vol.8, No.7/8, 1995;

[4] F.C. Morabito, G. Simone, "A Comparison of Feature Extraction Techniques for Remotely Sensed SAR Images Classification", *Proceedings of International Conference on Systems, Signals, Control, Computers (SSCC),* Vol.2, pp. 107-110, Durban (South Africa), Sept. 1998;

[5] A.Farina, F.C. Morabito, G.Simone, "Automated Lineament Detection in SAR Images based on the joint use of Wavelet and Hough Transforms", *accepted for publication Radar '99 International Conference*, Brest (France), May 1999;

[6] S. Barbarossa, L. Parodi, "SAR Image Classification by Wavelets", *Proc. of IEEE International Radar Conference*, 1995, pp. 462-467;

[7] Y.C.Tzeng, K.S.Chen, "A Fuzzy Neural Network to SAR Image Classification", *IEEE Transactions on Geoscience and Remote Sensing*, Vol.36, No.1, January 1998, pp.301-307;

[8] M.Ceccarelli, A.Farina, A.Petrosino, R.Vaccaro, F.Vinelli, "SAR Image Segmentation using Textural Information and Neural Classifiers", *L'Onde Electrique*, Vol.74, No.3, May 1994, pp.24-28.

Neural Networks for Spectral Analysis of Unevenly Sampled Data

Roberto Tagliaferri and Angelo Ciaramella
Dipartimento di Matematica ed Informatica,
Università di Salerno, and INFM unità di Salerno,
via S. Allende, 84081 Baronissi (SA) Italia
and IIASS "E. R. Caianiello", Vietri s/m Italia

Leopoldo Milano and Fabrizio Barone
Dipartimento di Scienze Fisiche, Università di Napoli "Federico II"
Istituto Nazionale di Fisica Nucleare, sez. Napoli,
Complesso Universitario di Monte Sant'Angelo,
via Cintia, I-80126 Napoli Italia

May 28, 1999

Abstract

In this paper we present a neural network based estimator system which performs well the frequency extraction from unevenly sampled signals. It uses an unsupervised Hebbian nonlinear neural algorithm to extract the principal components which, in turn, are used by the MUSIC frequency estimator algorithm to extract the frequencies.

We generalize this method to avoid an interpolation preprocessing step and to improve the performance by using a new stop criterion to avoid overfitting.

The experimental results are obtained comparing our methodology with the others known in literature.

1 Introduction

Periodicity analysis of unevenly collected data is a relevant issue in several scientific fields. Classical spectral analysis methods are unsatisfactory to solve the problem. In this paper we present a neural network based estimator system which performs well the frequency extraction from unevenly sampled signals. It uses an unsupervised Hebbian nonlinear neural algorithm to extract the principal components of the signal auto-correlation matrix, which, in turn, are used by the MUSIC frequency estimator algorithm to extract the frequencies [5], [11], [15]. We generalize this method to avoid an interpolation preprocessing step, which generally adds high noise to the signal, and improve the system performance by using a new stop criterion to avoid overfitting problems. The experimental results are obtained comparing our methodology with the others known in literature (see [6], [8], [10], [7], [14], [3]).

2 Evenly and unevenly sampled data

In what follows, we assume x to be a physical variable measured at discrete times t_i. $x(t_i)$ can be written as the sum of the signal x_s and random errors R: $x_i = x(t_i) = x_s(t_i) + R(t_i)$. The problem we are dealing with is how to estimate fundamental frequencies which may be present in the signal $x_s(t_i)$ [1], [6], [8].

If X is measured at uniform time steps (even sampling) there are a lot of tools to effectively solve the problem which are based on Fourier analysis [6], [8], [10]. These methods, however, are usually unreliable for unevenly sampled data [4]. For instance, the typical approach of resampling the data into an evenly sampled sequence, through interpolation, introduces a strong amplification of the noise which affects the effectiveness of all Fourier based techniques which are strongly dependent on the noise level.

To solve the problem of unevenly sampled data, we consider two classes of spectral estimators:

Spectral estimators based on Fourier Trasform (Least Squares methods);

Spectral estimators based on the eingevalues and eingevectors of the covariance matrix (Maximum Likelihood methods).

Classic Periodogram [6], [8], [10], Lomb's Periodogram [7], Scargle's Periodogram [14], DCDFT [3] are the methods of the first class that we use, while MUSIC [6], [8], and ESPRIT [12] belong to second class.

The methods based on the covariance matrix are more recent and have great potentiality. Starting by this consideration, we develop a method based on the MUSIC estimator. It is compared with classic methods to highlight the results.

3 The Neural Estimator

In the last years several papers dealed with learning in PCA neural nets [9], [13], [5], [11], [15] finding advantages, problems and difficulties of such neural networks. In what follows we shall use a robust hierarchical learning algorithm

$$\mathbf{w}_{k+1}(i) = \mathbf{w}_k(i) + \mu_k g(y_k(i))\mathbf{e}_k(i), \tag{1}$$

$$\mathbf{e}_k(i) = \mathbf{x}_k - \sum_{j=1}^{i} y_k(j)\mathbf{w}_k(j) \tag{2}$$

where $\mathbf{w}_k(i)$ is the weight vector of the $i-th$ output neuron at step k, $y_k(i)$ is the corresponding output, μ_k is the learning rate and $g(t) = \tanh(\alpha t)$ is learning function because it has been experimentally shown that it is the best performing one in our problem [11], [15].

Our neural estimator (ne) can be summarized as follows:

1 Preprocessing: calculate and subtract the average pattern to obtain zero mean process with unity variance.

Interpolate input data if it is the case.

2 Initialize the weight matrix and the other neural network parameters;

3 Input the $k-th$ pattern $\mathbf{x}_k = [x(k), \ldots, x(k+N+1)]$ where N is the number of input components.

4 Calculate the output for each neuron $y(j) = \mathbf{w}^T(j)\mathbf{x}_i \qquad \forall i = 1, \ldots, p.$

5 Modify the weights $\mathbf{w}_{k+1}(i) = \mathbf{w}_k(i) + \mu_k g(y_k(i))\mathbf{e}_k(i) \qquad \forall i = 1, \ldots, p.$

6 If convergence test is true then goto **STEP 8.**

7 $k = k+1$. Goto **STEP 3.**

8 End.

9 Frequency estimator: we use the frequency estimator MUSIC. It takes as input the weight matrix columns after the learning. The estimated signal frequencies are obtained as the peak locations of the function of following equation [6], [8], [15]:

$$P_{MUSIC} = \frac{1}{M - \sum_{i=1}^{M} |\mathbf{e}_f^H \mathbf{w}(i)|^2}$$

where $\mathbf{w}(i)$ is the $i-$th neural network weight vector after learning, and $\mathbf{e}_f^H$ is the pure sinusoidal vector. In the case of an interpolation preprocessing $\mathbf{e}_f^H = [1, e_f^{j2\pi f}, \ldots, e_f^{j2\pi f(L-1)}]^H$. In the generalization to non interpolated input data, $\mathbf{e}_f^H = [1, e_f^{j2\pi f t_0}, \ldots, e_f^{j2\pi f t_{(L-1)}}]^H$ where $\{t_0, t_1, ..., t_{(L-1)}\}$ are the first L components of the temporal coordinates of the uneven signal.

When f is the frequency of the $i-$th sinusoidal component, $f = f_i$, we have $\mathbf{e} = \mathbf{e}_i$ and $P_{MUSIC} \to \infty$. In practice we have a peak near and in corrispondence of the component frequency. Estimates are related to the highest peaks

Furthermore, to optimize the performance of the PCA neural networks, we stop the learning process when $\sum_{i=1}^{p} |\mathbf{e}_f^H \mathbf{w}(i)|^2 < M \; \forall f$, so avoiding overfitting problems. In fact leaving the stop condition used in the ne causes to the ne to find periodicities not present in the signal, while the new condition preserves it from this problem. A simple example is illustrated in figure 2 where we can see how the frequency identification varies depending on the stop condition (see next section for signal information). In fact, without the early stopping (see figure 2.b), Δf after 100 epochs remains at a value between 0.20 and 0.25 and we cannot know when the system reaches the best performance. The new stopping criterion, instead, permits to Δf to have a final value about 0.0 just after 50 epochs (see figure 2.a).

4 Experimental Results

Many experiments on synthetic and real signals were made, and in this paper we present the results obtained with one specific real signal, which highlights the main features of our problem.

The real signal is related to the Cepheid SU Cygni [2]. The sequence was obtained with the photometric technique UBVRI and the sampling made from June to December 1977. The light curve is composed by 21 samples, and has a period of 3.8^d, as shown in figure 1.

The first experiment is concerning the interpolation. In this case we apply three different methods by using the Signal Processing Mathlab ® Toolbox: linear, cubic and spline, because they are quite simple and the most used ones. In figure 3 there is a plotting of the interpolating functions and the frequency estimates obtained by the ne with the spline interpolated signal as input.

In this case, the parameters of the ne are: $N = 10$, $p = 2$, $\alpha = 20$, $\mu = 0.001$. The estimate frequency interval is $[0(1/JD), 0.5(1/JD)]$. The estimated frequency without interpolation is 0.260 (1/JD).

A comparison is made with the other methods cited in a previous section and the experimental results are shown in figure 4 and in table 1. Only the Lomb's Periodogram is in agreement with the right periodicity, but showing some spurious peaks. Furthermore, if we enlarge the frequency window for the two best performing methods, while the ne continues to work well, the Lomb's periodogram does not work at all as illustrated in figure 5.

5 Concluding Remarks

In this paper we have illustrated an improved technique based on PCA neural Networks and MUSIC to estimate the frequency of unevenly sampled data. It has been shown that it obtains good results on real data (here we used the SU Cygni light curve) compared with other well-known methods. In fact, it obtains a good estimate of the signal frequency also with few unevenly sampled inputs, it reduces the noise problems related to input data interpolation, it optimizes the convergence by introducing an early stopping criterion, and, finally, it is more resistant to the dimension of the frequency windows.

Future research lines regard the introduction of genetic algorithms to optimize the weight initialization of the PCA neural networks and to use filters to extract and identify one frequency at each time when dealing with multi-frequency signals.

References

[1] T. J. Deeming, Fourier analysis with unequally-spaced data, Astrophysics and Space Science, vol.36, pag.137-158, 1975.

[2] J. D. Fernie, Photometry of the Classical Cepheid SU Cygni. Astronomical Society of the Pacific, vol.91, pag.67-70, 1979.

[3] S. Ferraz-Mello, Estimation of periods from unequally spaced obsevations. The Astronomical Journal, vol.86, pag.619, 1981

[4] J. H. Horne, S. L. Balius, A prespriction for period analysis of unevenly sampled time series, The Astrophisical Journal, vol.302, pag.757-763, 1986.

[5] J. Kharhunen, J. Joutsensalo, Representation and separation of signals using nonlinear PCA type learning. Neural Networks, vol.7, pag.113-127, 1994.

[6] S. M. Kay, Modern spectral estimation: Theory and application. Englewood Cliffs: Prentice-Hall, 1988.

[7] N. R. Lomb, Least-squares frequency analysis of unequally spaced data. Astrophysics and Space Science, vol.39, pag.447-462, 1976.

[8] S. L. Marple, Digital spectral analysis with applications, Prentice-Hall, Englewood Cliffs, N.J., 1987.

[9] E. Oja, A simplified neuron model as principal component analyzer, Journal of Mathematical Biology, vol.15, pag.267-275,1982.

[10] A. V. Oppenhaim, R. W. Schafer, Digital Signal Processing, Prentince-Hall, 1965.

[11] M. Rasile, L. Milano, R. Tagliaferri, G. Longo, Periodicity Analysis of Unvenly Spaced Data by Means of Neural Networks, in Neural Nets WIRN Vietri 97, M. Marinaro and R. Tagliaferri Editors, Springer Verlag, pag.201-212, 1997.

[12] R. Roy, T. Kailath, ESPRIT- Estimation of Signal Parameters via Rotational Invariance Tecniques, in F.A. Grünbaum, J. W. Helton and P. Khargonear Editors, Signal Processing Part II: Control Theory and Applications, Springer Verlag, New York, pag.369-411, 1990.

[13] T. D. Sanger, Optimal unsupervised learning in a single-layer linear feedforward network, Neural Networks, vol.2, pag.459-473, 1989.

[14] J. D. Scargle, Studies in astronomical time series analysis; statistical aspects of spectral analysis of unevenly spaced data, The Astrophisical Journal vol.263, pag.835-853, 1982.

[15] R. Tagliaferri, A. Ciaramella, L. Milano, F. Barone, G. Longo, Spectral Analysis of Stellar Light Curves by Means of Neural Networks, accepted for publication, Astronomy & Astrophysics Supplement Series, vol.137, June, 1999.

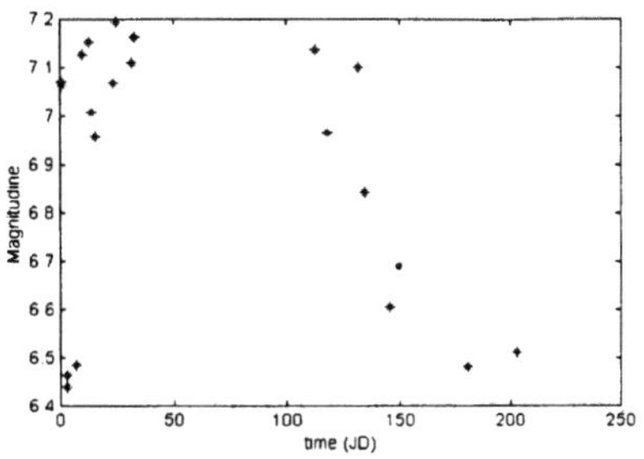

fig.(1) Light curve of SU Cygni.

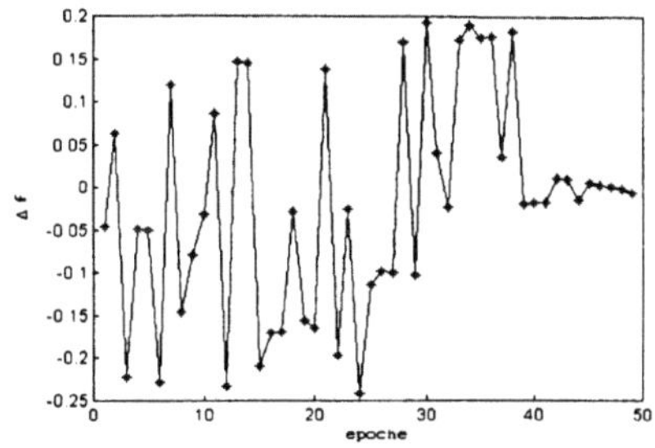

fig.(2.a) Learning with early stopping condition; Δf = (real frequency – estimate frequency).

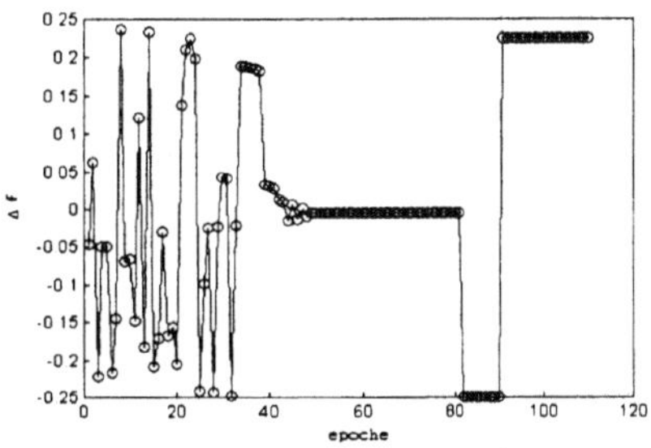

fig.(2.b) Learning without early stopping condition; Δf = (real frequency – estimate frequency).

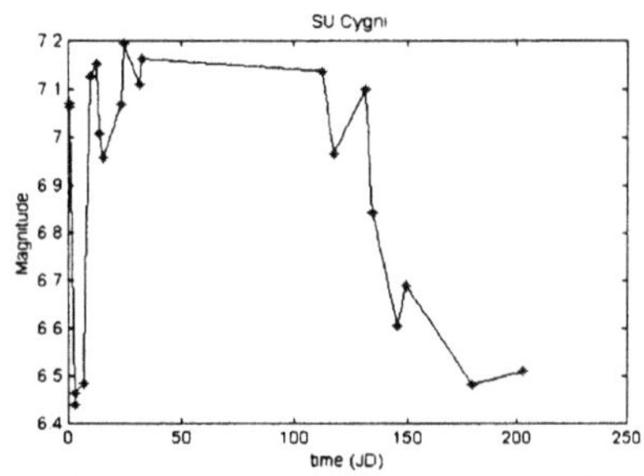

fig.(3.a) Linear interpolation of SU Cygni light curve.

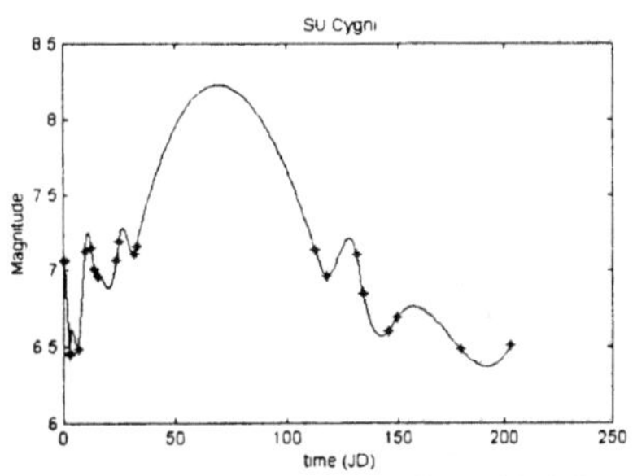

fig.(3.b) Spline interpolation of SU Cygni light curve.

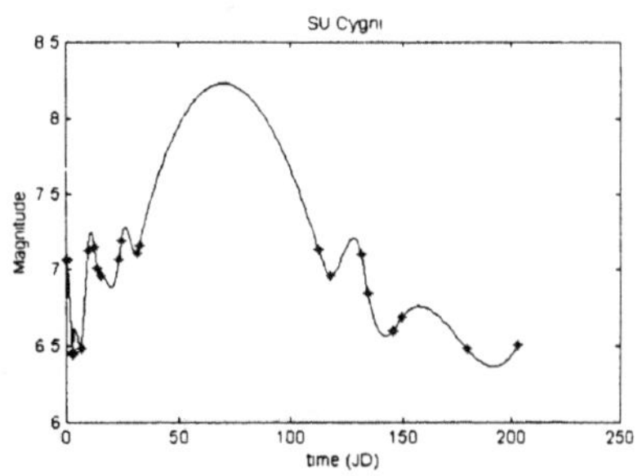

fig.(3.c) Cubic interpolation of SU Cygni light curve.

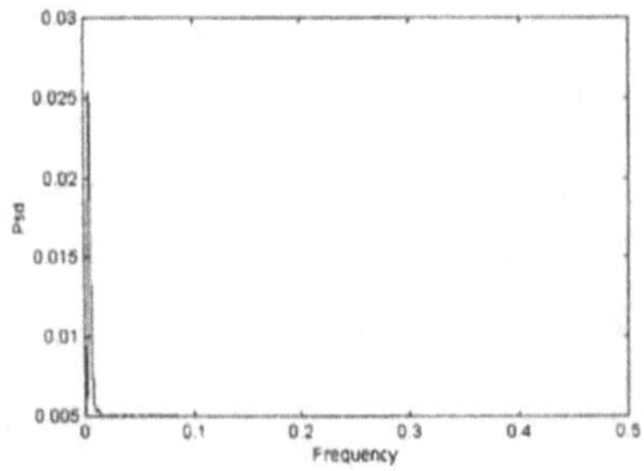

fig.(3.d) ne estimate of SU Cygni light curve with spline interpolation.

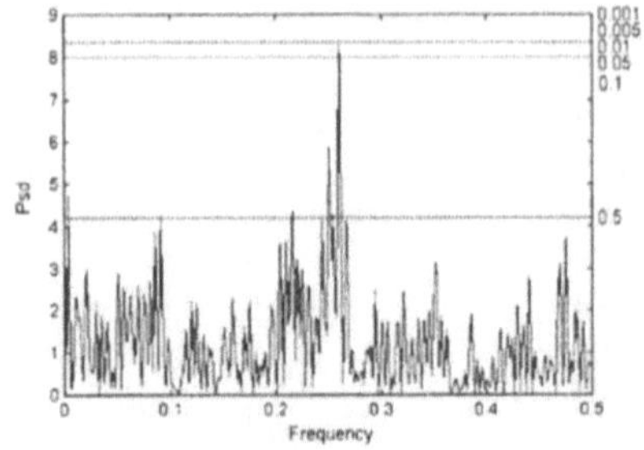

fig.(4.a) Lomb's Periodogram estimate of SU Cygni with confidence intervals.

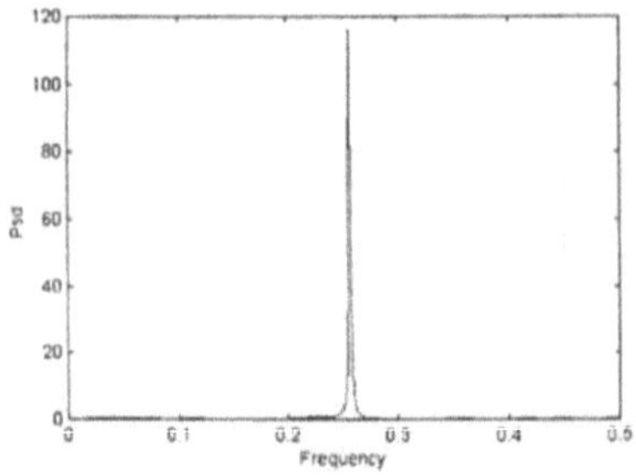

fig.(4.b) ne estimate of SU Cygni.

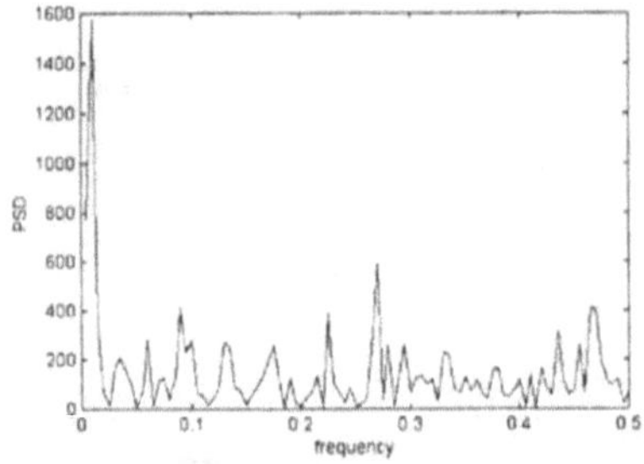

fig.(4.c) DCDFT estimate of SU Cygni.

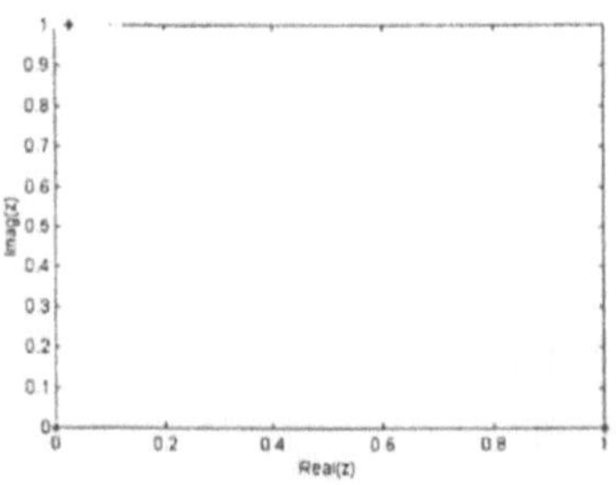

fig.(4.d) ESPRIT estimate of SU Cygni.

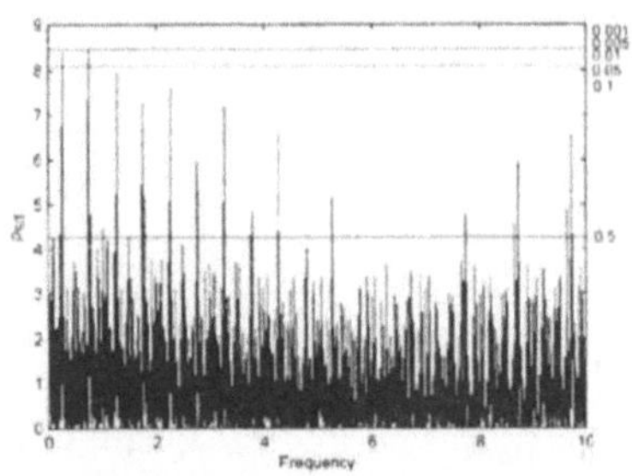

fig.(5.a) Lomb's Periodogram estimate of SU Cygni (with enlarged window) with confidence intervals.

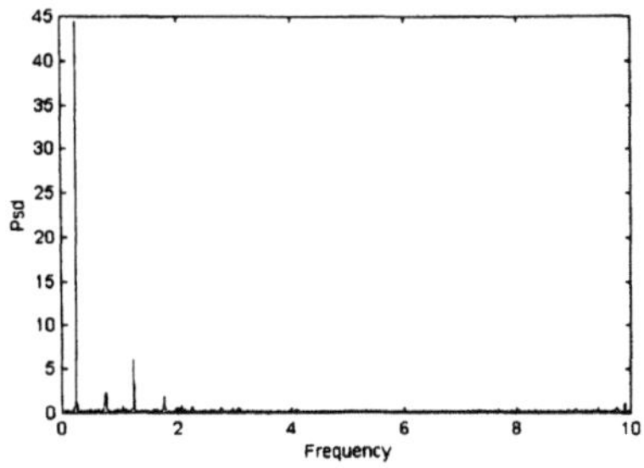

fig.(5.b) ne estimate of SU Cygni (with enlarged window).

Algorithm	Frequency Estimates	
Lomb's Periodogram	0.260	(1/JD)
DCDFT	0.015	(1/JD)
ESPRIT	0.25	(1/JD)
ne estimate (with interpolation)	0.015	(1/JD)
ne estimate (without interpolation)	0.259	(1/JD)

Table 1. Frequency estimates with the methods illustrated in the paper on the SU Cygni light curve.

Section 6
Architectures and Algorithms

Recursive Networks: An Overview of Theoretical Results

M. Bianchini, M. Gori, and F. Scarselli

Dipartimento di Ingegneria dell'Informazione
Università degli Studi di Siena
Via Roma, 56 — 53100 Siena, ITALY

Abstract

A naturally structured information is typical in symbolic processing. Nonetheless, learning in connectionism is usually related to poorly organized data, like arrays or sequences. For these types of data, classical neural networks are proven to be universal approximators.

Recently, *recursive networks* were introduced in order to deal with structured data. They also represent a universal tool to approximate mappings between graphs and real vector spaces. In this paper, an overview of the present state of the art on approximation in recursive networks is carried on. Finally, some results on generalization are reviewed, establishing the VC–dim for recursive architectures of fixed size.

1 Introduction

In several application domains, the information which is relevant for solving problems is encoded, often implicitly, into the relationships between the basic entities. Examples of structured domains have been found in medicine, chemistry, molecular biology, software engineering, and artificial intelligence [2]. The common feature shared by these tasks is that the computation should be based neither on simple arrays nor on sequences, but on dynamic data structures, incorporating also subsymbolic information. Of course, machine learning models conceived for dealing with sequences may be straightforwardly adapted for processing structured objects. For instance, a sequential representation of a binary tree, which may always be devised by a proper traversing of the structure, can be used to feed a recurrent network. However, such a preprocessing phase shows two major drawbacks: sequential representations of trees exponentially depend on their height, thus evidencing the long–term dependence problem of recurrent learning; moreover, fundamental information on the structure is definitely lost, thus making the generalization phase very difficult. Therefore, neural network models, capable of working directly on structured objects, are advisable, rather than classical models working on artificial information coding based on sequences.

In the last decade, some interesting connectionist models for the representation and processing of structured information have been proposed [1, 2, 6, 8, 9]. Like for classical neural models, many efforts have been mainly devoted to the problem of guaranteeing learning and approximation. In fact, while some theoretical results for recursive networks have been straightforwardly inherited from the previous work on static and recurrent models, there are important questions that are not meaningful and cannot been answered without considering the particular features of recursive processing. For example, the approximation of a function $f : T \to \mathbf{R}$, from the set of labelled trees T to $\mathbf{R}$, strongly depends on the norm which defines the distance between the elements of T. However, it is not obvious how a distance between trees with different labels, different structures, and different heights may be selected.

In this paper, we concentrate on the approach described in [2], proposing an overview of the most significant results on approximation in recursive networks. Moreover, we address the topic of generalization and show benefits and limits of the recursive approach, discussing existing results and open problems.

2 Approximation in recursive networks

2.1 Definitions and background topics

Let us suppose that the instances in the learning domain are labelled trees, with maximum *outdegree* k, that is each node has no more than k children. Let $\mathbf{U}_v$ be the label attached to the node v, where $\mathbf{U}_v$ is an element of the alphabet Σ. For each label $\mathbf{U}_v \in \Sigma$, we fix a representation $c(\mathbf{U}_v) \in \mathbf{R}^m$.

Definition 1 Let $\mathbf{Y}_F \in \mathbf{R}^l$ be an encoding for the empty tree. For a function $f : \mathbf{R}^{m+kl} \to \mathbf{R}^l$, the *induced function* $\tilde{f} : T \to \mathbf{R}^l$, being T the set of labelled trees, is recursively defined as

$$\tilde{f}(t) = \begin{cases} \mathbf{Y}_F & \text{if } t \text{ is empty,} \\ f(c(\mathbf{U}_v), \tilde{f}(t_1), \ldots, \tilde{f}(t_k)) & \text{otherwise,} \end{cases}$$

where v is the root of $t \in T$, and $t_1, \ldots, t_k$ are the k subtrees of v. A function $\mathcal{N}_r : T \to \mathbf{R}^n$ is calculated by a *recursive network* if there exist functions $f : \mathbf{R}^{m+kl} \to \mathbf{R}^l$ and $g : \mathbf{R}^l \to \mathbf{R}^n$, computed by an MLP, such that $\mathcal{N}_r \doteq g \circ \tilde{f}$.

A *recursive network* for approximation consists of two parts: a recursive component, induced by f, which is unfolded along the input structure and which encodes the input tree into a real l–dimensional vector, called *encoding network*, and an approximation feedforward component g, which realizes a mapping between the two real vector spaces $\mathbb{R}^l$ and $\mathbb{R}^n$ (see Fig. 1 for an example, where f and g are implemented by two–layered networks). If $k = 1$, the input is represented by a sequence and the recursive model reduces to a classical recurrent neural network.

Remark 1 The learning domain can be extended to include DOAGs, i.e. directed ordered acyclic graphs. A DAG D is ordered if, for each $v \in \text{vert}(D)$, a total order on the edges leaving from v is defined. Nevertheless, in the approximation setting, we shall require the DOAG either to be empty or to possess

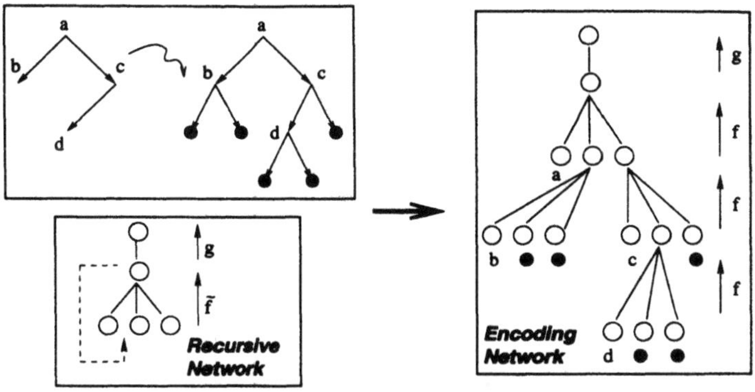

Figure 1: *Recursive* and *encoding* network associated with a given binary tree. The recursive network is unfolded through the structure of the tree.

a *supersource*, i.e. a vertex $s \in \text{vert}(D)$ such that every other vertex of D can be reached by a directed path starting from s. Note that if a DOAG does not have a supersource node, it is still possible to define a convention for adding an extra vertex s, with a minimal number of outgoing edges, such that s is a supersource for the expanded DOAG.

Recursive networks are an extension of common recurrent networks. In fact, the encoding function f plays the role of the transition function of recurrent networks, while the encoding of a tree $\tilde{f}(t)$ represents the internal state. However, in recurrent networks the current state is always fed to the network at the next time, whereas in recursive networks the encoding of the current tree will be fed only when the network will process the node which is the father of the current one. For this reason, the same kind of data can be treated by recurrent and recursive networks, but recurrent networks must process data in a sequential fashion, whereas recursive networks elaborate information following the underlinying structure.

Without restrictions on the dimension l of the encoding space (the internal state for recurrent networks) both models can approximate any continuous function $\overline{\mathcal{N}} : T_h \to \mathbf{R}^n$ defined on the set of trees T_h having height bounded by h and labels in a compact subset of $\mathbb{R}^m$. A simple intuitive explanation can be drawn in the case of recursive networks. In fact, notice that the encoding function $\tilde{f}$ could simply "*store*" the whole input tree into the real vector $\tilde{f}(t)$. For example, consider a binary tree that represents the expression $2+3\times(4+2)$ (see Fig. 2) and suppose to code any number a by the pair "$1, a$", and the symbols $+, \times, (,)$ by "2, 0", "3, 0", "4, 0", "5, 0", respectively. Then, the expression $2 + 3 \times (4 + 2)$ can be stored into $[1,2,2,0,1,3,3,0,4,0,1,4,2,0,1,1,2,5,0]$. In this way, the approximation function g reads the whole tree and computes the desired function $\overline{\mathcal{N}}$. Finally, f and g can be realized by two three–layered perceptrons.

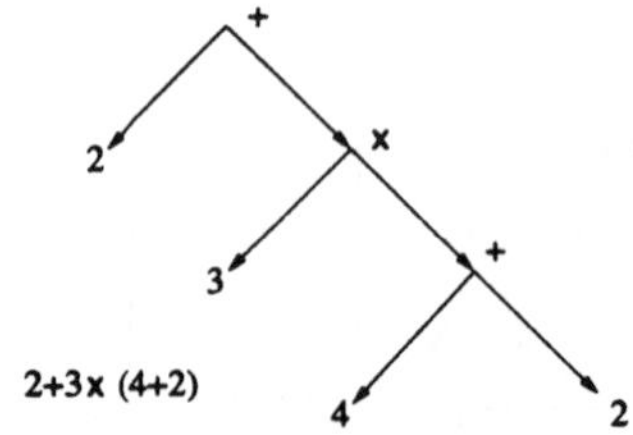

Figure 2: The binary tree that represents the expression $2 + 3 \times (4 + 2)$.

2.2 Recursive v.s. recurrent networks

When the dimension l of the encoding space has to be independent of the height of trees, the two models show meaningful differences. For example, consider the problem of realizing a *calculator*, i.e. a function $\overline{\mathcal{N}}_{Cal} : T \to \mathbb{R}^l$, for real expressions with operators $+, \times$ (see Fig. 2). It is easy to see that $\overline{\mathcal{N}}_{Cal} = I \circ \tilde{f}$, where $I : \mathbf{R} \to \mathbf{R}$ is the identity function,

$$f([a_1, a_2, a_3], t_1, t_2) = a_1 \times (t_1 + t_2) + a_2 \times t_1 \times t_2 + a_3 \quad (1)$$

and the vector $c(\mathbf{U}_v) = [a_1, a_2, a_3]$, which codifies the label $\mathbf{U}_v$, is $[0, 0, b]$ if $\mathbf{U}_v$ is a real b, $[0, 1, 0]$ if $\mathbf{U}_v = +$, and $[1, 0, 0]$ when $\mathbf{U}_v = \times$ holds.

Notice that the recursive network was constructed without any assumption on the length of the expression. According to [3], it can even be proved that f is approximable by an MLP with one hidden layer, where the number of hidden neurons does not depend on the desired degree of accuracy. Summing up, we get the following theorem.

Theorem 1 ***The function $\overline{\mathcal{N}}_{Cal}$ is approximable up to any degree of precision by a recursive network with the number of nodes that does not depend on the desired degree of accuracy.*** ■

Moreover, a similar result holds also when the set of operators include division and subtraction, provided that the network uses the inverse tangent activation function.

This independence from length is a nice property that could not be fulfilled without considering expressions as structured data. Processing an expression sequentially is a more complex task. The computation of the recursive networks can be based on the tree, at each time a product or a sum is carried out, whereas, a recurrent network must proceed in a sequential fashion and has to read a large part of the expression before being able to do any calculus.

2.3 Fixed encoding dimension

In [5], the case of recursive networks with fixed encoding dimension l is studied. It is proven that, even when $l = 1$, recursive networks can approximate, in probability, any function on trees T with labels in a finite set Σ.

Theorem 2 *Any function $\overline{\mathcal{N}} : T \to \mathbf{R}^n$ can be approximated in probability by a recursive network $\mathcal{N}_r : g \circ \tilde{f}$, where $f : \mathbf{R}^{m+k} \to \mathbf{R}$ and $g : \mathbf{R} \to \mathbf{R}^n$ are realized by two three layered perceptrons with a sufficient number of hiddens.* ■

Notice that the theorem works for approximation in probability, i.e. it is stated that, for any reals $\varepsilon > 0$ and $\delta > 0$ and any probability measure P on T, there is a recursive network $\mathcal{N}_r$ such that $P(\|\overline{\mathcal{N}}(\mathbf{x}) - \mathcal{N}_r(\mathbf{x})\| \geq \varepsilon) \leq \delta$ holds. Approximation in probability could be replaced by approximation w.r.t. the maximum norm, but the domain must be restricted to trees with bounded height T_h. Moreover, in [5] it is also proved that there are functions that cannot be approximated in maximum norm if trees are not restricted to have a bounded height. Finally, Theorem 2 can be extended also to trees with real labels, but it has to be assumed that the function to be approximated is measurable.

Unfortunately, Theorem 2 does not provide useful information on the relationship that links the function g with the encoding dimension l and the number of hidden neurons that must be employed in the networks. In fact, the encoding function f is realized independently from the function $\overline{\mathcal{N}}$. The example of Section 2.2 clearly showed that the recursive part of the network can become very simple, if f is adapted to the particular function we want to approximate. In this sense, a better analysis of this aspect could give also a deeper understanding of the differences between recurrent and recursive networks. Finally, another open question is the possibility to approximate, with respect to the maximum norm, trees having bounded lengths and real labels.

3 Recursive networks and generalization

In the classification as well as in the approximation setting, the VC–dimension constitutes a bound for the number of training patterns needed to make the training error representative of the error estimated on the whole input set [7].

Unfortunately, if no a priori information about the probability distribution of training examples is available, there are no guarantees for valid generalization. In fact, the VC–dimension [4] of recursive networks with sigmoidal neurons grows at least quadratically in the "*dimension*" of the patterns to be learnt (e.g. length of sequences, height of trees). Therefore, the VC–dimension of recursive networks becomes infinite if arbitrary inputs are allowed. In particular, the worst case analysis may be summarized in the following result [4].

Theorem 3 *Let T_h be the subset of trees with height h at most, $T = \bigcup_h T_h$. Let $\mathcal{F}_h = \{f|_{T_h}$ s.t. $f \in \mathcal{N}_r\}$ be, and define $\mathcal{N}_r$ as a recursive architecture with N neurons and W weights. Depending on the neuron activation function, Heavyside (H) or sigmoid (σ), then*

$$VC(\mathcal{F}_h^H) = \left\{ \begin{array}{l} \mathcal{O}(W \log(hN)) \\ \mathcal{O}(W(h + \log N)) \end{array} \right. , \quad VC(\mathcal{F}_h^\sigma) = \left\{ \begin{array}{l} \mathcal{O}(W^2N^2h^2) \\ \mathcal{O}(W^2N^2 2^h) \end{array} \right. , \quad \begin{array}{l} k=1 \\ k \geq 2 \end{array}$$

being k the arity of the tree. ■

In real–world learning tasks, a finite subset T' of the input space T may be chosen, being "***probabilistically representative***" of the whole space; on T' the "***risk***" can be controlled because of its finiteness. Therefore, an estimation on the difficulty of the whole learning problem can be also devised. Nevertheless, this does not give a theoretical bound for the training set dimension needed for accurate generalization in the worst case. For future research, it will be interesting to define a different measure to better capture the nature of generalization in recursive architectures.

4 Conclusions

Recursive networks are well suited for learning tasks involving structured data. The innovative aspects of recursive networks w.r.t. recurrent models were evidenced when dealing with trees and DOAGS, assessing their capacity of solving the long–term dependence problem due to sequentialization of structured objects. Their capability of approximating any function in probability was also reviewed [5], establishing the limitations when the infinite norm is employed. Finally, generalization issues have been discussed, showing that common VC–dimension measurements are inadequate to establish the generalization capability of recursive architectures.

References

[1] M. Bianchini, M. Gori, and F. Scarselli. Theoretical properties of recursive networks with linear neurons. Technical Report RT–DII–26/99, Dip. di Ingegneria dell'Informazione, Università di Siena, Siena, (Italy), 1999.

[2] P. Frasconi, M. Gori, and A. Sperduti. A general framework for adaptive processing of data structures. *IEEE Transactions on Neural Networks*, 9(5):768–786, September 1998.

[3] M. Gori, F. Scarselli, and A. C. Tsoi. On the clousure of set of functions that can be realized by a multilayer perceptron. *IEEE Transactions on Neural Networks*, 9(6):1086–1098, 1998.

[4] B. Hammer. Learning recursive data is intractable. Preprint 194, 7/1997, Osnabrücker Schriften zur Mathematik, Osnabrück, (Austria), 1997.

[5] B. Hammer. On the approximation capability of recurrent neural networks. In *NC'98, International Symposium on Neural Computation*, Vienna, (Austria), September 1998.

[6] A. Küchler and C. Goller. Inductive learning in symbolic domains using structure–driven recurrent neural networks. In G. Görz and S. Hölldobler, eds., *Advances in Artificial Intelligence*, pp. 183–197. Springer, Berlin, 1996.

[7] W. Maas. Vapnik–Chervonenkis dimension of neural nets. Preprint NC–TR–96–015, NeuroCOLT Technical Report Series, 1996.

[8] J. B. Pollack. Recursive distributed representations. *Artificial Intelligence*, 46(1–2):77–106, 1990.

[9] A. Sperduti. Labelling RAAM. *Connection Science*, 6(4):429–459, 1994.

A Persistent and Parallel Approach to Object Oriented ANNs Simulators

Antonio d'Acierno, Ivana Marra
IRSIP - CNR
via P. Castellino, 111 - 80131 Napoli - Italy
dacierno.a@irsip.na.cnr.it

Lucio Sansone
DIS - Universita' degli Studi di Napoli "Federico II"
via Claudio 21 - 80125 Napoli - Italy
sansone@unina.it

Abstract

This paper describes an *Object Oriented* software environment for ANN's simulation. A key feature of our software is that it uses *Persistence*, by means of an Object Oriented Data Base. Moreiver, the simulator is *Parallel*: it is being developed using multithreading, so that efficiency and portability have been obtained. Some preliminary results are shown.

1 Introduction

Do we need another neural networks simulator? Our idea is that, typically, researchers develop their own code; this is mainly due to the fact that, in the *Academia*, the main research is in the field of neural architectures and learning algorithms.

Quite different is the situation considering companies that *use* neural networks, or simply taking into account experiences of researchers that *apply* neural algorithms to solve challenging problems. Here, people typically uses freeware, shareware or commercial software; to have just an idea of the *pletora* of available products, see [6].

As far we know, no software is able to be (clearly at the same time):

- **General Purpose**: the system should provide as many as possible neural algorithms.
- **Extensible**: the software should be open, allowing users to modify algorithms.
- **Efficient in using the available hardware**: shared-memory multi-processor computers are today widely available (also at PC level); the available power is not always completely used.
- **Platform Independent**: Internet is the actual (global) platform to be considered.

- **Persistent**: data and structures have to be stored not simply using the file-system but with Data Base Management Systems (DBMS) able to provide data integrity and sharing.

This paper just describes preliminary steps towards these requirements.

2 Our Approach

To be extensible a software has to be designed and developed using Object Oriented (OO) technologies and languages; to be platform independent clearly we have to use Java while, to be fast, the software has to be parallel. In the following we will try to discuss such requirements, together with the interaction with a DMBS.

2.1 Object Oriented

In a neural environment, there are a lot of entities that have to be considered. First, clearly, we have the neural net; a net is a set of connected neurons and is characterized by a learning algorithm and by a retrieving algorithm. Two nets can differ both in topological sense (i.e. *how the neuron are connected*) as well as in mathematical terms (i.e. *how the neural structure learns and how the neurons fire*).

Another entity that has to be considered is the training set: we can have multi-dimensional training sets (used for classification in broad sense) as well as one-dimension training sets (mainly used for time series prediction and analysis). Training sets used for classification, moreover, can have an input part and a target part, or each input pattern can belong to a given class.

A net and a training set are connected in an experiment; an experiment has a lot of parameters such as initial conditions, number of patterns used for training and testing, and so on.

An application is the result of many experiments, whereas the same experiment can belong to different applications.

This scenario has been formalized using OMT (see figure 1). Once the system has been designed, its implementation using an OO language of course will provide, by means of inheritance, extensibility.

2.2 Persistent

To be used in a working environment, our software needs to be connected to a DBMS to store nets, weights, examples, experiments and so on; in other words it has to implement *persistence*.

Given the problem at hand, of course, standard relational DBMS cannot be used since complex objects cannot be easily described in terms of standard *tables*. At the same time an approach based on *BLOb's* (Binary Long Objects) is not really useful since, in most commercial systems, a BLOb cannot be queried.

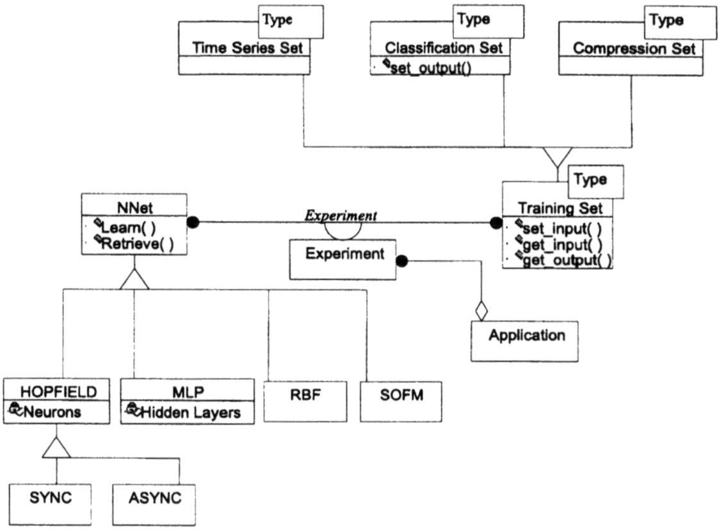

Figure 1: Classes and Relations in our software.

A possibility is to use an Extended Relational DBMS (ERDBMS); the problem here is that ERDBMSs are classical DBMSs and thus persistent data are external to and independent of applications [4]. Moreover, queries are performed joining tables and the more complex the relationships are, the more the application performance suffers. When no complex queries have to be performed more efficient (in terms both of execution time and developing time) approaches can be used.

From an OO perspective, we have that a program is simply a collection of objects that disappear when the program execution ends. Our approach is thus based on the use of an OODBMS that *simply* provides a way:

- to extend the lifetime of objects past the termination of the run that creates them;
- to efficiently retrieve persistent objects.

The product we are using in our project is ObjectStore that provides an unified programming interface to both persistently and transiently allocated data, through a tight integration with C++ and Java; in other words, the OODBMS and the programming language share the same data model and the resulting unified interface yields access speed for persistent objects. In fact, object access in ObjectStore means dereferencing what looks like normal C or C++ pointers. Performance for ObjectStore thus boils down to ensuring that the amount of time required to dereference pointers to persistent objects is as close as possible to that of transient objects. ObjectStore does this through the use of a unique Virtual Memory Mapping Architecture (VMMA) that involves sophisticated memory mapping, caching, and clustering techniques. Moreover,

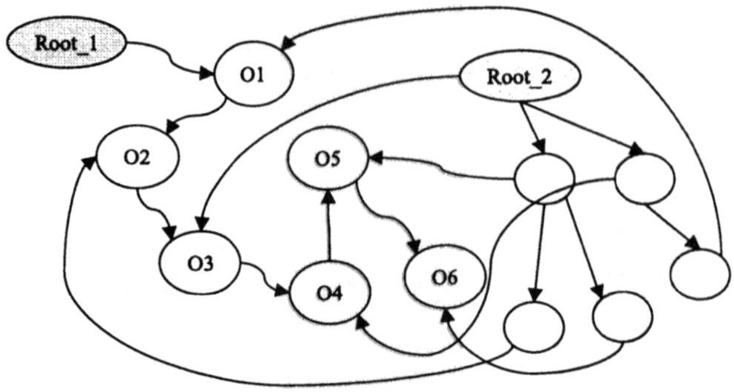

Figure 2: Navigation in *Objectstore*

ObjectStore provides the integration, through the Object Adapter module, into the distributed CORBA[1][7] environment so making client-server applications and heterogeneous computing easy to be implemented.

In the used system the key concept is *navigation* (figure 2); this does not seem a problem, since we do not need complex queries. Navigation simply means that a Data Base has some (two in figure 2) entry points (*roots*) and objects being retrieved by crossing pointers.

2.3 Parallel

As it is well known, the behavior of an ANN is determined by parameters denoted weights and a learning procedure is used to compute these parameters. Since such a learning procedure tends to be very time-consuming it is quite obvious to try to capitalize on the intrinsic parallelism of these systems in order to speed the computations up.

We are implementing a parallel code for the on-line Back-Propagation learning algorithm; since we are using, to obtain efficiency and portability, multi-threading, we have adopted, to obtain the parallel algorithm, the CLEPSYDRA mapping scheme [2]. That schema allows to describe the parallel code as a collection of processes, each simulating a Feed-Forward network, the only difference in the parallel algorithm being in the output layer (assuming that just an hidden layer exists). Here, in fact, a reduction operation has to be performed to evaluate the input value of output neurons.

Figure 3 shows very preliminary results. We tested an SGI ONYX NUMA[2] parallel computer equipped with P10000 processors and we used, as preliminary test-case, the *Nettalk* benchmark (203 input neurons and 23 output neurons). Given the mapping we are using, the efficiency just depends on the number of processors and on the number H of hidden neurons. For the case under study, we are able to use 4 processors efficiently (say with an efficiency ≥ 0.7) if H

[1]CORBA is the acronym for *Common Object Request Broker Architecture*.
[2]NUMA is the acronym for *Not Uniform Memory Access*.

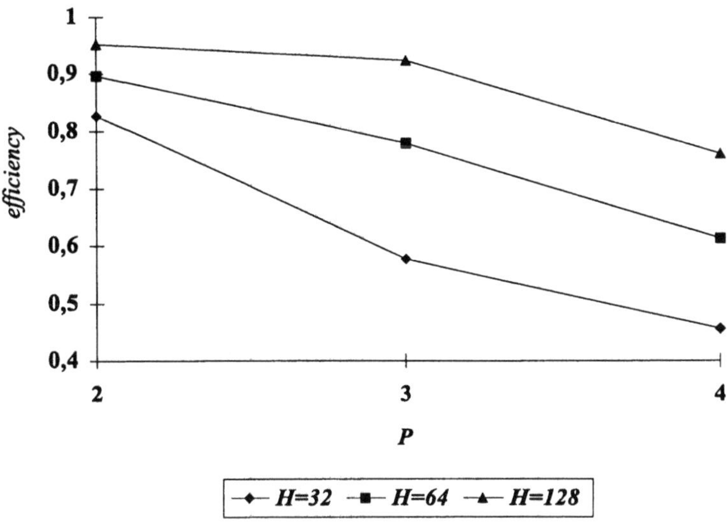

Figure 3: Preliminary results on the *Nettalk* benchmark. The efficiency as H vary.

is $\geq$ 128. Such a result is quite interesting, given the optimaztions not yet implemented.

The CLEPSYDRA schema is also being used for implementing the on-line training algorithm for RBF networks [3].

We are implementing a parallel code for Hopfield nets; as it is well known here the problem is the retrieving phase and we are using a parallel formulation [1] that allows to simulate both synchronous and asynchronous nets.

An open problem we are dealing with is the parallel implementation of learning algorithms for SOFMs. In this case, in fact, the trivial net partitioning cannot be used *as it is* since, while it improves the step of finding the neuron closest to the input pattern, it typically does not speed up the updating phase (near neurons being assigned to the same processors). A possible solution we are investigating consists in a scheme that, while still based on *global* communcation, allows to parallelize the updating phase by means of a circular mapping of weights.

3 Conclusions

In this paper we described the design of an Object Oriented software environment for Neural Networks Simulation we are developing. Our software is extensible; this is obtained using, in the design phase as well as in the implementation phase, object oriented techniques. Our software is parallel and, to ensure both efficiency and portability, we assume, as target machine, an SMP (Symmetric Multi processor) computer; this implies that multi-threading has to be used. Mapping schemes proposed elsewhere can easily meet the requirements of such constraints.

A key concept to be addressed is that our software will be *Open Source*; in fact, giving users of a product access to its source code, encourages natural product evolution and, even most importantly, reflects the trend toward networked collaboration [5].

Moreover by exploiting the OO architecture of the system, it seems a natural evolution to import it into the CORBA environment. Specifically, the migration towards a *state of art* client-server architecture will require the implementation, on the client side, of multi-threaded Java applications/applets able to query the C++ server. At the same time, the server has to be able to provide (besides persistence) efficient parallel methods and computing power for light clients.

References

[1] A. d'Acierno, R. Vaccaro, On Parallelizing Recursive Neural Networks on Coarse-Grained Parallel Computers: A General Algorithm, Parallel Computing, 20, 1994, pp. 245-256.

[2] A. d'Acierno, Back-Propagation Learning Algorithm and Parallel Computers: The CLEPSYDRA Mapping Scheme, Neurocomputing, To Appear.

[3] A. d'Acierno,Training On-Line Radial Basis Function Networks on Parallel Computers, Neurocomputing, Submitted.

[4] M. E. S. Loomis, Object DataBases: The Essentials, Addison-Wesley, Reading, Massachussetts, 1995.

[5] T. O'Reilly, Lessons from Open-Source Software development, Communications of the ACM, April 1999, pp. 33-37.

[6] PNNL, http://www.emsl.pnl.gov:2080/proj/neuron/neural/systems/.

[7] A. Pope, The CORBA Reference Guide, Addison Wesley, 1997.

An Experimental Comparison of Three PCA Neural Techniques

Simone Fiori *

Dipartimento di Ingegneria Industriale, Univesità di Perugia - Perugia, Italy
E-MAIL: SFR@UNIPG.IT

Abstract. We present a numerical and structural comparison of three neural PCA techniques: The GHA by Sanger, the APEX by Kung and Diamataras, and the ψ–APEX first proposed by the present author in [3].

1 Introduction

Performing the Principal Component Analysis (PCA) of stationary multivariate random data means computing the eigenvectors of the covariance matrix corresponding to the largest eigenvalues, and the projection of the data on the eigenvectors to obtain a number of principal components. Since the pioneering work of Prof. Oja and his research group, on-line estimation of principal components by linear neural networks has become an important research field (see for instance [1,9–11,15] and references therein) both for the interesting implications on unsupervised learning theory and applications to neural signal processing [4–7,2].

Over the recent years, several neural structures and learning rules for performing PCA have been proposed in the scientific literature. In this paper, we consider two of them: the well-known Generalized Hebbian Algorithm (GHA) by Sanger [14] and the Adaptive Principal-component Extractor (APEX) by Kung and Diamantaras [2], with the aim to compare their performances and structure complexity to those exhibited by an algorithm proposed by the present author in [3] as an improvement of APEX.

2 The GHA, APEX and ψ–APEX PCA Neural Techniques

2.1 The feed-forward network and GHA rule

The Generalized Hebbian Algorithm by Sanger [14] is one among the best known learning algorithms that allows a linear neural network to extract a selected number of principal components from a multivariate random process. It applies to a single-layered feedforward neural network that may be described with the

* This research was financially supported by the Italian MURST.

usual notation as $\mathbf{z}(t) = \mathbf{Q}^T(t)\mathbf{x}(t)$, where $\mathbf{x}(t) \in \mathcal{R}^p$, $\mathbf{z}(t) \in \mathcal{R}^m$, thus $\mathbf{Q}(t) \in \mathcal{R}^{p \times m}$. The GHA rule writes:

$$\mathbf{Q}(t+1) = \mu\mathbf{x}(t)\mathbf{z}^T(t) + \mathbf{Q}(t)(\mathbf{I}_m - \mu\mathtt{LT}[\mathbf{z}(t)\mathbf{z}^T(t)]) \ , \tag{1}$$

where μ is a positive learning stepsize, the operator $\mathtt{LT}[\cdot]$ returns the lower-triangular part of the matrix contained within, and $\mathbf{I}_m$ denotes the identity matrix of size m.

2.2 The laterally-connected network and APEX rule

Kung and Diamantaras developed a learning rule for Rubner-Tavan's Principal Component neural network [13] described by the following input-output relationships:

$$\mathbf{z}(t) = \mathbf{Q}^T(t)\mathbf{x}(t) \text{ and } \mathbf{y}(t) = \mathbf{z}(t) + \mathbf{H}^T(t)\mathbf{y}(t) \ . \tag{2}$$

The input vector $\mathbf{x}(t) \in \mathcal{R}^p$, the output vector $\mathbf{y}(t) \in \mathcal{R}^m$ (with $m \leq p$, arbitrarily fixed), the direct-connection $p \times m$ weight-matrix $\mathbf{Q}(t)$ and the lateral-connection $m \times m$ strictly upper-triangular weight-matrix $\mathbf{H}(t)$ are intended to be evaluated at the same discrete temporal instant t. The Kung-Diamantaras' APEX learning rule for the weight-matrix $\mathbf{Q}$ and the lateral inhibitory weight-matrix $\mathbf{H}$ may be recast to:

$$\mathbf{Q}(t+1) = \mu\mathbf{X}(t)\tilde{\mathbf{Y}}(t) + \mathbf{Q}(t)[\mathbf{I}_m - \mu\tilde{\mathbf{Y}}^2(t)] \ , \tag{3}$$

$$\mathbf{H}(t+1) = -\mu\mathtt{SUT}[\mathbf{Y}(t)\tilde{\mathbf{Y}}(t)] + \mathbf{H}(t)[\mathbf{I}_m - \mu\tilde{\mathbf{Y}}^2(t)] \ , \tag{4}$$

where μ is a positive learning rate, $\mathbf{X}$ is a $p \times m$ matrix, $\mathbf{Y}$ and $\tilde{\mathbf{Y}}$ are $m \times m$ matrices defined by:

$$\mathbf{X} \triangleq [\underbrace{\mathbf{x}\ \mathbf{x}\ \cdots\ \mathbf{x}}_{m}] \ , \ \mathbf{Y} \triangleq [\underbrace{\mathbf{y}\ \mathbf{y}\ \cdots\ \mathbf{y}}_{m}] \ , \ \tilde{\mathbf{Y}} \triangleq \mathrm{diag}(y_1, y_2, \ldots, y_m) \ ,$$

and operator $\mathtt{SUT}[\cdot]$ returns the strictly upper triangular part of the matrix contained within. Kung and Diamantaras proved the convergence of the above network, under some conditions and in the mean sense [2, Theorem 4.3], to a Principal Component analyzer.

2.3 The new ψ–APEX class

A PCA transformation is such that the transformed signals in $\mathbf{z}(t) = \mathbf{Q}^T(t)\mathbf{x}(t)$ are characterized by maximum variance; furthermore, any PCA vector $\mathbf{q}_k$ must be orthogonal with respect to each other and must possess unitary norm. These targets can be thought of as separate objectives to be attained by means of same Rubner-Tavan's laterally-connected neural network. In [3] we proposed to adapt the *direct-connection* weight-matrix $\mathbf{Q}$ in order to maximize the powers of the transformed signal by means of the following learning rule:

$$\mathbf{Q}(t+1) = \mu\mathbf{X}(t)\tilde{\mathbf{Y}}(t) + \mathbf{Q}(t)[\mathbf{I}_m - \mu\tilde{\mathbf{Y}}(t)\tilde{\mathbf{Z}}(t)] \ , \tilde{\mathbf{Z}} \triangleq \mathrm{diag}(z_1, z_2, \ldots, z_m) \ , \tag{5}$$

and to adapt the *lateral-connection* weight-matrix $\mathbf{H}$ only, in order to make them uncorrelated the components of vector $\mathbf{y}$, according to the following rule:

$$\mathbf{H}(t+1) = -\mu \mathtt{SUT}[\mathbf{Y}(t)\tilde{\mathbf{Y}}(t)] + \mathbf{H}(t)[\mathbf{I}_m - \mu\tilde{\mathbf{\Psi}}(t)] \; , \; \tilde{\mathbf{\Psi}} \triangleq \mathrm{diag}(\psi_1, \psi_2, \ldots, \psi_m), \quad (6)$$

where the ψ_k are arbitrary functions of the y_k, that at least guarantee the stability of the network [3].

The above learning equations derive from the gradient optimization of the criterion:

$$2N(\mathbf{Q},\mathbf{H}) \triangleq \mathrm{tr}[\mathbf{y}\mathbf{y}^T] + \mathrm{tr}[(\mathbf{Q}^T\mathbf{Q} - \mathbf{I}_m)\tilde{\mathbf{\Lambda}}] + \mathrm{tr}[(\mathbf{H}^T\mathbf{H})\tilde{\mathbf{\Psi}}] \; , \quad (7)$$

where $\tilde{\mathbf{\Lambda}}$ is a diagonal matrix of Lagrange multipliers whose optimal values can by found by solving $\mathbf{Q}^T \frac{\partial N}{\partial \mathbf{Q}} = \mathbf{0}$, and operator tr[$\cdot$] returns the trace of the matrix contained within. Then equations (5)-(6) arise by assuming $\Delta\mathbf{Q} = +\mu\frac{\partial N}{\partial \mathbf{Q}}$ and $\Delta\mathbf{H} = -\mu\frac{\partial N}{\partial \mathbf{H}}$ [3].

The choice of the free functions ψ_k has been discussed in [3] under the assumption that for each neuron the identity $\psi_k(y_k) = \psi(y_k)$ holds true, and in this paper some examples are given in the Section devoted to algorithms comparison. A learning rule with the structure (5)-(6) with a generic ψ is termed ψ–APEX. It is worth noticing that the y^2–APEX is very similar to, but is not the same algorithm as, the original APEX. However, as $\mathbf{H} \to \mathbf{0}$ also $\tilde{\mathbf{Z}} \to \tilde{\mathbf{Y}}$, thus these algorithms asymptotically behave in the same way, therefore we call *APEX-like* the former since it closely resembles the pair (3)-(4).

3 Numerical and Structural Comparison

3.1 Computer simulations

In this Subsection simulation results obtained by using Sanger's GHA [14], standard APEX [2] and new algorithms belonging to the ψ–APEX class, are shown and discussed. Such PCA algorithms have been run with a network input signal $\mathbf{x} = \mathbf{K}\mathbf{s}$, where $\mathbf{K}$ is a $p \times p$ orthonormal matrix ($\mathbf{K}^T\mathbf{K} = \mathbf{I}$) randomly generated, and $\mathbf{s}$ is an array of p mutually uncorrelated zero-mean random signals s_k endowed with different powers $\sigma_k^2 = E[s_k^2]$. Signals s_k are placed in $\mathbf{s}$ so that their powers appear in descending order, i.e. $\sigma_k^2 > \sigma_\ell^2$ if $k < \ell$. This implies that the first m Principal Components of $\mathbf{x}$ (with $m < p$) are the first m column-vectors of $\mathbf{K}$. Each algorithm starts from the same initial conditions: random subject to a normal distribution for $\mathbf{Q}$ and null values for $\mathbf{H}$. In order to compare the algorithms in term of convergence speed, we use as a measure of convergence an estimate of the averaged reconstruction error (r.e.) $||\mathbf{e}||^2 = ||\mathbf{x} - \mathbf{Q}\mathbf{Q}^T\mathbf{x}||^2$ as in [2]. As an estimate, at any time-step t we use the expression $\frac{1}{t}\sum_{\tau=0}^{t} ||\mathbf{x}(\tau) - \mathbf{Q}(\tau)\mathbf{Q}^T(\tau)\mathbf{x}(\tau)||^2$ that after convergence approximates $E[||\mathbf{e}||^2]$ if t is large enough.

The first two simulation results are shown for a neural network with $p = 16$ and $m = 4$. Twenty data sets containing 5000 samples each have been generated

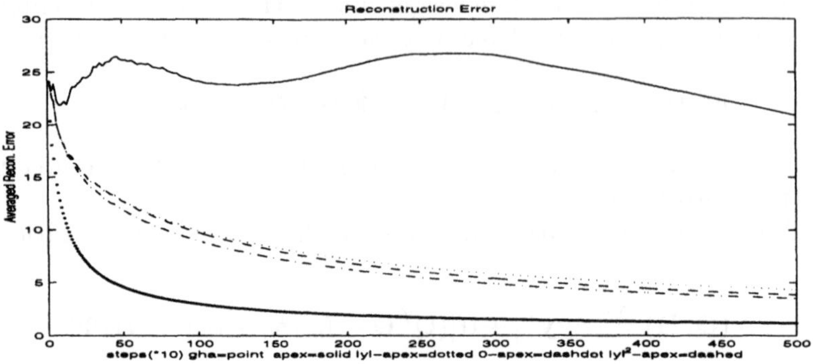

Fig. 1. Averaged r.e. in the good eigenvalue spread case.

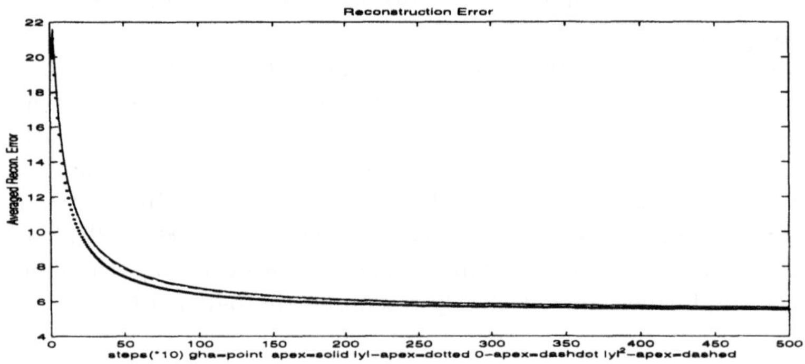

Fig. 2. Averaged r.e. in the poor eigenvalue spread case.

and ten values of learning stepsize μ have been used so that the network has been trained in 200 different situations. The first experiment covers the good eigenvalue spread case, where eigenvalues σ_k^2 are drawn from the law $\sigma_k^2 = 2^{2-k}$. Figure 1 shows the averaged r.e. for the GHA, APEX, $|y|$–APEX, 0–APEX and y^2–APEX algorithms. Figure 2 shows instead the result obtained in the poor eigenvalue spread case where the data set defined as in [2] has been used. Simulations show that in the good eigenvalue spread case the new algorithms behave better than the APEX even if they are slower than the GHA. In the poor eigenvalue spread case, that is a more difficult problem, the five algorithms exhibit almost the same performances. We noted that as the learning rate μ increases, the behaviors of GHA, $|y|$–APEX, y^2–APEX and 0–APEX tend to be more and more similar, while original APEX remains considerably slower.

In order to improve the performances of the APEX and APEX–like algorithms, it is possible to embed into the learning equations (3)-(4) and (5)-(6) a mechanism that varies the stepsize μ during the learning phase. Here we used the variable learning rate $\mu_k(t+1) = \mu_k(t)/(\gamma + \mu_k(t)y_k^2(t))$ as suggested by Kung and Diamantaras in [2]. Figure 3 shows the reconstruction error, in the good

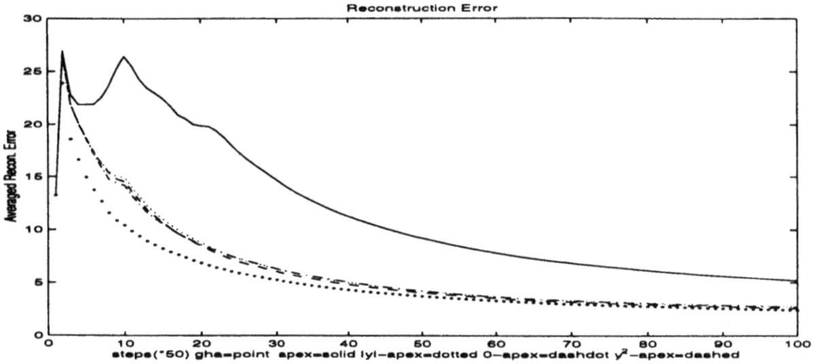

Fig. 3. Averaged r.e. in the good eigenvalue spread case, self-controlled learning step-sizes.

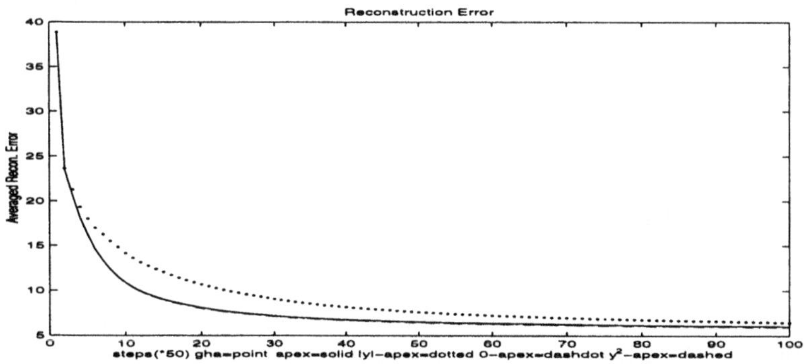

Fig. 4. Averaged r.e. in the poor eigenvalue spread case, self-controlled learning step-sizes.

eigenvalue spread case, averaged over 20 data sets. This result confirms that the use of a self-controlled learning rate dramatically enhances the performance of the algorithms. Figure 4 refers instead to the poor eigenvalue spread case. In this situation the use of a non-constant learning rate makes the speed of the ψ-algorithms comparable to that of the GHA.

3.2 Structure comparison

As a further element of this comparison, Table 1 provides an estimate of the architectural complexity of the neural networks in terms of the number of elementary operations required by the corresponding learning rules and networks' input-output relationships with respect to the input (p) and output (m) sizes. Here an "operation" is defined as a multiplication eventually followed by a sum. These three algorithms are listed in order of decreasing complexity. The 0–APEX one exhibits the lowest complexity degree in this comparison.

Algorithm	Operations
GHA	$2pm + 0.5(m^2 + m)(p + 1)$
APEX	$3pm + 2m^2 + m$
y^2–APEX	(Like APEX)
0–APEX	$3pm + 1.5m^2 + 0.5m$

Tab. 1. Complexity comparison.

4 Conclusion

The aim of this paper was to present a fair comparison among three neural PCA techniques: the GHA and APEX found in the literature and the ψ–APEX first proposed by the present author. Numerical simulations and structure complexity evaluations show the members of the ψ–APEX class exhibit interesting features. Prospective applications of the new algorithms are on image compression [14] and adaptive filtering [2]. Extensions of the proposed methods to non-linear PCA [10] for Blind Source Separation are also under investigation: Preliminary results are to appear [8,12].

5 Acknowledgments

The author wishes to thank Professor R.W. Newcomb (who is with the University of Maryland) that in a personal communication suggested a way to recast the basis of the ψ–APEX learning theory presented in [3] in a more clever and elegant form, leading to the formulation of the criterion (7).

References

1. P.F. Baldi and K. Hornik, *Learning in neural networks: A survey*, IEEE Trans. on Neural Networks, Vol. 6, No. 4, pp. 837 – 858, July 1995
2. K.I. Diamantaras and S.Y. Kung, *Principal Component Neural Networks: Theory and Applications*, J. Wiley & Sons, 1996
3. S. Fiori, A. Uncini, and F. Piazza, *A new class of APEX–like PCA algorithms*, Proc. of Int. Symposium on Circuits and Systems (IEEE-ISCAS), Vol. III, pp. 66 - 69, 1998
4. S. Fiori and A. Uncini, *A unified approach to laterally-connected neural nets*, Signal Processing IX - Theories and Applications (EUSIPCO), Vol. I, pp. 379 – 382, 1998
5. S. Fiori, A. Uncini, and F. Piazza, *Application of the MEC network to Principal Component Analysis and Source Separation*, Artificial Neural Networks, pp. 571 – 576, 1997, Springer-Verlag
6. S. Fiori and F. Piazza, *Orthonormal strongly-constrained neural learning*, Proc. of International Joint Conference on Neural Networks (IEEE-IJCNN), Vol. 2, pp. 1332 – 1337, 1998
7. S. Fiori, A. Uncini, and F. Piazza, *neural learning and weight flow on Stiefel manifold*, X Italian Workshop on Neural Nets (SIREN-WIRN), pp. 325 – 333, 1998

8. S. Fiori, A. Uncini, and F. Piazza *Neural blind separation of complex sources by extended APEX algorithm*, To appear on the Proc. of Int. Symposium on Circuits and Systems (IEEE-ISCAS), May 30 - June 2, 1999, Orlando (Florida - USA)
9. K. Hornik and C.-M. Kuan, *Convergence analysis of local feature extraction algorithms*, Neural Networks, Vol. 5, pp. 229 – 240, 1992
10. J. Karhunen, *Optimization criteria and nonlinear PCA neural networks*, Proc. of International Conference on Neural Networks (ICNN), pp. 1241 – 1246, 1994
11. F. Palmieri, J. Zhu, and C. Chang, *Anti-Hebbian learning in topologically constrained linear networks: A tutorial*, IEEE Trans. on Neural Networks, Vol. 4, No. 5, pp. 748 – 761, Sept. 1993
12. E. Pomponi, S. Fiori, and F. Piazza, *Complex independent component analysis by nonlinear generalized Hebbian learning with Rayleigh non-linearity*, Proc. of International Conference on Acoustics, Speech and Signal Processing (IEEE-ICASSP), Vol. 5, pp. 2507 – 2510, 1999
13. J. Rubner and P. Tavan, *A self-organizing network for Principal-Component Analysis*, Europhysics Letters, Vol. 10, No. 7, 1989, pp. 693 – 698
14. T.D. Sanger, *Optimal Unsupervised Learning in a Single-Layer Neural Network*, Neural Networks, Vol. 2, pp. 459 – 473, 1989
15. A. Weingessel and K. Hornik, *SVD algorithms: APEX-like versus Subspace Methods*, Neural Processing Letters, Vol. 5, pp. 177 – 184, 1997

Scale-Based Clustering Optimization via Gravitational Law Imitation

F.M. Frattale Mascioli, A. Rizzi, G. Scrocca, and G. Martinelli
INFO-COM Dpt., University of Rome "La Sapienza",
Via Eudossiana 18, 00184 Rome, Italy
mascioli@infocom.ing.uniroma1.it

Abstract

Clustering optimization is pursued in the present work by relying on a simple imitation of a physical law: the gravitational attraction. The resulting algorithm belongs to the category of hierarchical clustering. It is based on a merging approach in a scale-space, where the given data are represented. Its performance depends on a suitable quality index that indirectly measures the compactness and isolation of the clusters. This index suggests that the "natural" partition of clusters should be stable over a considerable scale parameter interval (lifetime). The effect of this stability index on the proposed algorithm is illustrated by some clustering examples.

1 Introduction

Cluster analysis arises in several fields, as vector quantization, classification, function approximation, outlier elimination, fuzzy rule determination. The efficient implementation of clustering is always a basic prerequisite to the success of algorithms devoted to the solution of the above problems. For this reason, an abundant literature can be found at this regard and several approaches have been pursued. The preliminary aspects to be fully considered in developing a clustering algorithm are:

a) definition of what we mean for data clustering;
b) strategy to be followed for deriving an algorithm which complies with the given definition;
c) introduction of an appropriate performance index for evaluating how much the resulting partition meets our expectation.

Let us examine these points in detail and in relation to the solutions proposed in the present paper.

a) The traditional definition of cluster, with reference to the space where the data are represented as patterns, is [1]: "A cluster is a set of patterns close among them and separated from the patterns of other clusters". This definition is not precise and is strictly dependent on the scale used for representing the data. At a very fine scale each data pattern can be viewed as a cluster, while at a very coarse scale the entire data set can be seen as a single cluster. The proper use of the scale notion can be borrowed from the computer vision community. The structures of an image are significant only if they survive to a large variation of its scale. Consequently, we complete the previous definition by adding the requirement that clusters must be *stable* over a considerable range of the scale of the space where the data are represented.

b) The usual optimization strategy underlying a clustering algorithm consists in minimizing a relative index measuring simultaneously the *compactness* and *isolation* of the clusters, obtained by partitioning the data set. The main problem regarding the use of a relative index is the frequent presence of spurious results, such as solutions in which each pattern is a cluster or the entire data set is contained in a single cluster. This problem can be overcome by applying algorithms based on a scale parameter [2, 3], together with a proper validity index. This index should select the partition that remains unchanged over the widest scale parameter interval (in the following: *stability* index). However, algorithms described in [2, 3] do not follow strategies strictly tailored to such type of index. From this point of view, what we need is the introduction of a fictitious mutual action among the patterns of the data set. This is useful in order to implement a fusion (or splitting) criterion that generates a set of partitions of the original data set, tailored to the application of a stability index. A simple way for obtaining this result is to imitate the physical laws that originate the natural clusters. In particular, we consider the gravitational attraction. We suggest an intuitive strategy to mimic the above physical force with intensity proportional to a scale parameter. Namely, the clustering process is modulated by the scale parameter: starting from a low value of this parameter and varying it iteratively up to a large value, we obtain a succession of partitions (one for each parameter sample) of the data set of interest. The clusters, as the scale increases, are subjected to a fusion process, which reduces their number and increases their size and population. More precisely, when two or more clusters are fused in a new one, the resulting cluster contains all their patterns and only them. For this characteristic, we can define the resulting algorithm as *hierarchical constructive*. The scale parameter corresponds to the minimum distance between two centroids required to perform cluster fusion. The proposed algorithm will be denoted as *gravitational* and will be discussed in Sect. 2.

c) In the hierarchical constructive algorithm here discussed, a valid partition is selected by means of the stability index. It takes into consideration the measure of scale parameter intervals in which a partition remains unchanged (no fusion between clusters). The stability index is strictly connected to the scale-based approach [2] and to the imitated physical force. More precisely, since the gravitational force increases proportionally to the scale parameter (from another point of view, this fact corresponds to a space contraction), the more a partition is stable, the more its clusters are "natural" with respect to the imitated force. The partitions that are considered acceptable by the stability index, are also characterized by the best compactness and isolation of clusters. In Sect. 3, the characteristics of the stability index are explained more in detail. Examples regarding the use of the proposed algorithm will be presented in Sect. 4.

2 Gravitational Clustering Algorithm

The physical law we mimic is gravity. We assume that the data patterns have a fictitious unitary mass and are subjected to a gravitational force proportional to a scale parameter. Increasing the scale, the patterns collapse toward characteristic points (centroids) of the denser regions. The influence of the data presentation order is overcome by starting with one cluster per pattern of the data set. After initialization, the scale parameter is increased iteratively; consequently, clusters sufficiently close begin to merge. The scale parameter is increased at each iteration by an increment value chosen on the basis of the minimum distance (Euclidean or not) among data patterns. Since the measurement of distance is sensitive to the ranges of the pattern fields, each of them is preliminarily normalized. It is evident the constructive nature of the algorithm: each cluster increases its attraction force proportionally to the scale parameter value, and when a suitable threshold is attained, two or more clusters merge. Therefore, new clusters originate, increasing their size and population. As a consequence of the fusion, the number of clusters decreases until all the data are contained in a single cluster. In the following, the proposed algorithm will be shortly denoted as GCA (Gravitational Clustering Algorithm).

The clustering model underlying GCA can be expressed, from a scale-space point of view, as follows:

$$C(\underline{x};p) = h : m^{(h)}\left(\underline{x};\underline{W}^{(h)}\right) = \operatorname*{Min}_{h=1}^{N_c(p)}\left\{m^{(h)}\left(\underline{x};\underline{W}^{(h)}\right)\right\} \tag{1}$$

where $\underline{x} \in R^M$ is the input pattern, $h \in N$ is the cluster label assigned to $\underline{x}$, $N_c(p)$ is the number of clusters determined during training for a given value of the scale parameter p, and $m^{(h)}\left(\underline{x};\underline{W}^{(h)}\right)$ is the distance between the actual pattern $\underline{x}$ and the centroid $\underline{c}^{(h)}$ of the h-th cluster. Each function $m^{(h)}$ depends on a set of parameter $\underline{W}^{(h)}$ to be determined on the basis of the training set T and the parameter p, i.e. $\underline{W}^{(h)} = \underline{W}^{(h)}(T;p)$. In the case of Euclidean distance, $m^{(h)} = \sqrt{\left(\underline{x}-\underline{c}^{(h)}\right)^T \cdot \left(\underline{x}-\underline{c}^{(h)}\right)}$ and $\underline{W}^{(h)} = \underline{c}^{(h)}$. Different definitions of distance can be used in specific applications.

The iterative procedure consists in the following steps:

1. Initialization: let $i=0$; create as many clusters as the number of pattern in the data set; let the centroid of each cluster be equal to the corresponding pattern; compute the minimum distance $d_{min}^{(0)}$ between centroids of each pair of clusters; let the scale parameter p_0 be equal to $d_{min}^{(0)}$.
2. Fusion process: in i-th iteration, two clusters can be merged if the distance between their centroids is less than or equal to the value p_i. Merge them determining new clusters and compute their centroids.
3. Partition validity: compute the value of the performance index with regard to the data set partition resulting by the fusion process.
4. Stop criterion: stop the procedure if a one-cluster partition has been found, otherwise continue.
5. Scale parameter upgrade: let $i=i+1$; compute the minimum distance $d_{min}^{(i)}$ between centroids of each pair of clusters in i-th partition; let $p_i = d_{min}^{(i)}$. Go to step 2.

The characteristic aspects of the algorithm concern mainly the fusion strategy. In general, it may happen that more than two clusters can be fused; in this case, they are merged simultaneously in a unique cluster. The strategy to be followed during each iteration for a multiple fusion is the following:

i. calculate the mutual distances among centroids;
ii. compute for each centroid the number n_d of distances from all the other centroids with a value less than or equal to p_t;
iii. select the centroid with the highest n_d and merge it with all the other centroids that have a distance less than or equal to p_t; if there are more centroids in this situation, calculate for each one the average of the distances under threshold and select the centroid with the smallest value (in this way, more compact clusters are privileged).

The population of the new cluster coincides with the union of the populations of the fused clusters. Therefore, the new centroid $\underline{c}^{(t)}$ is the weighted linear combination of the centroids $\underline{c}^{(j)}$ of the fused clusters, with weights coincident with the ratios $\overline{n}_j$ between their populations n_j and the population of the resulting cluster $\underline{c}^{(t)}$, i.e.: $\underline{c}^{(t)} = \sum_{j=1}^{f} \overline{n}_j \underline{c}^{(j)}$; with f equal to number of fused clusters, $\overline{n}_j = \frac{n_j}{n_t}$, and $n_t = \sum_{j=1}^{f} n_j$. It is easy to show that the centroid $\underline{c}^{(t)}$ is always the center of gravity of the new cluster. The fusion can be carried out following different alternatives. However, the previous one yielded the best results in connection with several tests.

GCA generates a hierarchy of partitions; among these partitions we have to find the most valid one by means of a specific performance index. At this regard, the fitness of a partition is evaluated, during the constructive process, on the basis of a proper stability index. It will be explained more in detail in the following section.

3 Stability Performance Indices

The validity of a data set partition operated by a clustering algorithm must be ascertained in order to select the optimal solution. The quantitative value of a relative performance index is significant only in relation to the same data set we are clustering. Consequently, we can use it for determining the optimal partition of the data set with respect to a parameter, such as the scale, but it cannot be used for comparing partitions of different data sets. However, the quantitative measurement of the partition fitness is highly heuristic. For this reason, numerous performance indices have been introduced in the technical literature. In general, they are based on the estimate of the compactness and isolation of the clusters in a given partition of the data set. The determination of the most appropriate index to a specific case is difficult and highly subjective [4].

Stability indices are a particular class of performance indices. The definition of these indices is related to clustering algorithms that depend on a scale parameter. In the case of GCA, each partition is evaluated in terms of its invariance with respect to the whole range of scale parameter values. The use of stability indices consequently requires the determination of the largest scale parameter interval where the partition is unchanged. There are several ways to express the difference between two partitions; in the present case, we assume that two partitions P_h and P_k are different if $N_c(h) \neq N_c(k)$, where $N_c(.)$ is the number of clusters of a partition. We use this measure to compare couple of partitions obtained for consecutive value of the scale parameter. In particular, in correspondence to *i*-th partition, the stability index used in conjunction with GCA is equal to $SI(i)=p_i-p_{i-1}$. Index $SI(i)$ is an estimate of the *lifetime* of *i*-th partition. As a consequence of the GCA mechanism, the longer is the lifetime of a partition, the better is compactness and isolation of its clusters.

4 Simulation Tests

We have constructed three artificial data set, which permits to clearly follow the capacity of our algorithms in overcoming specific clustering problems. The cases considered are limited to a max of two dimensions, in order to compare the resulting partition with our intuition. However, GCA can be applied also to multidimensional problems. The considered tests are:

1. 1-D gaussian distributions;
2. fractal structures;
3. anomalies.

4.1 Monodimensional Gaussian Distributions

The data set is constituted by 4 sets of patterns, each containing 35 values, generated by gaussian distributions with centers in correspondence to 0.1, 0.3, 0.7, and 0.9. The overall data set is shown in Fig.1.

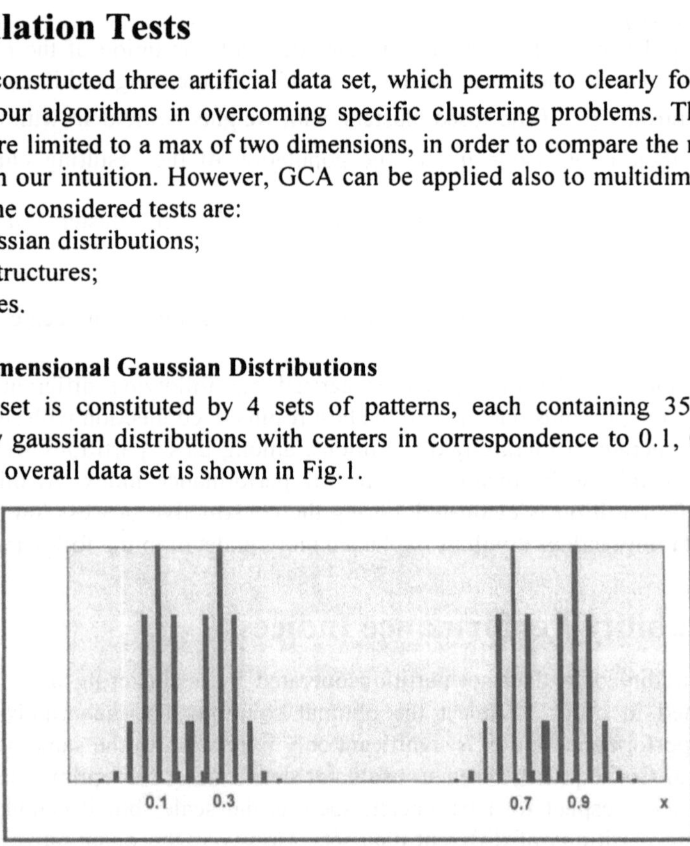

Fig. 1: Problem 4.1 data set histogram.

From the direct inspection of this figure, we deduce that two principal clusters are present; each of them is obtained by overlapping two contiguous clusters. This result is clearly detected by the use of the lifetime index *SI*. Fig. 2 shows the number of clusters N_c versus the scale parameter p (i.e., the *lifetime diagram*).

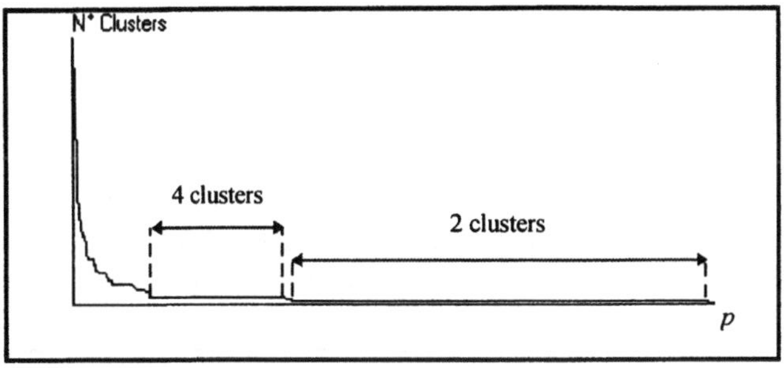

Fig. 2: lifetime diagram (N_c versus p).

There are two stable regions corresponding to 2 and 4 clusters. The former is larger than the latter. Partition hierarchy is well represented by the *dendrogram* in Fig. 3.

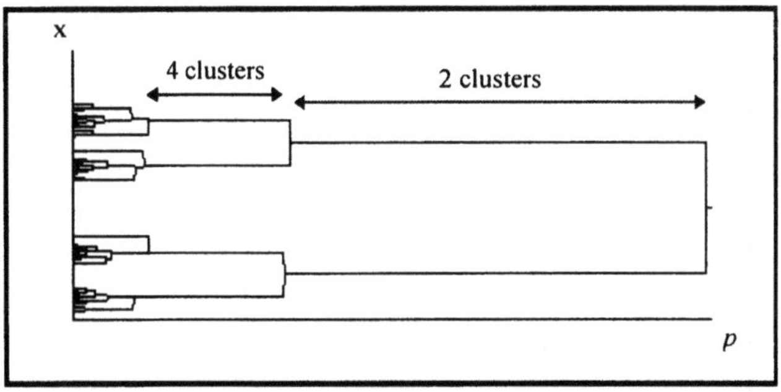

Fig. 3: dendrogram.

This figure clearly shows the formation of the successive partitions, starting from the initial condition of having one cluster per pattern. Cluster evolution is also depicted in the *catastrophe diagram* (Fig. 4).

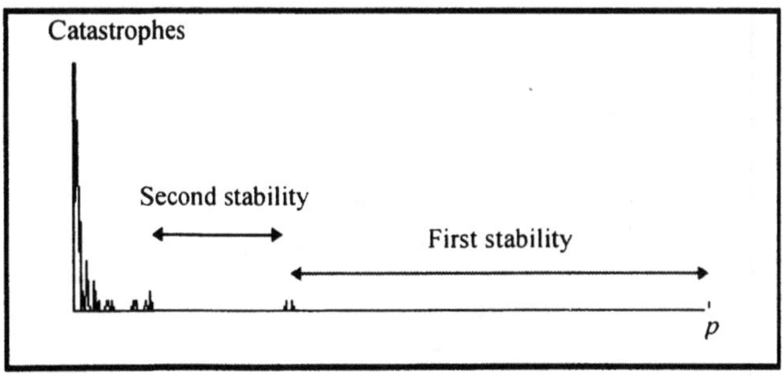

Fig. 4: catastrophe diagram.

In this diagram we report the difference between $N_c(i-1)$ and $N_c(i)$. In the stable intervals there are not catastrophes. Moreover, the presence of some contributions in correspondence to the border between the two stable regions points out that the number of clusters does not pass directly from 4 to 2, but takes on the value 3 too.

4.2 Fractal Structures

The fractal nature of a data set is characterized by several recursive levels of auto-similar structures. Consequently, looking at the data set with different resolution (scale), these structures can be perceived. The data set of fractal nature is shown in Fig. 5: it contains two recursive levels of pentagonal structures (125 patterns).

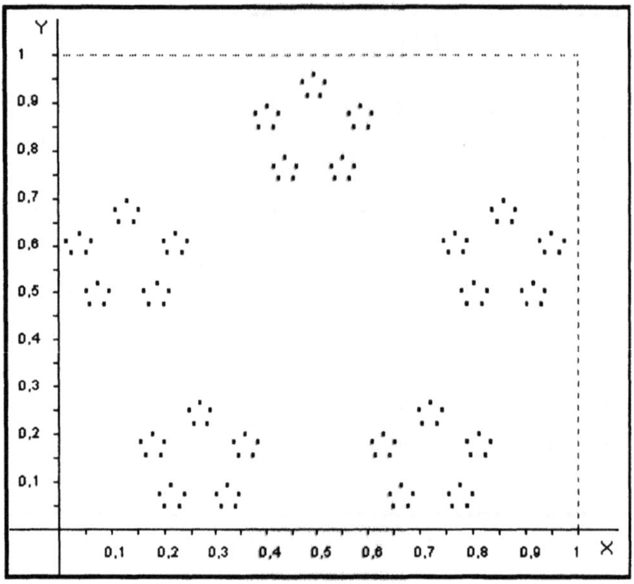

Fig. 5: auto-similar pentagonal structures.

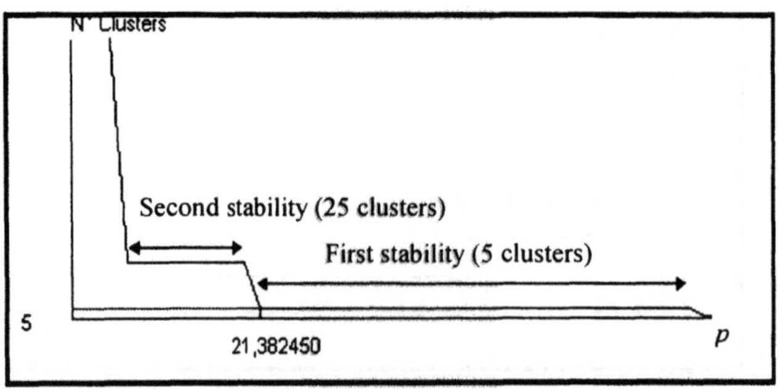

Fig. 6: lifetime diagram for problem 4.2.

We apply GCA with lifetime index. As expected, two stable intervals are individuated (Fig. 6), the larger one corresponds to 5 clusters, while the second one to 25 clusters. In Fig. 7 the two stable partitions are shown.

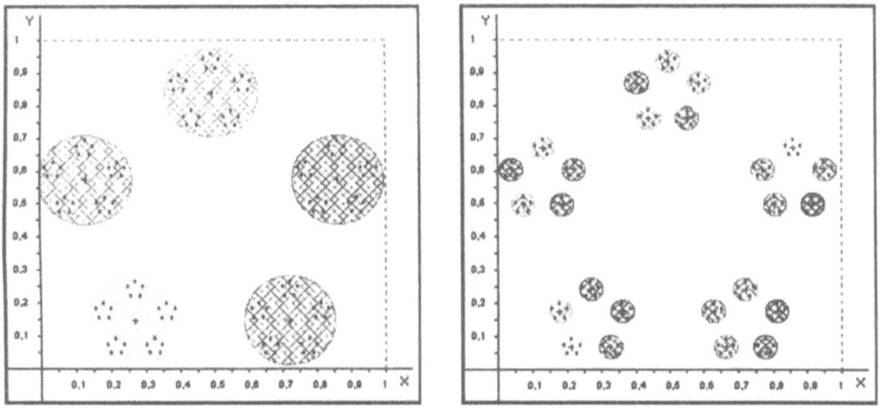

Fig. 7: partitions obtained by GCA for the considered fractal data set.

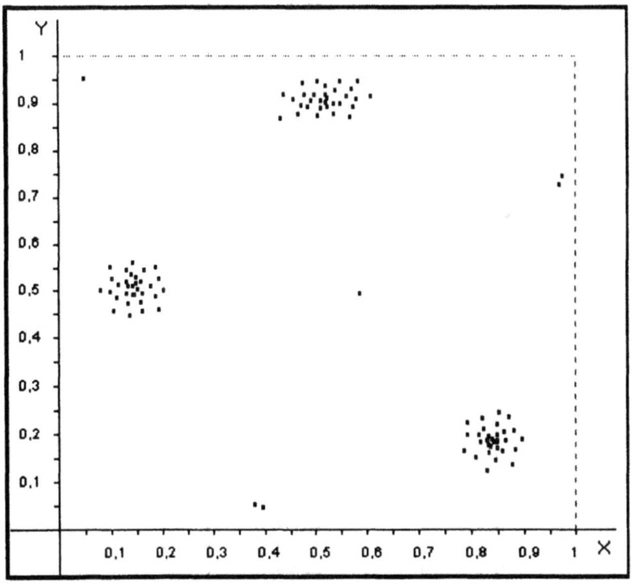

Fig. 8: problem 4.3 data set.

4.3 Anomalies

A cluster is anomalous if it contains a very reduced number of patterns, much less than that of the other clusters of the partition. Anomalous clusters contain data with characteristics very different from the average. They can be individuated on the basis of their reduced population and then eliminated if necessary. As an example, we consider the data set of Fig. 8. The application to this case of GCA with lifetime index easily determines 7 clusters, as shown in the lifetime diagram of Fig. 9. The 4 anomalous clusters are characterized by a population of 1 or 2 patterns (Fig. 10).

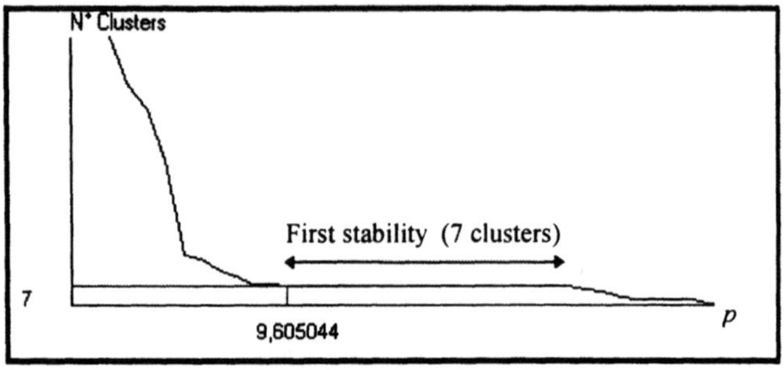

Fig. 9: lifetime diagram for problem 4.3.

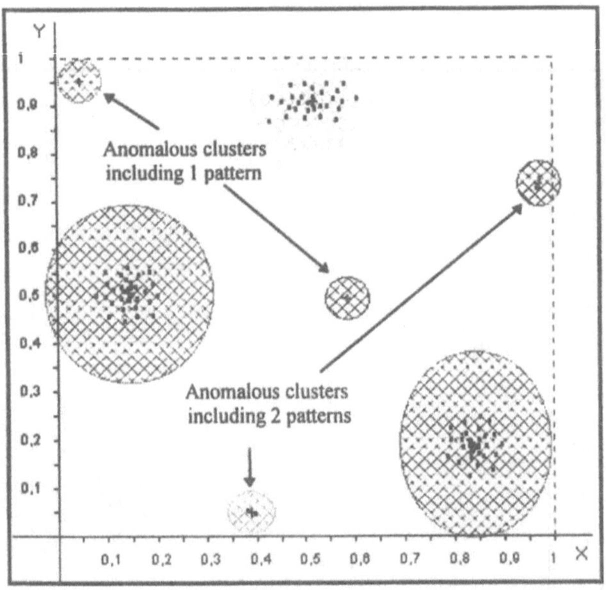

Fig. 10: anomalous clusters detected by GCA.

5 Discussion

The idea of imitating the way physical clusters originate is the guideline of the clustering algorithm proposed in the present paper. The physical law taken into account is the gravitational attraction. The best results obtained by applying the proposed algorithm were always reached in connection with a performance index of stability type: the lifetime index.

The examples presented in the paper were artificially constructed with the goal of representing in an intuitive way some interesting situations. The application of the proposed algorithm to real problems, also with multidimensional data set, yields good results. An important issue to be further investigated is the computational cost. It grows with the square of the cardinality of the data set to be clustered. A drastic reduction of this cost can be achieved by applying in a preprocessing step a classic *C*-Means procedure, with *C* much less than the cardinality of the data set. Successively, the gravitational algorithm can be applied replacing each cluster determined by *C*-Means with its centroid and assigning to it a mass equal to its population.

References

[1] A.K. Jain, R.C. Dubes. Algorithms for clustering data. Prentice Hall, Englewood Cliffs, 1988.

[2] Y.F. Wong. Clustering data by melting. *Neural Computation*, Vol. 5, No. 1, pp. 89-104, 1993.

[3] S.V. Chakravarthy, J. Ghosh. Scale-based clustering using the radial-basis function network. *IEEE Trans. on Neural Networks*, Vol. 7, No. 5, pp. 1250-1261, Sept. 1996.

[4] J.C. Bezdek, W.Q. Li, Y. Attikiouzel, and M. Windham. A geometrical approach to cluster validity for normal mixtures. *Soft Computing*, Vol. 1, No. 4, pp. 166-179, 1997.

A General Assembly as implementation of a Hebbian rule in a Boolean Neural Network

F.E. Lauria, M. Milo, R. Prevete, S. Visco
Department of Scienze Fisiche, University of Naples INFM, Pad. 19,
Mostra d'Oltremare, 80125 Napoli (Italy).
E-mail: lauria(milo, prevete, visco)@na.infn.it
http://www.na.infn.it/Gener/Cyber/nnet.htm

Abstract

Usually the Hebbian learning spontaneously seems to produce associative memory behavior in the network where they are applied. The unsupervised learning performed by the Hebbian rule, automatically creates associations into the network as soon as the responses to the inputs are computed. The paradigm we are discussing here is different from the classical unsupervised learning paradigm, and it is a quite general solution for the implementation of a a Hebbian rule in a Boolean neural network. Our system may not be seen as an associative memory only, it is both a controller and a classifier.

1 Introduction

In this paper we discuss how Boolean neural network can learn its own internal representations by a Hebbian learning rule, to perform sequential task compositions. The network computes input sequences, randomly distributed into a noisy environment, and initially it has only primitive capabilities (or primitive rule) activated by a subset of network inputs. Initially the network has a *low level* learning capability, performed by the computation of the activities of the hidden nodes. Therefore the network can be structured to perform *high level* learning capabilities by evaluating the randomly incoming input sequences. Furthermore, it is trained to control the execution of the primitive tasks, since the learned sequences are composed of primitive tasks. Thus the network can be trained to learn new capabilities (or new rules) with continuos and coherent activity. In [4], we showed how the discussed learning procedure is performed by a network structured into suitably chosen subnets, *the assemblies*, [4,5].
We call *Adaptable Boolean Neural Network*, or ABNN, the Boolean neural network structured in assemblies executing the discussed learning procedure.

2 The network model

The system is a Boolean neural network composed of a set of N+m deterministic and synchronous nodes, connected by directed and delayed output connections, [1,3,5]. The nodes have a time dependent binary activation and output value. The network is composed of two different set of nodes: a set of m sensory input nodes, receiving no input connections and a set of N hidden and output nodes. The activation function of the hidden or output node i, at the time t, where τ is the smallest indivisible time bit, is given by:

$$f_i(t)=\theta(\sum_{j=1}^{N+m}\sum_{d=1}^{D} C_{ijd}\cdot f_j(t-d\cdot\tau)-T)$$

where θ is the Heaviside step function. C_{ijd} is the weight of the d-delayed output connection, transmitting the output signal from the node j to the node i. D is the maximum delay in the network and $T > 0$ is the threshold of the node i, with $i\in\{1,..,N\}$. We consider a low-level learning rule in a Boolean Neural Network (BNN), such that the weights, or coupling coefficients, can be either time independent or time dependent, and the evolution of the time dependent weights is controlled by the rule itself. We assume as low-level learning rule a suitable Hebbian learning rule [2].

The time dependent coupling coefficients can have two *tags* (or *properties*): *Hebb_in* and *Hebb_out*. We assume that each node can be addressed with:

- no tags
- either the tag *Hebbian_in* or the tag *Hebbian_out*
- both the tags.

If a node has the tag *Hebbian _in* we will call it *Hebbian _in* node, analogously a *Hebbian _out* node will be called *Hebbian _out* node. Moreover, let H_{in} and H_{out} be the finite sets containing the indices of all *Hebbian _in* and *Hebbian _out* nodes, respectively. At each time t for each i and j, with $i\in H_{in}$ and $j\in H_{out}$, we consider the associated coupling coefficients $C_{ijd}(t)$ as time dependent. Moreover, we call *Hebbian connection* the monodirectional delayed connection going from j to i with delay d. If a time dependent coupling coefficient becomes greater than a suitable threshold, said *Hebbian threshold*, its value becomes constant and it is said *frozen*. The time dependent coupling coefficient are regulated by a suitable Hebbian rule[2].

3 The General Assembly

We define some neural subnets, said *Iterable Boolean Neural Subnets*, or *IBNSs*. We use the IBNSs to obtain a network able to perform an high-level learning. We call Adapdable Boolean Neural Network (ABNN) a Boolean Neural Network composed of IBNSs.

The IBNSs must be activated by the ABNN inputs. Therefore, we connect each IBNS to the ABNN inputs with a suitable connection so that one and only one ABNN input is associated to one IBNS. E.g., let the binary code for j be the j-th IBNS *identifier*. Then, we connect the j-th IBNS to the ABNN input such that if j is the ABNN input subvector value, the j-th IBNS input node becomes active. Actually, in an ABNN we need two sorts of IBNSs, namely the *assemblies* and the *primitives*. Whereas the assemblies are IBNSs playing the role of both the input and the hidden nodes in a BNN, the output of the primitive is the ABNN output as they are in a one-to-one correspondence with the ABNN output nodes. So, if the input node of a primitive j is active then j causes a suitable ABNN output. If the input node of a primitive j is active we say j is active.

If S' and S'' are finite sequences of primitive tasks an ABNN can be trained to control the sequence S ≡ S'S", i.e., the sequence S is composed of the sequence S' followed by the sequence S". As S' and S" are sequences of primitives, either one or both of them can be composed of just one primitive. An ABNN controls the sequences S, S' and S'' by means of the assemblies. That is, let A' and A'' be the assemblies controlling the output of the sequences S' and S'', respectively, then an ABNN can be trained so that an assembly A is able to control the output of S ≡ S'S". We call A the *father*, or *parent*, assembly, A' the *first son*, or *first offspring*, A" the *second son*, or *second offspring*. An assembly can be *parent* or *offspring*. A' and A" must be different from A, even if they can be coincident within each other. That is, an assembly cannot be its own father but it can have finitely many identical sons.
The *assembly* can be seen as a sufficient condition on the weights of the BNN in which is implemented a low-level learning rule.
To avoid that a sequence S is of the type S≡ S'S, i.e. S contains itself, we propose a new assembly model called *general assembly*.
In the version here discussed an general assembly has one Hebbian-in and two Hebbian-out nodes. For simplicity we call *Hebbian-in connections* and, respectively, *Hebbian-out connections* the Hebbian connections terminating on the Hebbian-in nodes and, respectively, the Hebbian connections stemming out of the Hebbian-out nodes.
In an general assembly there are the following subnets: *input control subnet*, *output control subnet*, *reset subnet*, *identifier subnet*, *or subnet* and *store subnet* [4].

3.1 The General assembly Functioning

The general assembly functioning can be distinguished in two phases. To satisfy the necessary condition for the coherent implementation of the Hebbian rule, see [4], the Hebbian-out connections of an general assembly must be frozen before its Hebbian-in connections. Accordingly, if we want a coherent ABNN, the general assembly functioning must change as soon as its Hebbian-out connections are frozen. Initially, the ABNN is set on by activating just one component of its input subvector. Thus in every general assembly a controlling node *g* becomes active. The node *g* is always active and inhibits the general assembly Hebbian-in nodes. That is, until the node *g* is inhibited, the Hebbian-in nodes cannot be activated. In this way, no hebbian connection between a general assembly and itself can be frozen, then no sequence of the type S≡ S'S can be learned. The general assembly algorithm is divided in two phases: the *first phase of the algorithm* when the Hebbian-out connections are not yet frozen; and the *second phase of the algorithm,* when the Hebbian-out connections are frozen.

3.1.1 The first phase of the algorithm

If at the time t the identifier of the general assembly A is the ABNN input subvector value, at the next time step (t+ τ) the relative IBNS is activated. There is a one to one correspondence between the identifiers and the IBNSs, so no more than one identifier at the time can be present at the ABNN input subvector value. Then, the general assembly A executes the following algorithm:

a) it inhibits its own input nodes;
b) via the output control it executes the following steps: 1) it sets its own *store subnet* so to accept an IBNS identifier: 2) it activates its own *reset subnet* 3) it activates its own *identifier subnet*;
c) its identifier subnet, broadcasts the A identifier to all the other assemblies;
d) its first Hebbian-out connection stemming out of the output control, becomes active;
e) after a constant delay D, namely the delay in the connection of the reset subnet, (e. g. a delay equal to the greatest delay D), if the "or" subnet of A does not receive another general assembly identifier, then A activity is terminated. Otherwise, it stores the identifier received in its store subnet.
f) As the general assembly A receives an identifier equal to the one it has stored (i.e., as the parent receives for the second time the first offspring identifier) then A executes the following algorithm:
a') it resets its own output control subnet, then, first, it sets its own store to accept an IBNS identifier, second, it activates the resets subnet;
b') its second Hebbian-out connection stemming out of the output control becomes active;
c') it executes once more the procedure described in steps e).
g) As A receives an identifier equal to the new one it has stored (i.e., as the parent receives the second offspring identifier) then A executes the following algorithm: first, its store subnet activity is terminated, second, it activates the identifier subnet to broadcast the A identifier to all the other assemblies, third, it sends a resetting signal to the input control subnet, so resetting A to the initial conditions.

Comment. If the general assembly A' identifier is stored in the store subnet of A at the time t', then at the time t'-D the identifier of the general assembly A' has been the ABNN input subvector value.
Comment. Because of the step b), in a ABNN the general assembly A is the only general assembly storing another general assembly A' identifier. Once it has done so, A waits until A' is reset to the initial conditions , i.e. the offspring broadcasts its identifier for the second time.
Comment. If the general assembly A" is the second offspring of A, broadcasting its identifier, because of the step c'), A is the only general assembly in the ABNN, ready to store it. Then, A waits until A'' broadcasts its own identifier the second time, i.e., until the offspring is reset to the initial conditions.
Comment. If this first phase of the algorithm is repeated a sufficient number of times and A' and A'' are either primitives or assemblies with both Hebbian-out connections frozen, then A Hebbian-out connections are frozen. So, the A activity is always followed by the A' activity and, as soon as the A' activity is terminated, by the A'' activity.

3.1.2 The second phase of the algorithm

The second phase of the algorithm starts when the general assembly A has both its Hebbian-out connections frozen. This second phase allows the two Hebbian-in nodes of the assemblies A' and A" to be activated, if the general assembly A is active.

The general assembly A executes the following algorithm:

a) it inhibits both its own input nodes and Hebbian-out nodes of the other assemblies if at least one active Hebbian-out node belonging to a general assembly different from A is active after five τ. We call this alternative the *learning alternative*;

b) it is divided into cases: 1) if the learning alternative is satisfied for at least one general assembly different from A; 2) the learning alternative is not satisfied or it is satisfied for A only.
In the case 1) the general assembly A executes the following algorithm: the A output control inhibits its own Hebbian-out nodes and sends one signal to the A identifier subnet and, after two τ, executes the resetting step described in the first phase of the algorithm;
In the case 2) A executes the following algorithm: A predisposes its own store to accept an IBNS identifier; it activates the resets subnet; it activates its identifier subnet;

c) A identifier subnet broadcasts the A identifier to all the other assemblies;

d) A first Hebbian-out control node becomes active;

e) after the constant delay D if the "or" subnet of A does not receive another general assembly identifier, is reset to the initial conditions. Otherwise it stores the received identifier in its store subnet.

f) As a general assembly A receives an identifier equal to the one it has stored (i.e., as the parent receives for the second time the first offspring identifier) then A executes the following algorithm:

a') it resets its own output control subnet and again, first, it predisposes its own store to accept an IBNS identifier, second, it sets on the reset subnet;

b') its second Hebbian-out output connection stemming out of the output control becomes active;

c') it executes once more the procedure described in step e).

g) As A receives an identifier equal to the new one it has stored (i.e., as the parent receives the second offspring identifier) then A executes the following algorithm: 1) its store subnet activity is terminated; 2) it activates the identifier subnet to broadcast the A identifier to all the other assemblies; 3) it sends a resetting signal to the input control subnet, so resetting A to the initial conditions; 4) it inhibits its own g node.

4 Conclusions

As it is easy to verify, in a ABNN, the activity of one Hebbian-out node is followed, at most, by the activity of one Hebbian-in node. And they are connected by a Hebbian connection. Accordingly, once the training has been suitably completed, i.e., once some of the Hebbian connections have been frozen, just one primitive at the time can be active.

In our simulations we have verified the adaptable Boolean neural net behaviour under different input conditions, as when it interacts with environments characterised by different noise to signal ratios, in other words, with different mixes of random and correlated events, where the correlated events are the ones the net has to learn. For input streams composed by up to 80% of random events and 20% of correlated events, the number of training sessions increases up to three times, with respect to the number of training sessions necessary when the net input is composed of the correlated events, only.

Acknowledgements

This research was supported in part through the INFM-MPI 40% fund, the MPI 60% funding.

References

[1] D.E.Rumelhart, J.L.McClelland and the PDP Research Group - "Parallel distributed processing", vol 1, Foundations" MIT Press, Cambridge (Mass.) (1986).

[2] D.O.Hebb - "The organization of the behavior. A neurophsychological theory", J. Wiley & Sons, New York (NY) USA (1949).

[3] Mc Culloch, W. S. and W. Pitts - "A logical calculus of the ideas immanent in the nervous activity", Bull. Math. Bioph. 5 (1943) 115-143.

[4] F.E.Lauria, M.Sette, S.Visco - "Adaptable Boolean neural networks", Fridericiana Scientia, Liguori (Napoli)(1996) 212, ISBN 88-207-2676-9, http://www.na.infn.it/Gener/Cyber/nnet.htm.

[5] Caianiello, E. R. - "Outline of a theory of thought processes and thinking machines", J. Theor. Biol. 2 (1961) 204-235.

Training Semiparametric Support Vector Machines

Davide Mattera*, Francesco Palmieri*, and Simon Haykin**

* Dipartimento di Ingegneria Elettronica e delle Telecomunicazioni, Università degli Studi di Napoli Federico II, Via Claudio, 21, 80125, Napoli, Italy

** Communications Research Laboratory, McMaster University, 1280 Main Street W., Hamilton, Ontario, Canada, L8S 4K1

Abstract

The semiparametric Support Vector Machine (SVM) has recently been introduced as a generalization of the classical SVM to the case in which some *a priori* knowledge about the considered problem is available. Semiparametric SVM training requires that we solve an optimization problem very similar (it only imposes a larger number of equality constraints) to that to be solved for classical SVM training. In both cases training is usually performed by means of existing software packages. Since this black-box approach may be undesirable, with reference to the classical SVM, some simple and explicit algorithms, difficult to extend to the semiparametric case, have recently been proposed. In this paper we introduce a simple iterative algorithm for semiparametric SVM training which compares well with some typical software packages, can be simply implemented and has minimal memory requirements.

1 Introduction

Given the set $X \subseteq \mathcal{R}^d$, an unknown real function $\overline{f}$ defined on X and a set of ℓ training examples $\{(\boldsymbol{x}_i, y_i),\ \boldsymbol{x}_i \in X, y_i \in \mathcal{R}\}_{i=1,\ldots,\ell}$, SVM theory [12] suggests of approximating the unknown function $\overline{f}$ with the function $f(\boldsymbol{x}) = \boldsymbol{w}^T\Phi(\boldsymbol{x}) + b$, by minimizing, over $\boldsymbol{w}$ and b, the regularized cost function:

$$\frac{1}{2}||\boldsymbol{w}||^2 + C\sum_{i=1}^{\ell} c_{\varepsilon_i}[y_i - \boldsymbol{w}^T\Phi(\boldsymbol{x}_i) - b], \tag{1}$$

where $\Phi(\boldsymbol{x})$ is a fixed nonlinear function of $\boldsymbol{x}$, $c_{\varepsilon_i}(\xi) \stackrel{\Delta}{=} \max(0, |\ \xi\ | - \varepsilon_i)$ is the cost function, and ε_i is a small positive constant, which specifies the required accuracy, usually set to the same value for all i. The constant C controls the amount of regularization; the smaller C is, the stronger is the regularization. It can be shown that $\boldsymbol{w} = \sum_{i=1}^{\ell} u_i\Phi(\boldsymbol{x}_i)$ and, therefore, $f(\boldsymbol{x}) = \sum_{j=1}^{\ell} u_j\mathcal{K}(\boldsymbol{x}, \boldsymbol{x}_j) + b$, where $\mathcal{K}(\boldsymbol{x}, \boldsymbol{x}_j) \stackrel{\Delta}{=} \Phi^T(\boldsymbol{x})\Phi(\boldsymbol{x}_j)$. The coefficients u_j are obtained by solving the following optimization problem [1] [12]:

$$\begin{cases} \min_{u_i} \frac{1}{2}\sum_{i=1}^{\ell}\sum_{j=1}^{\ell} u_i u_j \mathcal{K}(\boldsymbol{x}_i, \boldsymbol{x}_j) - \sum_{i=1}^{\ell} y_i u_i + \sum_{i=1}^{\ell} \varepsilon_i \mid u_i \mid , \\ \sum_{i=1}^{\ell} u_i = 0, \\ -C \le u_i \le C \qquad i = 1, \ldots, \ell \ . \end{cases} \tag{2}$$

At optimum only some u_i are nonzero and the corresponding examples are named *support vectors*. The value of b is equal to $b = y_i - \sum_{j=1}^{\ell} u_j \mathcal{K}(\boldsymbol{x}_i, \boldsymbol{x}_j) - \varepsilon_i \, sign(u_i)$ for all i such that $0 <\mid u_i \mid< C$.

Very recently the following semiparametric extension of the SVM has been proposed [11]. It is suggested to consider the following class of functions

$$f(\boldsymbol{x}) = \boldsymbol{w}^T \boldsymbol{\Phi}(\boldsymbol{x}) + \boldsymbol{w}_1^T \boldsymbol{\Psi}(\boldsymbol{x}), \tag{3}$$

where $\boldsymbol{\Psi}(\boldsymbol{x}) \triangleq [\psi_1(\boldsymbol{x}), \ldots, \psi_m(\boldsymbol{x})]^T$ is a vector of known functions. The approximation (3) to the unknown functions $\overline{f}$ is determined by minimizing, over $\boldsymbol{w}$ and $\boldsymbol{w}_1$, the regularized function:

$$\frac{1}{2}||\boldsymbol{w}||^2 + C\sum_{i=1}^{\ell} c_{\varepsilon_i}[y_i - \boldsymbol{w}^T \boldsymbol{\Phi}(\boldsymbol{x}_i) - \boldsymbol{w}_1^T \boldsymbol{\Psi}(\boldsymbol{x}_i)]. \tag{4}$$

Note that the classical SVM can be obtained by choosing $m = 1$ and $\psi_1(\boldsymbol{x}) \equiv 1$. The semiparametric SVM can be, therefore, considered an extension of the SV method to the case in which some *a priori* knowledge about the unknown target function is available; this allows us to choose a vector of functions $\boldsymbol{\Psi}(\boldsymbol{x})$ more suited to the considered problem. Let us note that some components of $\boldsymbol{\Psi}(\boldsymbol{x})$ can be set equal to the input vector $\boldsymbol{x}$. This represents the more general extension of the classical linear filtering techniques. It appears to be well suited for some engineering problem where the linear filter has already shown to be effective [3] but its approximation performance needs to be improved with some nonlinear extension.

Let us note that the objective function (4) is somewhat restrictive. For instance, only $\boldsymbol{w}$ is regularized while $\boldsymbol{w}_1$ remains unregularized in (4). Therefore, the number m of components of $\boldsymbol{\Psi}(\boldsymbol{x})$ needs to be chosen sufficiently small. The choice of ε_i-insensitive cost function allows us to reduce the number of components u_i different from zero at optimum. The linear behavior of the classical ε_i-insensitive function is introduced for rendering the method robust to the presence of outliers [4]. The reduction in the number of components in the nonlinear expansion can be also obtained by applying other methods of subset selection while the choice of a robust cost function renders the optimization problem worthlessly more complicated when the noise, contaminating the given examples, is known to have non-impulsive characteristics.

[1] Problem (2) is usually written, by expressing $u_i = \alpha_i^* - \alpha_i$ and $\mid u_i \mid= \alpha_i^* + \alpha_i$, as a quadratic problem with 2ℓ variables.

However, when the cost function (4) is adopted, we are faced with a convex optimization problem whose solution is given [11] by $f(\boldsymbol{x}) = \sum_{j=1}^{\ell} u_j \mathcal{K}(\boldsymbol{x}, \boldsymbol{x}_j) + \boldsymbol{w}_1^T \boldsymbol{\Psi}(\boldsymbol{x})$, where the coefficients u_i can be found by solving the following optimization problem:

$$\begin{cases} \min_{u_i} \frac{1}{2} \sum_{i=1}^{\ell} \sum_{j=1}^{\ell} u_i u_j \mathcal{K}(\boldsymbol{x}_i, \boldsymbol{x}_j) - \sum_{i=1}^{\ell} y_i u_i + \sum_{i=1}^{\ell} \varepsilon_i \, |u_i| \, , \\ \sum_{i=1}^{\ell} u_i \psi_j(\boldsymbol{x}_i) = 0 \qquad j \in \{1, \ldots, m\}, \\ -C \leq u_i \leq C \qquad i = 1, \ldots, \ell \; . \end{cases} \tag{5}$$

The components of $\boldsymbol{w}_1$ are usually determined [11] as the Lagrangian multipliers corresponding to the equality constraints.

All methods so far proposed in the literature require the usage of standard software packages which can be employed directly for the whole problem or to solve a number of subproblems iteratively [8]. Simple and explicit training algorithms have been proposed with reference to the classical SVM [6, 7, 10, 9, 5]. Our purpose here is to propose an explicit algorithm for solving (5) that permits a very simple implementation. This may be very useful in comparison to using pre-written software packages which may be costly to use, may lack transparency, flexibility and control over memory usage and computational complexity. We first consider the case in which the dimension of the vector $\boldsymbol{\Phi}(\boldsymbol{x})$ is so large that we need to work in the kernel domain. Then, we consider the case in which the vector $\boldsymbol{\Phi}(\boldsymbol{x})$ is smaller and we can work directly to find the optimum vector $\boldsymbol{w}$.

2 The proposed algorithm

We propose an algorithm for the case of generic m that requires to solve different times the same optimization problem without equality constraints ($m = 0$).

2.1 A particular case

Consider first the case in which $m = 0$. The resulting optimization problem

$$\begin{cases} \min_{u_i} \frac{1}{2} \sum_{i=1}^{\ell} \sum_{j=1}^{\ell} u_i u_j \mathcal{K}(\boldsymbol{x}_i, \boldsymbol{x}_j) - \sum_{i=1}^{\ell} y_i u_i + \sum_{i=1}^{\ell} \varepsilon_i \, |u_i| \, , \\ -C \leq u_i \leq C \qquad i = 1, \ldots, \ell, \end{cases} \tag{6}$$

can be solved with the method proposed in [6, 7] that consists in cycling on all the variables applying the following updating rule:

$$u_i = \min \left\{ C\Theta(|e_i| - \varepsilon_i), \frac{||e_i| - \varepsilon_i|}{\mathcal{K}(\boldsymbol{x}_i, \boldsymbol{x}_i)} \right\} sign(e_i), \tag{7}$$

4 Conclusions

We proposed a new method for performing semiparametric SVM training. Its principal advantages are simplicity and minimal memory requirement. We pay their simplicity, often necessary in real-time adaptive problems, with an increase in the number of iterations required for reaching convergence.

References

[1] D. P. Bertsekas. *Constrained Optimization and Lagrange Multiplier Methods*. Academic Press, Inc., San Diego, CA, 1982.

[2] K. Fukunaga. *Introduction to statistical pattern recognition*. Academic Press, San Diego, California, 1990. Second Edition.

[3] S. Haykin. *Adaptive Filter Theory*. Prentice-Hall, New York, 1996.

[4] P. J. Huber. *Robust Statistics*. John Wilcy and Sons, New York, 1981.

[5] O. L. Mangasarian and D. R. Musicant. Successive overrelazation for support vector machines. Technical Report 98-12, Computer Sciences Department, University of Wisconsin, USA, 1998.

[6] D. Mattera. *Nonlinear Modeling from Empirical Data: Theory and Applications [in italian]*. PhD thesis, Università degli Studi di Napoli Federico II, National Italian Libraries of Rome and Florence, Feb 1998.

[7] D. Mattera, F. Palmieri, and S. Haykin. An explicit algorithm for training support vector machines. submitted to IEEE Signal Processing Letters, August 1998.

[8] E. Osuna, R. Freund, and F. Girosi. An improved training algorithm for support vector machines. In *Neural Networks for Signal Processing VII — Proceedings of the 1997 IEEE Workshop*, pages 276 – 285, 1997.

[9] J. Platt. Fast training of support vector machines using sequential minimal optimization. In B. Scholkopf, C. Burges, and A. Smola, editors, *Advances in Kernel Methods - Support Vector Learning*. MIT Press, Cambridge, MA, USA, 1998.

[10] A. Smola and B. Schölkopf. A tutorial on support vector regression. Technical Report TR-1998-030, NEURO-COLT, Royal Holloway College, 1998.

[11] A. J. Smola, T. Friess, and B. Schölkopf. Semiparametric support vector and linear programming machines. In *Advances in Neural Information Processing Systems*. MIT Press, 1999.

[12] V. Vapnik. *The Nature of Statistical Learning Theory*. Springer Verlag, New York, 1995.

Building Neural and Logical Networks with Hamming Clustering

Marco Muselli
Istituto per i Circuiti Elettronici - CNR
via De Marini, 6 - 16149 Genova, Italy

Abstract

The solution of binary classification problems is obtained by employing a new learning method, called Hamming Clustering (HC). It is able to build in a constructive way a two-layer perceptron with binary weights, which can be easily implemented by means of conventional logical ports.

This technique generalizes the information contained in the given training set by combining input patterns that are close each other according to the Hamming distance. The output class is assigned in a competitive way, thus allowing the treatment of ambiguous samples.

The application of HC to the signal prediction in genomic sequences shows its ability to determine regularities in complex problems.

1 Introduction

Many real-world classification problems belonging to different application fields can be described through Boolean variables. In these cases the employment of general methods, like the back-propagation algorithm [1], which are expressly developed for the treatment of real inputs, increases the complexity of the problem and leads to higher execution times that limit the applicability range of the considered approach.

It is thus important to develop suitable training algorithms for binary classification problems, which exploit their peculiar characteristics to make easier the learning process and to improve the properties of the resulting connectionist model [2, 3]. Particular attention must also be devoted to the implementation on physical supports; the need of dealing with only two-state variables can indeed reduce the complexity of the devices to be realized.

A new learning technique of this kind is the procedure of Hamming Clustering (HC): it generates two-layer perceptrons which can be directly implemented through a combinatorial logical network, containing only AND, OR and NOT ports. The learning task proceeds by pattern: a sample of the training set is randomly chosen at each iteration and one or more clusters in the input space are generated starting from that element. Since the method can subsequently work on samples belonging to different classes, the construction of the logical network is based on a particular kind of competitive learning.

During the training HC is able to determine redundant inputs, which do not affect the behavior of the resulting configuration. A final pruning phase allows a further reduction of the network complexity and can yield a significant improvement in the corresponding generalization ability.

Some theoretical results [4] ensure that HC has a polynomial computational cost $O(n^2cs + nc^2)$, where n is the dimension of the input space, s is the size of the given training set, whereas c is the number of hidden units in the resulting neural network. Also the amount of memory required is small and shows an asymptotic behavior $O(n(c+s))$.

The application of HC to the prediction of relevant signals in genomic sequences shows the good performances of the networks generated. Furthermore, it is straightforward to extract from these two-layer perceptrons a collection of templates that code the behavior of the underlying classifier. This property suggests a possible application of HC as a method for extracting understandable rules from a given set of samples. The analysis of this aspect is the subject of a complementary publication [5].

2 The Hamming Clustering procedure

Consider a training set S containing s samples $(\mathbf{x}_j, y_j)$, $j = 1, \dots, s$, generated by a Boolean function $f : \{-1,+1\}^n \longrightarrow \{-1,+1\}$, eventually in the presence of noise. The n components x_{ji}, $i = 1, \dots, n$, of the input patterns $\mathbf{x}_j$ and the corresponding output y_j are binary-valued, coded by the integers -1 and $+1$. Such a function f subdivides the vertices of the hypercube $\{-1,+1\}^n$ into two subsets H^+ and H^-, according to the corresponding output:

$$H^+ = \{\mathbf{x} \in \{-1,+1\}^n \mid f(\mathbf{x}) = +1\}, \quad H^- = \{-1,+1\}^n \setminus H^+ \qquad (1)$$

In the same way, S can be subdivided into two subsets S^+ and S^-:

$$S^+ = \{\mathbf{x}_j \in \{-1,+1\}^n \mid (\mathbf{x}_j, y_j) \in S,\ y_j = +1\}, \quad S^- = S \setminus S^+$$

Note that $H^+ \cap H^- = \emptyset$, but $S^+ \cap S^- = \emptyset$ only if the training set S is *consistent*, i.e. it does not contain *ambiguous* input patterns $\mathbf{x}$, for which both $(\mathbf{x}, +1)$ and $(\mathbf{x}, -1)$ belong to S.

The procedures of *Hamming Clustering* (HC) determine the unknown function f or, equivalently, the sets H^+ and H^-, by combining the elements of $\{-1,+1\}^n$ in a competitive way. The technique adopted by HC can be better described by denoting with I_n the set $\{1, 2, \dots, n\}$ of the first n positive integers and by introducing the following definition.

Definition 1 *The set* $C_l(\mathbf{x}, L) = \{\mathbf{z} \in \{-1,+1\}^n \mid z_i = x_i$ *for every* $i \notin L\}$, *where* $\mathbf{x} \in \{-1,+1\}^n$ *and* $L \subset I_n$ *has size* $l = |L|$, *will be called* l-cube *(or simply* cube*) with* vertex $\mathbf{x}$ *and* generator L.

The well known definition of Hamming distance between two binary vectors can then be easily extended to the case of two cubes:

Definition 2 *The Hamming distance* d_H *between an* l*-cube* $C_l(\mathbf{x}, L)$ *and a* k*-cube* $C_k(\mathbf{z}, K)$, *with* $0 \le l, k \le n$, *is given by* $d_H = \sum_{i \notin L \cup K} |x_i - z_i|/2$.

HAMMING CLUSTERING

1. Choose at random a sample $(\mathbf{x}, y)$ in the training set S.
2. Build one or more cubes having a vertex in $\mathbf{x}$. Remove $(\mathbf{x}, y)$ from S. If the construction is not complete, go to Step 1.
3. Prune the set(s) of cubes generated.

Figure 1: General procedure followed by Hamming Clustering.

The goal of HC is to obtain from the training set S two collection of cubes C^+ and C^-, which approximate the sets H^+ and H^-. A two-layer perceptron that provides output $+1$ for all the input patterns belonging to C^+ (C^-) can be directly realized. It is sufficient to build a layer of hidden threshold units performing AND operations that provide output $+1$ only when the input patterns contained in each cube $C_l(\mathbf{x}, L) \in C^+$ (C^-) are presented. The two-layer perceptron is completed by a final neuron performing a logical OR among the output of the hidden units. Such a configuration will be called *single-class HC network*.

A different architecture can be employed to realize a classification similar to that performed by the nearest-neighbor algorithm. This digital network will be called *majority HC network* and is formed by two consequent blocks. The first of them performs the computation of the Hamming distances between the input pattern and each of the elements belonging to C^+ and C^-, while the second one chooses the class containing the cube having minimum Hamming distance. Both the single-class HC and the majority HC network contain only binary weights and can be directly implemented by using only logical AND, OR and NOT ports [4].

The algorithm followed by HC is reported in Fig. 1; Steps 1 and 2 are repeatedly executed and form the basic iteration of the method.

2.1 Cluster generation

The computational kernel of HC is given by the generation of cubes having a vertex in the sample $(\mathbf{x}, y)$ previously selected (at Step 1 of Fig. 1). Suppose without loss of generality that $\mathbf{x} \in S^+$: the separation of the sets C^+ and C^- is ensured if each of the l-cubes $C_l(\mathbf{x}, L)$ generated in the current iteration satisfies the following two conditions:

$$C_l(\mathbf{x}, L) \cap S^- \subset S^+ \,, \quad C_l(\mathbf{x}, L) \cap C^- = \emptyset \tag{2}$$

The procedure employed by HC for the generation of cubes having a vertex in $\mathbf{x} \in S^+$ is reported in Fig. 2. Minor changes are necessary if the training set S is not consistent. The wide freedom in the choice of the generator L

METHOD FOR THE GENERATION OF CUBES IN HC

1. Compute the s^- Hamming distances d_j between $\mathbf{x}$ and the patterns $\mathbf{x}_j \in S^-$ by applying Def. 2. Let D_j be the subset of I_n containing the d_j indices of the components that differ in $\mathbf{x}$ and in $\mathbf{x}_j$:
$$D_j = \{i \in I_n \mid x_i \neq x_{ji},\ \mathbf{x}_j \in S^-\} \quad , \quad \text{for } j = 1, \ldots, s^- .$$
2. Extract a subset $L \subset I_n$ such that $L \cap D_j \neq D_j$ for every $j = 1, \ldots, s^-$. The l-cube $C_l(\mathbf{x}, L)$, where $l = |L|$, satisfies conditions (2).
3. For every cluster $C_k(\mathbf{u}, K) \in C^-$, if $C_l(\mathbf{x}, L) \cap C_k(\mathbf{u}, K) \neq \emptyset$, remove from L an index $i \notin K$ such that $x_i \neq u_i$.

Figure 2: Procedure followed by HC for the generation of an l-cube.

allows us to define additional criteria that can influence the generalization ability of the resulting logical network. Two of them are:

Largest cube criterion: it requires that the size of the generator L chosen at Step 2 is maximum. To this end the index $i \in L$ that is included in the greatest number of sets D_j associated with the input patterns $\mathbf{x}_j \in S^-$ is repeatedly removed from L until conditions (2) are satisfied.

Maximum covering cube criterion: it maximizes the number of samples of S included in the current cluster and is mostly employed in the construction of majority HC networks. This criterion needs the construction of the sets $D_j^+ \subset I_n$ containing the indices of the components that differ in $\mathbf{x}$ and in the elements $\mathbf{x}_j \in S^+$. Then L is built by choosing the indices which are contained in the greatest number of sets D_j^+.

2.2 Pruning

In general the collection of cubes C^+ and C^- generated by Steps 1-2 of HC contain redundant elements which can be removed without changing the behavior of the resulting network. Thus, a pruning phase can be used to optimize the complexity of the final configuration. To describe the main pruning methods employed in HC we introduce the following definition:

Definition 3 *The set* $Q = \{\mathbf{x} \in \{-1, +1\}^n \mid (\mathbf{x}, y) \in S,\ \mathbf{x} \in C_l(\mathbf{u}, L)\}$ *will be called* cover set *of the l-cube* $C_l(\mathbf{u}, L)$. *The number q of input patterns contained in Q will be named* covering *of* $C_l(\mathbf{u}, L)$.

Minimal pruning: the easiest way to reduce the size of C^+ and C^- is to find a minimum subset of cubes that correctly classify all the input patterns in S. This can be realized by employing a near-optimal greedy procedure that extracts the clusters with maximum covering one at a time. In this way the resulting collections C^+ and C^- cannot contain redundant clusters.

Threshold pruning: in some classification problems it can be advantageous to maintain in C^+ and C^- only the cubes whose covering exceeds a fixed threshold τ. Normally, the value of τ is proportional to the maximum covering $q_{\max}$ found in the set of clusters to be pruned (e. g. $\tau = 0.1 \cdot q_{\max}$).

3 Application of HC to the signal prediction in genomic sequences

The properties of the two-layer perceptrons generated by HC have been tested on the problem of locating regulatory signals in genomic sequences [6]. In particular, HC has been applied to the identification of poly(A) signals on an inhomogeneous set of human DNA sequences. It is known that a fixed six base pattern such as AATAAA does not determine univocally the presence of a poly(A) signal; consequently, it is interesting to establish if a greater number of adjacent bases is sufficient to predict its precise position.

To this end we have considered a window containing a fixed number b of locations that slides on the genomic sequence and captures corresponding bases. Since HC works on binary input patterns, a four bit coding for the DNA bases has been adopted. In this way it is possible to associate every cube generated by HC with a template which can be used for the recognition of poly(A) signals.

For each value of $b = 6, 7, \ldots, 12$ about 1000 sequences containing poly(A) signals have been extracted by the EMBL database [7]; half of them (S^+) have been employed by HC to perform the training of binary neural networks together with a high number of base sequences corresponding to gene regions that do not contain the target signal (S^-). The resulting training sets S contain more than 100,000 samples, thus preventing the application of time-consuming classification methods. On the contrary, the low computational cost of HC allows us to obtain within a reasonable time a complete set of templates to be used for the analysis of new patterns.

Tab. 1 presents the results of the classification performed on ten genomic sequences randomly extracted from the test set. For each of them it is reported the accession number in the EMBL database, the total number of bases, the number of poly(A) signals found experimentally and the number of possible targets located by HC. An asterisk near this last value denotes that one target has been missed by the resulting network.

As one can note, the number of presumed poly(A) signals decreases with the window length b; this shows that the flanking region plays an important role as supposed. In several test sequences the redundancy of the prediction vanishes when $b = 9$ or $b = 10$; consequently, these are the optimal window lengths to be used in the recognition of these signals.

Genomic sequence			Window length						
Accession n.	N. bases	N. poly(A)	6	7	8	9	10	11	12
d12676	2329	1	14	6	4	3	2	3	6
j00184	2685	1	1	1	1	1	1	1	1
j05582	4139	1	4	2	1	1	1	1	1
l11706	3250	1	2	2	1	1	1	1	2
m26679	3079	1	8	6	3	2	1	1	3
m33518	6349	1	6	3	2	2	2	2	2
m59040	1794	1	9	6	2	1	2	*2	3
m63154	1621	2	6	5	2	2	3	3	5
m69245	1503	1	8	6	1	2	1	*1	4
m91083	1539	1	1	1	1	1	1	1	1

Table 1: Number of presumed poly(A) signals obtained by HC.

References

[1] D. E. RUMELHART, G. E. HINTON, AND R. J. WILLIAMS, Learning internal representations by error propagation. In *Parallel Distribute Processing*, D. E. Rumelhart and J. L. McClelland, eds., Cambridge, MA: MIT Press, 1986, 318–362.

[2] D. L. GRAY AND A. N. MICHEL, A training algorithm for binary feedforward neural networks. *IEEE Transactions on Neural Networks*, **3** (1992), 176–194.

[3] H. M. A. ANDREE, G. T. BARKEMA, W. LOURENS, A. TAAL, AND J. C. VERMEULEN, A comparison study of binary feedforward neural networks and digital circuits. *Neural Networks*, **6** (1993), 785–790.

[4] M. MUSELLI AND D. LIBERATI, Training digital circuits with Hamming Clustering. Submitted to *IEEE Transactions on Circuit and Systems*, (1998).

[5] M. MUSELLI AND D. LIBERATI, Inferring understandable rules through digital synthesis. Accepted as a contribution to *WIRN'99 - The 11-th Italian Workshop on Neural Nets*, (1999).

[6] E. E. SNYDER AND G. D. STORMO, Identification of coding regions in genomic DNA sequences: An application of dynamic programming and neural networks. *Nucleic Acid Research*, **21** (1993), 607–613.

[7] C. M. RICE, R. FUCHS, D. G. HIGGINS, P. J. STOEHR AND G. N. CAMERON, The EMBL data library. *Nucleic Acid Research*, **21** (1993), 2967–2971.

Inferring Understandable Rules through Digital Synthesis

Marco Muselli and Diego Liberati
Istituto per i Circuiti Elettronici, Consiglio Nazionale delle Ricerche
via De Marini 6, 16149 Genova, Italy

Abstract

The extraction of a set of rules underlying a classification problem is performed by applying a new algorithm reconstructing the AND-OR expression of any Boolean function from a given set of samples.

The basic kernel of the method, called Hamming Clustering (HC), is the generation of clusters of input patterns that belong to the same class and are close each other according to the Hamming distance. Inputs are identified and neglected, which do not influence the final output, thus automatically reducing the complexity of the final set of rules.

The performances of HC are evaluated through artificial and real-world benchmarks: its application to the breast cancer prognosis leads to the derivation of a small set of rules solving the associated classification problem.

1 Introduction

Training algorithms for connectionist methods exhibit the typical ability of inferring relationships from the given examples. A drawback of this approach is that these relationships, implicitly embedded in the trained neural network, are not explicitly available to the human user. Neural networks know how to do, but not how to explain.

It would be important also for connectionist models to reconstruct the set of procedural relations typical of the expert system approach, whose drawback, on the other side, is the long and complex work of interdisciplinary experts required. Several attempts have been performed to achieve this goal [1, 2, 3], but their application is not general and often leads to poor results.

In this paper we propose the use of a new technique, called Hamming Clustering (HC), for the solution of classification problems with binary inputs. It is essentially a method for the synthesis of digital circuits, which is able to directly infer the explicit rules in the `if-then` form typical of the expert systems, while achieving performances comparable to those of artificial neural networks, in terms of both efficiency and efficacy, and allowing a direct implementation on a physical support, since it does not require variable weights with floating point precision.

In fact, any binary classification problem can be solved by a proper Boolean function f, which can always be written in AND-OR form, i.e. a logical sum of AND operations [4]. Each of these AND operations can be viewed as an `if-then` rule, whose antecedents and consequent are its inputs and output, respectively. The combination of these rules univocally defines the unknown function f.

Not all the available inputs are necessarily used to determine the output of each AND operation: it is important to select which inputs are useless in order to simplify the resulting set of rules. Moreover, there can be redundant features whose rejection does not influence the behavior of the Boolean function f: a good method for rule extraction must be able to perform such task on the basis of the given training set.

The goal of classical algorithms for the synthesis of digital circuits is exactly the determination of the AND-OR expression of a Boolean function f starting from a portion of its truth table. Thus, at least in principle, they could be used to extract rules from a binary training set. Unfortunately, fast techniques yields only rough approximation of the correct AND-OR expression [5, 6], since their target is indeed to build a logical network which only satisfies the given constraints.

On the other side, some of the available methods require a high computational time to achieve an useful result [4, 7]. On the contrary, theoretical results [8] ensure that HC has a polynomial computational cost $O(n^2cs+nc^2)$, where n is the dimension of the input space, s is the size of the training set, and c is the number of AND ports in the resulting digital circuit. Also the small amount of memory required shows a polynomial asymptotic behavior $O(n(c+s))$.

A detailed description of the procedure followed by HC to build a two-layer perceptron or an equivalent logical network from a given training set forms the subject of a complementary contribution [9]. Here we are interested in comparing the performances of HC, in terms of generalization ability and computational cost, with those of other widely used learning algorithms and methods for the synthesis of digital circuits. Furthermore, we want to analyze the capacity of HC to infer understandable rules from a collection of samples associated with a given classification problems.

In the following sections the results obtained on artificial benchmarks concerning the reconstruction of Boolean functions and the so-called Monk's problems [10] are presented. Moreover, the real-world classification problem codified in the Wisconsin Breast Cancer Database [11] is considered and a set of rules underlying this prognostic process is extracted and analyzed. The execution times reported refers to a DEC ALPHAStation 500/400 running under operating system DIGITAL UNIX 4.0A.

2 Reconstruction of Boolean functions

The first group of tests has the aim of analyzing the ability of HC to reconstruct Boolean functions from examples. In every trial an AND-OR expression is randomly built and the truth table for the corresponding function is generated. A fraction r $(0 < r < 1)$ of the total 2^n samples is then employed as a

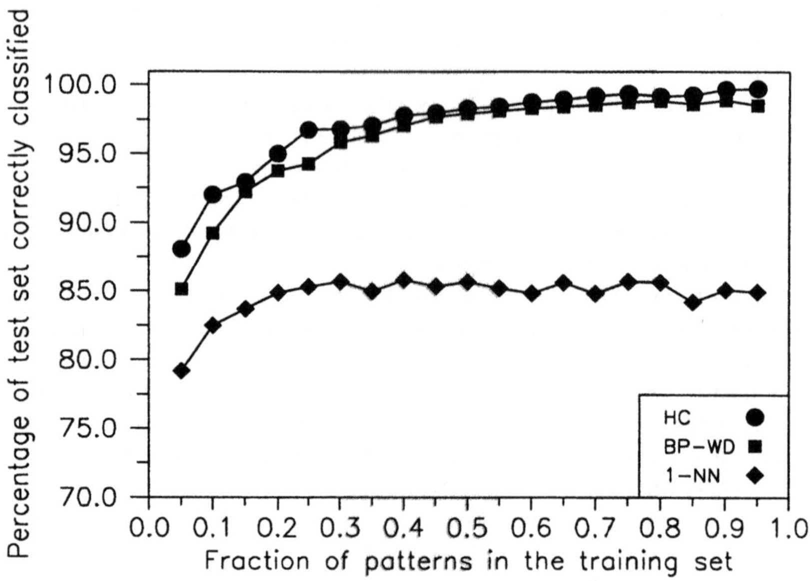

Figure 1: Percentage of input patterns correctly classified by HC (Hamming Clustering), BP-WD (Back-Propagation with Weight Decay) and 1-NN (Nearest-Neighbor method), in the test phase for the reconstruction of Boolean functions.

training set for the algorithms considered, while the remaining $2^n(1-r)$ input patterns form the test set for the evaluation of the generalization ability.

In every run the AND operations used to realize the unknown Boolean function are exactly n and are generated by choosing at random (with uniform probability) the number of terms in each of them and subsequently the relative operands. To each operand the logical operation NOT is applied with probability 0.5 before its insertion in the AND-OR expression.

We have set $n = 10$ in every trial, while the fraction r of samples in the training set assumes the values $r = 0.05, 0.1, 0.15, \ldots, 0.9, 0.95$. Fig. 1 compares the generalization ability achieved by HC, by the Nearest-Neighbor method (1-NN) [12] and by the Back-Propagation algorithm with Weight Decay (BP-WD) [13, ch. 9.2]. The results obtained with computationally expensive techniques for the synthesis of digital circuits, such as the Quine-McCluskey procedure [4], are similar to those obtained with HC and are not reported for the sake of simplicity.

On the other side, the performance of fast methods, like ESPRESSO [5], is too low (about 35%) to be included in the figure, ESPRESSO being not suited for the solution of classification problems. As one can see, the generalization ability of HC and BP-WD increases with the size of the training set, always outperforming 1-NN.

Concerning the computational costs of training, it is interesting to note that the execution time of HC (0.7 seconds on the average) is slightly higher than that of ESPRESSO, suitably designed to be fast. The greater computational burden of HC is due to the reduction of complexity of the logical networks generated. BP-WD requires a learning time much higher (about two orders of magnitude for $r > 0.4$).

3 Monk's problems

A classical set of benchmarks, widely used in the artificial literature, is Monk's problems [10]. These are classification problems defined in a "robot" domain, described by six discrete attributes. With this characterization 432 different robots can be obtained. The aim of each proposed benchmark is the determination of a general classification rule starting from a limited set of samples. The first two Monk's problems are noise-free, whereas in the third case the outputs of the training set can undergo small changes from their correct value. This last test can therefore provide a measure of the robustness of HC.

Since the inputs are discrete but not binary, a proper transformation must be applied to allow the generation of an AND-OR expression. The resulting binary input patterns contain 15 components, each of which is associated with the presence of a particular attribute in the corresponding robot. The result obtained with HC on the Monk's problems are the following:

Problem #1: Within an execution time of 0.06 seconds HC has been able to obtain the exact expression describing the unknown classifier. The method extracting rules from neural networks described in [2] obtained 12 more hidden units, each representing a different redundant rule.

Problem #2: The logical complexity of the underlying classifier is equivalent to that of the parity problem. In this case patterns with low Hamming distance have different output with high probability; for this reason HC is in great trouble in finding the exact set of rules. Nevertheless, HC generates an AND-OR expression containing 32 logical products, less redundant than that achieved by [2] with 43 hidden units.

Problem #3: From the total 432 examples, 122 are randomly selected, 5% of whom are misclassified, simulating the action of noise in the aquisition of the training set. The Boolean function found by HC contains 11 logical products and achieves a generalization error of 3%, partially recovering the effect of noise and reaching the same performance of the best neural methods tested in [10].

4 Breast Cancer Prognosis

A classical real-world classification problem is the Winsconsin Breast Cancer Database [11], where breast cancer is to be prognosed on the basis of 9 features, assuming integer values in the range $1, \ldots, 10$: clump thickness (x_1), uniformity of cell size (x_2), uniformity of cell shape (x_3), marginal adhesion (x_4), single epithelial cell size (x_5), bare nuclei (x_6), bland chromatin (x_7), normal nucleoli (x_8), mitoses (x_9). The available input-output pairs are 699, of which 458 (65.5%) are benign and 241 (34.5%) malignant.

Since 16 input patterns include missing data, the corresponding pairs have been removed. The remaining 683 have been subdivided into training (372 patterns) and test (311 patterns) set. Since the application of HC requires binary inputs, a thermometer code with length 9 has been used to translate the integer values in the database, thus obtaining input patterns with 81 binary components.

The execution of HC with minimal pruning yields the generation of 6 rules, two of which describe only one example each. The required CPU time is 0.25 seconds. The four significant rules producing a benignant prognosis are

- $x_1 \leq 8$ AND $x_4 \leq 7$ AND $x_5 \leq 5$ AND $x_6 \leq 2$ AND $x_7 \leq 9$ AND $x_8 \leq 8$ AND $x_9 \leq 3$
- $x_1 \leq 8$ AND $x_3 \leq 2$ AND $x_4 \leq 7$ AND $x_5 \leq 5$ AND $x_6 \leq 3$ AND $x_8 \leq 8$
- $x_3 = 1$ AND $x_4 \leq 9$ AND $x_8 \leq 8$
- $x_1 \leq 8$ AND $x_4 \leq 7$ AND $x_5 \leq 5$ AND $x_6 \leq 4$ AND $x_7 \leq 2$ AND $x_8 \leq 8$ AND $x_9 \leq 3$

The direct application of these four rules yields a generalization error of about 2% (14 misclassified patterns). It is interesting to consider a result of Drago and Ridella [14] concerning a measure of the relative input saliency obtained via the application of the method described in [15]. The joint use of the two more relevant features, namely the sixth and the first ones, allowed them to achieve an error percentage of 3.95% through the training on the whole data set.

The search for a similar rule among the four listed above produces two simple disjoint quantitative thresholds for the variables $x_1 \leq 8$ and $x_6 \leq 4$. This leads to a generalization error of 7.3% when the learning is performed on the training set only. However, this single rule is not the best possible one; the addition of the three most recurrent other conditions $x_4 \leq 9$, $x_5 \leq 5$ and $x_8 \leq 8$ reduces the error percentage to 4.4%.

References

[1] R. M. Goodman, C. M. Higgins, J. W. Miller, and P. Smyth, Rule-based neural networks for classification and probability estimation. *Neural Computation*, **4** (1992), 781–804.

[2] K.-P. Huber and M. R. Berthold, Building precise classifiers with automatic rule extraction. in *Proceedings of the IEEE International Conference on Neural Networks*, Perth, 1995, III–1263–1268.

[3] G. Towell and J. W. Shavlik, Interpretation of artificial neural networks: Mapping knowledge-based neural networks into rules. in *Advances*

in Neural Information Processing Systems 4, J. E. Moody, S. J. Hanson, and R. P. Lippmann, eds., San Mateo, CA: Morgan Kaufmann, 1992, 977–984.

[4] D. L. DIETMEYER, *Logical Design of Digital Systems (third edition)*. Boston, MA: Allyn and Bacon, 1988.

[5] R. K. BRAYTON, G. D. HACHTEL, C. T. MCMULLEN, AND A. L. SANGIOVANNI-VINCENTELLI. *Logic Minimization Algorithms for VLSI Synthesis*, Hingham, MA: Kluwer Academic Publishers, 1984.

[6] S. J. HONG, R. G. CAIN, AND D. L. OSTAPKO, MINI: A heuristic approach for logic minimization. *IBM Journal of Research and Development*, **18** (1974), 443–458.

[7] H. W. GSCHWIND AND E. J. MCCLUSKEY, *Design of Digital Computers*, New York: Springer-Verlag, 1975.

[8] M. MUSELLI AND D. LIBERATI, Training digital circuits with Hamming Clustering. Submitted to *IEEE Transactions on Circuit and Systems*, (1998).

[9] M. MUSELLI, Building neural and logical networks with Hamming Clustering. Accepted as a contribution to *WIRN'99 - The 11-th Italian Workshop on Neural Nets*, (1999).

[10] S. THRUN ET AL., A performance comparison of different learning algorithms, *Technical Report CMU-CS-91-197*, Pittsburgh, PA: Department of Computer Science, Carnegie Mellow University, 1991.

[11] C. J. MERZ AND P. M. MURPHY, UCI repository of machine learning databases [http://www.ics.uci.edu/~mlearn/MLRepository.html]. Irvine, CA: University of California, Department of Information and Computer Science, 1996.

[12] R. O. DUDA AND P. E. HART, *Pattern Classification and Scene Analysis*, New York: John Wiley and Sons, 1973.

[13] C. M. BISHOP, *Neural Networks for Pattern Recognition*, New York: Oxford University Press, 1995.

[14] G. P. DRAGO AND S. RIDELLA, Pruning with interval arithmetic perceptron, *Neurocomputing*, **18** (1998), 229–246.

[15] J. M. STEPPE AND K. W. BAUER, Improved feature screening in feedforward neural networks, *Neurocomputing*, **13** (1996), 47–58.

From Spiking Neurons to Dynamic Perceptrons

Francesco Palmieri* Antonella Luongo * Andrew Moiseff**

* Dipartimento di Ingegneria Elettronica e delle Telecomunicazioni
Università degli Studi di Napoli Federico II
Via Claudio, 21, 80125, Napoli, Italy

** Department of Physiology and Neurobiology,
The University of Connecticut,
Storrs, CT 06269-4156

Abstract

This paper begins a systematic validation for a simple and reliable artificial neural network model that can be directly related to the main behaviour of biological neural networks. The sigmoid-plus-linear filter appears to be a promising candidate if the sigmoidal function is calculated in reference to the pulse generation refractory effects. We directly compare a classical spiking neuron model with a scheme based on a sigmoidal function plus a linear filter. The filter is computed as the best least squares fit to the output of the spiking model. The results seem to confirm that FIR and IIR neural networks may be able to represent the essence of the signal processing performed by biological neurons.

Introduction

In Artificial Neural Networks (ANN), the classical linear combination-plus-sigmoid neuron model provides a reasonable link to the average firing rate of pulsed neurons. Unfortunately it does not include time and it seems to be too extrapolated to be directly applicable to modeling measured responses from biological neurons. The question is then: is there a reasonably simple model for ANNs that can capture most of the essence of the information processing typical of Biological Neural Networks (BNN) ? For example: do neural spikes carry information about a distribution or are a fine code of time-varying information? [4]. What is the scale at which such information is transported? Given two different spike patterns scoring the same average firing rate, do they carry the same information signal?

We have started considering the classical description of a pulsing neuron with a threshold and a refractory effect and related it to the emergence of the sigmoidal function in the ANN counterpart when excited with a step function. ANNs with sigmoids and linear filters have been proposed as a likely model for BNNs [10], but a direct link with the firing mechanism, to our knowledge,

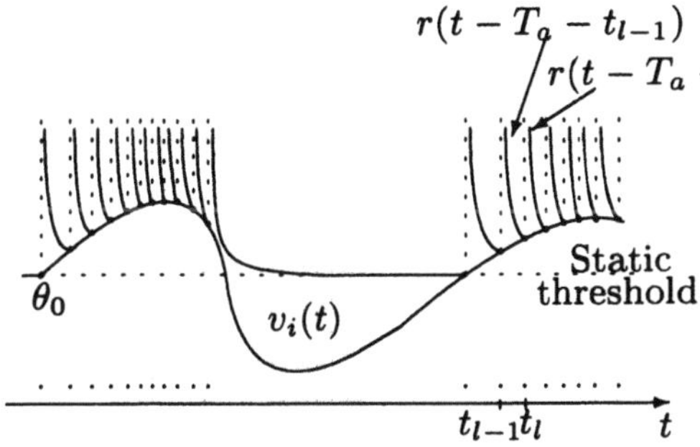

Figure 1: The spiking process with the refractory effect

has never been explicitly considered. Also, FIR and IIR neural networks have been proposed in the ANN literature [1, 6] proving to be a very promising generalization of static perceptrons specially in applications in which time plays a critical role (speech recognition, time series prediction, etc.). Our work points into this direction showing how such a model may very well be also a rather faithful way of capturing the essence of the information processing carried out by pulsing BNNs.

Neural Sampling

Suppose that we are modeling the action of synaptic connection between neuron i and neuron j. Let us name neuron i membrane potential $v_i(t)$ and $u_{ij}(t)$ its contribution to membrane potential $v_j(t)$.

The spike initiation process is such that when, starting from a resting state, $v_i(t)$ goes above a threshold θ_0, an action potential is generated. The neuron then enters an *absolute refractory period* T_a within which any further spike generation is inhibited. Subsequently, a *relative refractory phase* is initiated with a varying threshold following a pre-defined time evolution $r(t)$. Hence, if the potential $v_i(t)$ is greater than $\theta_0 + r(t - T_a)$ a new spike is generated and the whole process repeats. The model is depicted in Figure 1. The result is a spike train with occurrence times described by a sequence $\{t_l\}$ that carry the information about $v_i(t)$. Such a simple model has been originally proposed by Lapique that considered only an absolute refractory interval T_a and subsequently improved with a number of different forms for the refractory function (see Figure 2, first five rows and first two columns) [11]. Such a scenario seems to be a reasonable description of what is measured from biological neurons. Suppose now that voltage $v_i(t)$ is a step function $v_i(t) = v\ u(t)$. If $v > \theta_0$ the neuron will fire regularly with a period T solution of the equation $\theta_0 + r(T - T_a) = v$. Since $r(t)$ is a monotonic function we solve to get the firing rate:

$$\frac{1}{T} = \frac{1}{T_a + r^{-1}(v - \theta_0)} = \varphi(v). \tag{1}$$

This is the *sigmoidal function* $\varphi(v)$ that relates the input voltage v to the firing rate $1/T$. The first five rows of Figure 2 show the sigmoidal functions corresponding to various classical choices of the refractory function $r(t)$. They all begin from the static threshold θ_0 and approach asymptotically the maximum firing rate $1/T_a$. We have also calculated what would be the refractory function if we had already chosen a sigmoid $\varphi(v)$. The relation is

$$r(t) = \varphi^{-1}\left(\frac{1}{t+T_a}\right) - \theta_0. \tag{2}$$

The last two rows of Figure 2 show the refractory function that would correspond to a logistic and to a piecewise linear sigmoidal function. Note how the logistic refractory function appears to be quite unrealistic since tends to $-\infty$ as t increases. This would allow firing even if the voltage v is below the static threshold θ_0. Looking at Figure 1 we note that the refractory effect generates a non uniform sampling of $v_i(t)$. The pulses are denser in the intervals in which $v_i(t)$ is larger. We refer to this type of nonuniform coding of $v_i(t)$ as *neural sampling*. This because, if $v_i(t) > \theta_0$ from knowledge of the refractory function, and the pulse occurrence times $\{t_l\}$, exact reconstruction of $\{v_i(t_l)\}$ is possible. The existence of a reconstruction strategy for $v_i(t)$ is obviously limited to the time intervals in which $v_i(t) > \theta_0$ and to a criterion on maximum allowed variation for the signal. The modulation process is highly non linear and shows a saturation effect by allowing a minimum spacing between pulses equal to T_a. A reasonable heuristic criterion to asses what is preserved of $u(v_i(t) - \theta_0)$ after sampling, could be to assume that $u(v_i(t) - \theta_0)$'s bandwidth B is below the maximum resolution of the sampler, i.e. $B < \frac{1}{2T_a}$. This does not seem to be too far fetched because $v_i(t)$ is supposedly the result of other integrated spikes that must satisfy the same time resolution constraint. A different scenario may emerge if $B < \frac{1}{2T_a}$ in which neural sampling does not capture all the variations of $u(v_i(t) - \theta_0)$ (aliasing) even though the samples maintain exact time information. This may be a likely model for some neurons of the first stages of the auditory system in which phase information is preserved even though the time resolution would allow only an aliased (decimated) version of $u(v_i(t) - \theta_0)$. We disregard this scenario in this paper leaving it for somewhere else, and concentrate on the former one.

It is typical in analog pulse modulation (i.e. Pulse Duration Modulation, PDM) studied in communication systems [2], to use a low-pass filter to get back a reasonable version of the modulating signal. Synaptic integration at the $i-j$ junction could play such a role to form the contribution of neuron i to $v_j(t)$. In the simplest scenario such a contribution could be just a scaled version of the saturated $u(v_i(t) - \theta_0)$. However, the type of synaptic function and its branch configuration, simply through delay compositions, could also perform linear filtering.

This leads to the scheme of Figure 3 (upper branch) in which the contribution of neuron i to neuron j could be globally described as $u_{ij}(t) = \sum_l h_{ij}(t - t_l)$, where $h_{ij}(t)$ includes pulse shape, axon transmission, synaptic integration and post-synaptic propagation.

Refractory funct. $r(t-T_a)$	$r(t)$	Sigmoidal function $\varphi(v)$	$\varphi(v)$
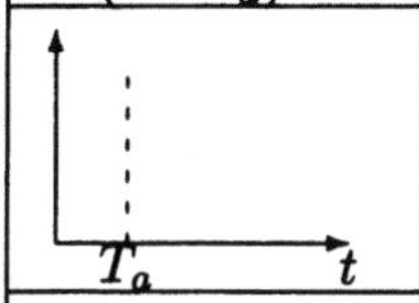	Lapicque	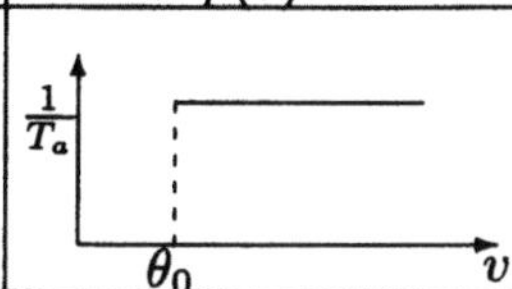	$u(v-\theta_0)$
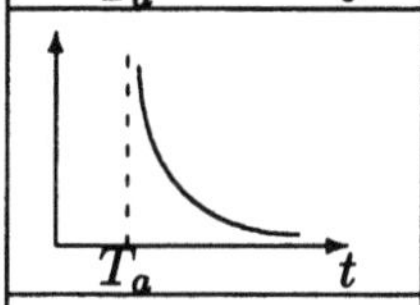	Buller $\frac{A}{t}\,u(t)$	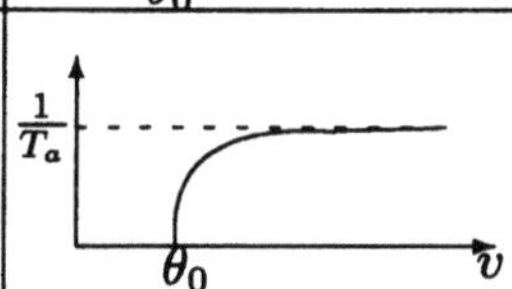	$\frac{1}{T_a+\frac{A}{v-\theta_0}}\,u(v-\theta_0)$
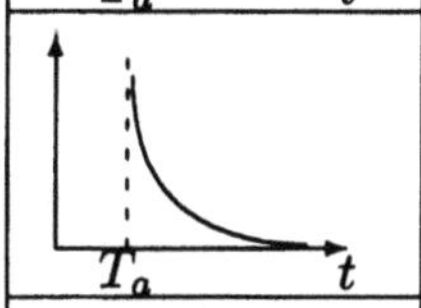	Geisler and Goldberg $\frac{A}{e^{t/\tau}-1}\,u(t)$	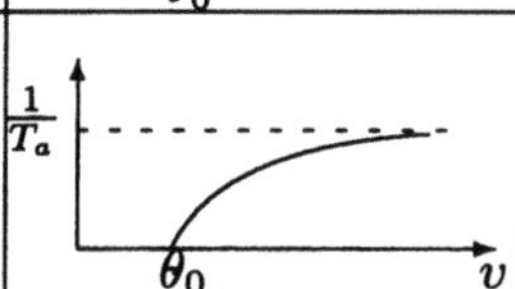	$\frac{1}{T_a+\tau\, ln(1+\frac{A}{v-\theta_0})} \cdot u(v-\theta_0)$
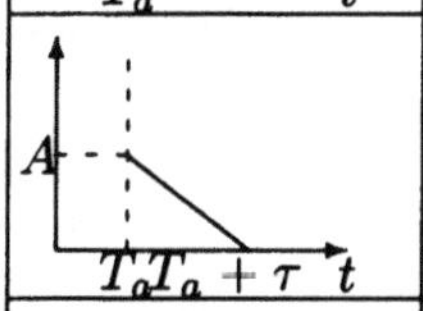	Calvin and Stevens $A(1-\frac{t}{\tau})(u(t)-u(t-\tau))$	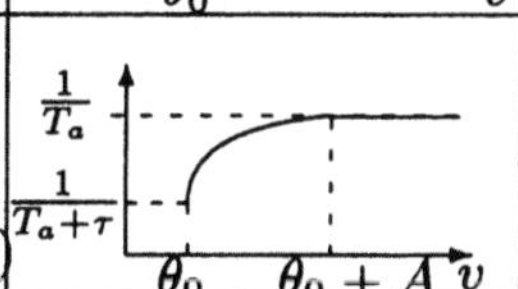	$\begin{cases} 0, & v \le \theta_0 \\ \frac{1}{T_a+\tau\,(1-\frac{v-\theta_0}{A})}, & v \in (\theta_0, \theta_0+A) \\ \frac{1}{T_a}, & v \ge \theta_0+A \end{cases}$
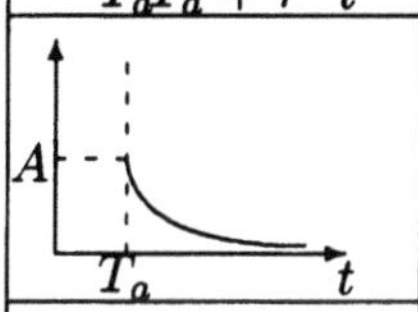	Weiss $A\,e^{-t/\tau}u(t)$	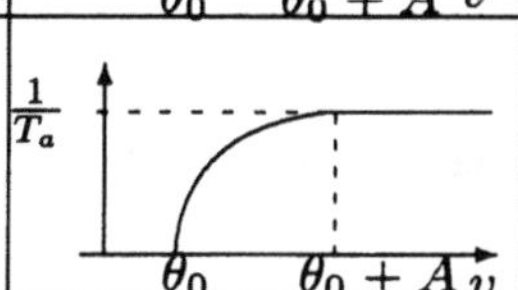	$\begin{cases} 0, & v \le \theta_0 \\ \frac{1}{T_a+\tau ln(\frac{A}{v-\theta_0})}, & v \in (\theta_0, A+\theta_0) \\ \frac{1}{T_a}, & v \ge A+\theta_0 \end{cases}$
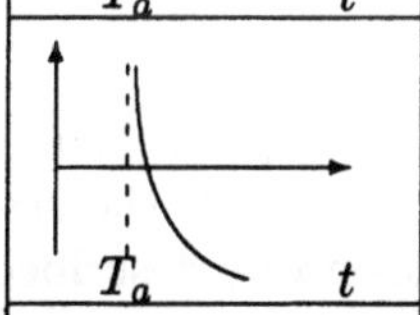	logistic $A\,ln(\frac{T_a}{t})u(t)$	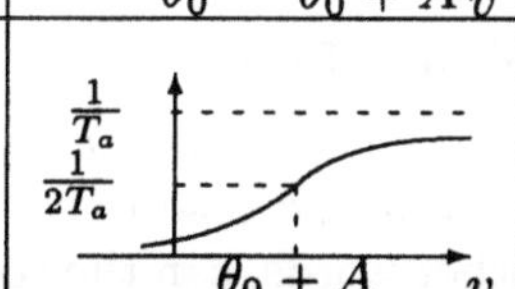	$\frac{1}{T_a\left(1+e^{-\frac{v-\theta_0}{A}}\right)}$
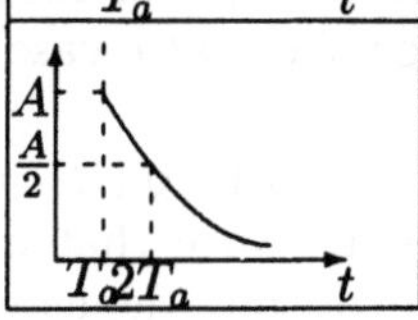	piecewise-linear $A\,\frac{T_a}{t+T_a}\,u(t)$	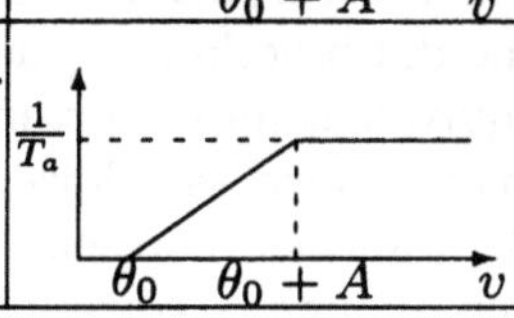	$\begin{cases} 0, & v \le \theta_0 \\ \frac{1}{T_a}\frac{v-\theta_0}{A}, & v \in (\theta_0, \theta_0+A) \\ \frac{1}{T_a}, & v \ge \theta_0+A \end{cases}$

Figure 2: The correspondence between refractory functions and sigmoids

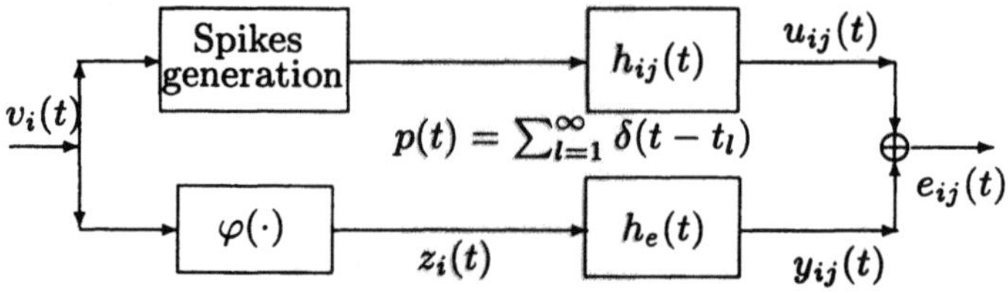

Figure 3: Block diagram showing the spike transmission and the equivalent model.

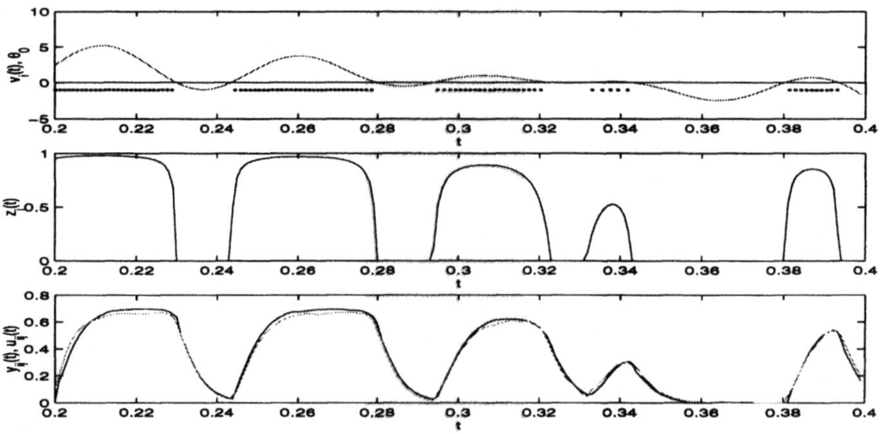

Figure 4: The simulation results.

A Dynamic Equivalent Model

The question is now, can we draw an equivalence with a simple dynamic system that includes the static case (with constant firing rate) with the sigmoidal function? In other words, can the contribution of neuron i be described by a memoryless sigmoidal function φ (the same derived for the static case) followed by a linear equivalent impulse response $h_e(t)$?

Such an hypothesis is drawn as a lower branch in Figure 3 with $y_{ij}(t) = \varphi(v_i(t)) * h_e(t)$. Both branches represent a nonlinear system. Note that in the lower branch, even though a convolution appears, the output depends nonlinearly on the input magnitude. Given an equivalent integral response $h_{ij}(t)$, can we find an $h_e(t)$ to get an output $y_{ij}(t)$ sufficiently close to $u_{ij}(t)$? This would validate the hypothesis that the effect of neuron j to neuron j can be described by a sigmoidal function plus a linear filter also in a time-varying scenario.

We have performed extensive simulations with various input signals ranging from pure sinusoids to bandpassed noise taking as a typical example $h_{ij}(t) = Kt\, e^{\frac{-t}{\tau}} u(t)$ and found the least squares equivalent filter $h_e(t)$. The results show consistently a quite good agreement between the two branches. It is crucial in this equivalence that the sigmoidal function φ is the one computed from the refractory function used in the generation of the pulse trains. Figure 4 shows a 0.2 sec segment of one of our simulations based on a synthetic signal $v_i(t)$ of about 2 secs. The neural spikes have been generated taking $\theta_0 = 0$, Buller's refractory function and $T_a = 1$ msec. Figure 4a shows the original signal and the corresponding firing points. Figure 4b reports $z_i(t) = \varphi(v_i(t))$. Note the rectification and the saturation effects. Figure 4c shows the output of our best linear filter $h_e(t)$ applied to $z_i(t)$ and the signal $u_{ij}(t)$. The agreement between the two is excellent both for saturated and midrange values. The simulations have been performed on sampled signals. The sampling interval for $v_i(t)$ is $T_{ss} = 10^{-4}$ sec. With this resolution the spikes $\{t_l\}$ are generated and convolved with $h_{ij}(t)$ with $\tau = 3$msec. The convolution is obtained as the exact superposition of the impulse responses $h_{ij}(t)$ placed on the computed points $\{t_l\}$. The result is sampled at intervals of $T_s = 1$ *msec*. In the lower branch $z_i(t)$ is also sampled with $T_s = 1$ *msec* and it is passed into an FIR filter with 30 taps. The FIR coefficients have been obtained by solving a standard normal equation [5].

References

[1] A.D. Back and A.C. Tsoi. Fir and iir synapses: a new neural network architecture for time-series modeling. *Neural Computation*, 3:375–385, 1991.

[2] A.B. Carlson. *Communication Systems.* Prentice-Hall, 1986. Third Edition.

[3] C.E. Carr and M. Konishi. A circuit for the detection of intereaural time differences in the brain stem of the barn owl. *J. Neuroscience*, 10:3227–3246, 1990.

[4] D. Ferster and N. Spruston. Cracking the neuronal code. *Science*, 270:756–757, November 1995.

[5] S. Haykin. *Adaptive Filter Theory.* Prentice-Hall, New York, 1996. Third Edition.

[6] S. Haykin. *Neural Networks : A Comprehensive Foundation.* Prentice-Hall, New York, 1998. Second Edition.

[7] C. Koch and I. Seger. *Methods in Neural Modeling.* MIT Press, Cambridge, MA, 1989.

[8] W. Maas. Networks of spiking neurons: the third generation of neural network models. *Neural Networks*, 10(9):1659–1671, 1997.

[9] A. Moiseff and M. Konishi. Binaural characteristics of units in the owl's brainstem auditory pathway: Precursors of restrictive spatial receptive fields. *J. Neuroscience*, 3:2353–2362, 1983.

[10] S.A. Shamma. Spatial and temporal processing in central auditory systems. In C. Koch and I. Seger, editors, *Methods in Neural Modeling*, pages 247–289. MIT Press, Cambridge, MA, 1989.

[11] H.C. Tuckwell. *Introduction to Theoretical Neurobiology: Nonlinear and Stochastic Theories*, volume 2. 1988.

Using the Hermite Regression Algorithm to Improve the Generalization Capability of a Neural Network

G.Pilato, F.Sorbello and G.Vassallo*

Department of Electrical Engineering, University of Palermo
Palermo, Italy

**Centro di Ricerche Elettroniche in Sicilia, CRES*
Monreale, Italy

Abstract

In this paper it is shown that the ability of classification and the ability of approximating a function are correlated to the value (in the training points) of the gradient of the output function learned by the network.
It has been designed a feedforward neural network ("αNet") that makes use of the Hermite function regression formula to approximate its hidden unit activation functions; the Hermite algorithm leads to a smooth function approximation, hence it is obtained a low value of the gradient of the output function of each output unit.
Experimental results, concerning the 5-parity and the two-spiral problems, show the better skill of classification and interpolation of αNet with respect to a traditional feedforward network that uses sigmoids as activation function of the hidden units.

1. Introduction

The traditional feedforward network architecture uses, for its hidden units, sigmoidal activation function; the sigmoid is a monotone non-decreasing differentiable function with a very simple first derivative form, however it neither interpolate nor extrapolate efficiently in a lot of regression applications[2,3]. To overcome this problem a lot of studies have been done to improve the choice of nonlinear activation functions[5,6,7].

In the field of handwritten characters, Le Cun et al[8] have demonstrated that the recognition capability can be improved keeping low the gradient of the output functions along the curve corresponding to the rotation applied to the training set patterns.

In a previous work[1] it has been devised a method to keep low the value of the gradient of the output function associated at any neural unit of the output layer: this technique has shown to increase the recognition rate significantly.

A technique capable to smooth approximate unknown functions is that one used in the Projection Pursuit Learning Network[2,3]; in that work to better the statistical performance of the PPL[9], a orthogonal polynomial approximation is used leading to a smooth approximation function.

In this paper, starting from PPLN, it has been designed a feedforward neural architecture, αNet, in which the activation function of the hidden units is chosen before starting the learning phase as the approximation of well-known activation functions; this approximation is realized using the regression algorithm that makes use of the orthonormal Hermite polynomial functions.

The choice of this particular algorithm is due to the fact that from the obtained function with the Hermite regression algorithm it is possible to calculate easily and accurately the values of the approximated function and the first derivative, but the most important thing is that it leads to a smooth interpolation[2,3].

In the next section it is described the proposed solution to obtain a low gradient of the output function of any unit of the output layer in the nearby of the training samples; experimental results and conclusions are then given.

2. The proposed solution

The regularity of the output functions plays an essential role for the generalization capability of a neural network[1]; furthermore, since it is reasonable to guess, as a general assumption, that at very close inputs should correspond very close outputs, it can be supposed that the approximation capability is strictly related to the classification capability.

Let **O** be the input-output relationship learned by a feedforward neural network; let us consider two multidimensional points identified by the vectors **p** and **p'** where **p** is the pattern used for the network training and **p'** is a new pattern to classify (or whose output, obtained applying the function **O**, has to be approximated): it is supposed that **p'** is very close to **p**; in this case, considering the closeness of **p** and **p'** and supposing to linearize the function **O** in the nearby of **p**, it can be defined the error of generalization as:

$$\Delta E \equiv \nabla \mathbf{O}\big|_{\mathbf{p}} \cdot (\mathbf{p} - \mathbf{p}')$$

where: $\nabla \mathbf{O}|_{\mathbf{p}}$ is the value of the learned output function gradient computed in the point **p**; (**p-p'**) is the difference vector between the two vectors **p** and **p'**; "·" denotes the inner product.

To obtain a good classification ability it is necessary therefore to minimize the generalization error ΔE, however the distance $\|\mathbf{p}-\mathbf{p}'\|$ (that is correlated to the inner product) cannot be reduced, since the pattern **p** was used for the network training and the pattern **p'** is the element to classify (or whose output, obtained applying the function **O**, has to be approximated); as a consequence it is necessary to reduce the quantity $\nabla O|_p$, i.e. to keep the gradient of the output function near the training point as low as possible.

It has been designed a feed-forward network ("αNet") with a good generalization capability keeping low value of the gradient, in the nearby of the training points, of the output function of any unit belonging to the output layer[1].

Quality Factors

In order to measure the number of connections between the network units it can be defined the *Computational Cost of the Production Phase* $\mathbf{C_C}$:

$$C_C \equiv H \cdot (N + M)$$

where H is the number of hidden units, N is the number of the input units and M is the number of output units of the network.

In the same way, to measure the average gradient, in the nearby of the training points, of the output function of any unit belonging to the output layer it can be defined the *Gradient Cost* $\mathbf{C_G}$:

$$C_G \equiv \frac{1}{P}\sum_{r=1}^{P}\left(\sum_{k=1}^{M}\left|\nabla u_{kl}\right|_{\mathbf{x}_r}^{2}\right)$$

where u_k is the k-th output of the network, $\mathbf{x}_r$ is the r-th training pattern and P is the number of input-output patterns of the training set,

Clearly, it is necessary to keep the quantities C_C and C_G as low as possible. Hence it has been chosen to use an architecture where its hidden units have the activation function $f^*(x)$ given by the following formula[2,3]:

$$f^*(x) = \sum_{r=1}^{R} c_r^* h_r(x)$$

where h_r are the first R orthonormal functions of Hermite calculated in the point x and the coefficients c_r^* are the expansion coefficients of the Hermite approximation algorithm[2,3]; the weights c_r^* are chosen before starting the training phase so that $f^*(x)$ results the approximation of a well-known activation function.

It has been used this approximation function instead of another approximation algorithm in order to obtain a smooth representation of the output function of any unit belonging to the output layer; besides the Hermite orthonormal functions and their first derivatives can be easily calculated in a recursive manner[2,3].

3. Experimental Results

To test the skill of classification and interpolation of αNet, a full C-code simulator has been developed and in this paragraph the results obtained with two well-known benchmarks are reported: the 5-parity problem and the two-spirals problem.

For each experiment two networks have been compared. The former is a traditional feedforward network with sigmoid as activation functions of the hidden units that uses the CGRD algorithm with the restart conditions of Powell[10] to reach minimum of the error over the training set. The latter is the αNet, using the Hermite regression algorithm to approximate the activation functions belonging to the hidden layer; in particular the following activation functions have been approximated using the first 6 Hermite polynomials:

- a gaussian function with zero mean and unitary variance
- a sinus function

the network uses the same algorithm to reach minimum of the error (i.e. the CGRD algorithm with the restart conditions of Powell).

It can be defined the error obtained over the training set as:

$$Q_E \equiv \frac{1}{P}\sum_{i=1}^{P}\sum_{j=1}^{M}\left(t_j^{(i)} - o_j^{(i)}\right)^2$$

where P is the number of patterns that form the training set, $t_j^{(i)}$ is the j-th element of the i-th output vector of the training set and $o_j^{(i)}$ is the j-th output of the network corresponding to the i-th pair of input-output patterns of the training set.

The value of Q_E should be close to zero for both the networks that have to be compared, then the factors C_C and C_G are evaluated.

Experiments have also be done using the "true" gaussian and sinus function; however, the "Hermite approximated" and "true" functions are different, hence they are not comparable each other. The obtained results confirm the above sentence. Nevertheless the choice of the Hermite polynomials allows us to "compose" the hidden activation functions.

3.1 The 5-Parity Problem

The 5-parity problem consists in determining the minimum number of hidden nodes to let the hidden network learn the 2^5=32 possible configurations of the training set; in general, for the N-bit parity function, no less than (1+N/2) hidden-unit sigmoid functions are required[11].

The following results, based on an average of 10 starting configurations of the connection weights between the units of each network, have been obtained:

	Q_E	C_G	C_C
Traditional network 5-6-1	0	297.97	36
αNet 5-4-1 with Gaussian Approximation	0	44.33	24
αNet 5-3-1 with Sinus Approximation	0	69.44	18

Lower values of C_G using the αNet architecture having Gaussian or Sinus approximated hidden activation function have been obtained; αNet, besides, is able to learn the input/output relationship with just 4 or 3 hidden unit instead of the 6 units required by the traditional network using fixed sigmoids.

3.2 The Two Spirals Problem

The problem of the two-spirals consists in classifying the points in the x-y plane belonging to one of two interlocked spirals. The training set consists of 192 elements half of which are classified as "o" and the other half are classified as "x" (see fig.1A) This benchmark is not linearly separable and is extremely hard to resolve for a traditional multi-layer backpropagation network [12].

Here, to point out the correlation between interpolation and classification ability of a neural network, the problem is treated as interpolation: the output can assume the values +1 if the input pair belongs to the "x" spiral and -1 if the input pair belongs to the "o" spiral.

The following results, based on an average of 10 starting configurations of the connection weights between the units of each network have been obtained:

	Q_E	C_G	C_C
Traditional network 2-60-1	0.02	327.9	180
αNet 2-40-1 with Gaussian Approximation	0.02	102.63	120
αNet 2-36-1 with Sinus Approximation	0.02	67.31	108

αNet is able to learn the input/output relationship with just 40 or 36 hidden unit instead of the 60 units required by the traditional network using fixed sigmoids. Besides, lower values of C_G using αNet than using a traditional network with sigmoids have been obtained. This result leads to the supposition that αNet should have a better skill in classifying the two spirals than the traditional architecture; this assumption is confirmed looking at fig 1(B-D) where three samples of results obtained with both architectures are shown.

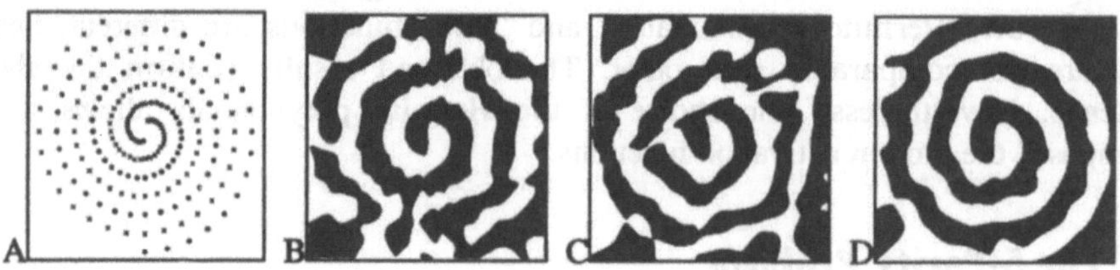

Figure 1: In A it is represented the training set for the two-spirals problem, in B it is shown the result obtained with a traditional feedforward network 2-60-1; in C it is represented the result obtained with the αNet 2-40-1 network with gaussian approximation while in D it is represented the result obtained with the αNet 2-36-1 network with sinus approximation

4. Conclusions

In this work it has been proposed that the ability of classification and the ability of approximating a function are correlated by the value (in the training points) of the gradient of the output function learned by the network after the training phase.

A feedforward neural network ("αNet") has been designed in which the activation function of the hidden units is chosen before starting the learning phase as the approximation, via the Hermite function regression formula, of one activation function well-known in literature.

Experimental results show that αNet has a better skill of classification and interpolation requiring less hidden units of a traditional feedforward network that uses a sigmoid as activation function of the hidden units. Besides, the possibility to "compose" the activation function of the hidden units allows focusing further optimization techniques.

References

[1] G.A.M.Gioiello, A.Tarantino, F.Sorbello, G.Vassallo, Simple techniques for an efficient recognition of handwritten characters using a MLP, The Journal of Intelligent Systems Vol 6 No 3/4 199-221

[2] J. N.Hwang, S. R.Lay, M.Meachler, D.Martin, J.Schimert, Regression Modeling in Back-Propagation and Projection Pursuit Learning, IEEE Trans. Neural Networks Vol.5 No 3 pp342-353, 1994

[3] J. N.Hwang, S.You, S. R.Lay, I.C.Jou, What's Wrong with a Cascaded Correlation Learning Network: A Projection Pursuit Learning Perspective, ftp.cis.ohio.state.edu/pub/neuroprose/

[4] K.Hornik, Multilayer Feedforward Networks are Universal Approximators, Neural Networks, 1989, **2**, 359-366

[5] B.Kosko, Neural Networks and Fuzzy Systems:A Dynamical Systems Approach to Machine Intelligence, Prentice Hall Englewood Cliffs NJ, 1991

[6] R.D.Jones, Y.C.Lee, C.W.Barnes , G.W.Flake, K.Lee, P.S.Lewis, S.Qian, Functional approximation and time series prediction with neural networks, Proc Int'l Jont Conf. on Neural Networks pp.649-665, 1990

[7] J.Moody, C.J.Darken, Fast Learning in Networks of Locally Tuned Processing Units, Neural Computations Vol 1 No 1 PP 281-294, 1989

[8] P.Simard, Y.Le Cun, J.Denker, B.Victorri, An Efficient Algorithm for Learning Invariances in Adaptative Classifiers, Proc of 11th IAPR International Conference on Pattern Recognition, 651-655, 1994

[9] J.H.Friedman, W.Stuetzle, Projection Pursuit Regression, Journal of the American Statistical Association 76 (376) 817-823, 1981

[10] M.J.D.Powell, Restart Procedures for the Conjugate Gradient Method, Mathematical Programming Vol 12 pp 241-254

[11] Minor, Parity With Two Layer Feed-Forward Nets, Neural Networks Vol 6 No 5 pp 705-707, 1993

[12] S.E.Fahlman, C.Lebiere, The Cascade-Correlation Learning Architecture, Advances in Neural Information Processing Systems, D.S. Touretzky ed. vol. 2 pp.524-532, San Mateo Calif: Morgan Kaufmann 1990

Development of Selectivity Maps in a BCM Network Using Various Connectivity Schemes

D. Remondini
Physics Department, Bologna University
Bologna, Italy

G. C. Castellani
Physics Department, Bologna University
Bologna, Italy

A. Bazzani
Physics Department, Bologna University
Bologna, Italy

R. Campanini
Physics Department, Bologna University
Bologna, Italy

F. Bersani
Physics Department, Bologna University
Bologna, Italy

Abstract

This paper is centered on the analysis of lateral connections in a network of neurons following BCM synaptic modification theory. This is a non-supervised algorithm, based on properties inspired by the behaviour of real neurons in hyppocampus and visual cortex, that allows synaptic strength to increase or decrease according to a time-varying threshold based on the previous "history" of neural activity. From a statistical point of view it represents a Projection Pursuit (PP) method based on an Objective Function that seeks "interesting" projections in input space, along those directions presenting statistical properties far from normal (gaussian). After a short introduction to BCM theory, we test various schemes of lateral connectivity between neurons receiving the same inputs, using assumptions inspired by the anatomical and functional properties of visual cortex of evolved mammals (like primates and cats). Using computer simulations, the different schemes (uniform, random, gaussian, coulombian, linearly and exponentially decaying connections) were compared in relation to their ability to improve selectivity and to form a "metric", namely to preserve the correspondence between close inputs and the topology of the corresponding activated neurons, in the case of linearly dependent inputs.

Unsupervised neural networks and feature extraction

Neural networks play a very important role in tasks like data clustering, classification and feature extraction.
Extracting features means searching for a restricted number of parameters (a space with very low dimensionality compared to the input space) that keep some relevant characteristics of the inputs, allowing a discrimination between them. Many statistical methods pursue this task in a very efficient way, with the so called "supervised methods" that try to loose as little information as possible. Unfortunately, in presence of very high dimensional input spaces they are subject to the problem known as "curse of dimensionality" (Intrator and Cooper, 1992). Moreover, supervised methods, during the training process, can fall in local minima far from the best solution, right because they keep too much information, even less "interesting" details.
A key point is to search for what is "interesting" to analyze for our purposes, even if we loose a faithful representation of our data. In this context unsupervised neural networks provide a very powerful tool. The BCM algorithm, when used for extracting features (Intrator and Cooper, 1992), belongs to a class of unsupervised methods called Exploratory Projection Pursuit (Huber, 1985; Friedman, 1987), that seeks for interesting projections in the input space. The notion of interesting projection derives from considerations on high dimensional spaces: due to Central Limit Theorem, most low-dimensional projections in this space are approximately normal. This suggests that most of the important features in our inputs lie in those directions whose single dimensional projected distribution is far from Gaussian. The extraction of non Gaussian characteristics is well performed by high order polinomial moments, although they have the drawback to be more sensitive to outliers. In the next section BCM rule will be shown to have an objective function based on high order moments.

BCM rule and its statistical properties

BCM neurons are characterized by the rule for updating weights, as firstly proposed in (Bienenstock, Cooper, Munro, 1982) .
Given an input vector $\vec{d}$ and a weight vector $\vec{m}$, the output of the neuron is given by:

$$C = \sigma(\vec{m} \cdot \vec{d})$$

where $\sigma(x)$ is a sigmoidal function, or simply by the scalar product:

$$C = \vec{m} \cdot \vec{d}$$

The weight modification rule is given by:

$$\dot{m}_i(t) = \Phi(C(t), \Theta(t)) d_i(t);$$

$\Phi(C(t), \Theta(t))$ a function such that $\Phi(0, \Theta(t)) = 0$ and $\Phi(\Theta(t), \Theta(t)) = 0$, $\Phi < 0$ for $c < \Theta$ and $\Phi > 0$ for $c > \Theta$. This function can be represented in a simple form by a second order polinomial:

$$\Phi = C(C - \Theta);$$

where$\Theta(t) = \frac{1}{\tau} \int_{-\infty}^{t} e^{\frac{s-t}{\tau}} C(s)^2 ds$, that represents the memory of the previous neuron activity weighted by a time exponential decay term.

This rule for weight updating can be seen as a method for risk minimization. Given an Objective Function $\hat{\Phi} = C^2 - \frac{1}{2} C\Theta$, with $\Theta = E[C^2]$, where E[] means "expected value of", we consider a Loss Function for every presented input $\vec{d}$:

$$L_m(\vec{d}) = -\mu \int_0^{(\vec{m}\cdot\vec{d})} \hat{\Phi}(s) ds = -\mu\{\frac{1}{3}(\vec{m} \cdot \vec{d})^3 - \frac{1}{4} E[(\vec{m} \cdot \vec{d})^2](\vec{m} \cdot \vec{d})^2\};$$

(where L_m represents a function depending on $\vec{m}$). The Risk Function is given by the expected value of the Loss Function for all the inputs:

$$R_m = -\mu E[L_m(\vec{d})] = -\mu\{\frac{1}{3} E[(\vec{m} \cdot \vec{d})^3] - \frac{1}{4} E^2[(\vec{m} \cdot \vec{d})^2]\}$$

Now the optimal projections in our input space (with respect to the initial objective function) are obtained minimizing the risk function via a gradient descent method (with respect to $\vec{m}$):

$$\dot{m}_i = -\frac{\partial}{\partial m_i} R_m = \mu E\{\Phi(C, \Theta) \cdot d_i\},$$

which is an averaged version of BCM rule. It can be demonstrated that the two algorithms converge to the same fixed points. Analogous results can be obtained when considering a sigmoidal transfer function: its effect is a reduction of sensitivity to outliers that could in fact be enhanced by the high order of polynomial moments used in the BCM algorithm.

For BCM algorithm, under general hypoteses, we can obtain the fixed points and determine their stability (Intrator & Cooper, 1992; ,Castellani et al., 1999). We can assume ergodic conditions and replace the time average for calculating Θ (if performed over a sufficiently long period) with the inputs expectation value. Given a set of linearly indepedent inputs, we can demonstrate that the stable states are those with maximum selectivity, where each neuron presents maximum response for only one input.

Role of connectivity: models and simulation

The previous discussion referred to the case of a single neuron. When more neurons receiving the same signals are considered we can obtain different projections in input space. How can we influence this multiple parallel search in a way that makes it more efficient?

The problem can be addressed introducing lateral connections, that allow comunication between neurons belonging to the same "layer" (in analogy with the connectivity structures found in animal visual cortex). The effect of lateral connections in a BCM network has been studied via computer simulations (Remondini, 1997),in a mean field approximation (Cooper and Scofield, 1988) or analytically (Castellani et al., 1999). For neurons receiving linearly independent inputs, we are able to describe the effect of lateral connections on the fixed points and their stability. An important assumption is that we deal with connections producing a little effect compared to the afferences, that is:

$$C_i(t) = \vec{m}(t) \cdot \vec{d}(t) + \sum_j L_{ij} \cdot C_j(t), \quad L_{ij} << 1.$$

The stable states of such network are the same as in the case of single neuron. Only the boundaries between basins of attraction result translated with respect to the origin. However, this shift has a relevant effect in network learning, since we always start with weights close to the origin ($m_i << 1$), so that different configurations are reached depending on the initial conditions. Excitatory connections between two neurons will lead them to the same stable state (associative behaviour), whereas inhibitory connections will produce competition and consequently a differentiation of tuning.

In the present paper we deal with linearly dependent inputs, in the form of strongly overlapping gaussian signals. A behaviour similar to that of independent inputs is expected, because we can think to our gaussian inputs as δ-like inputs with the addition of "tails" that make them overlap (for weak overlapping we could think to linear dependence as a "perturbation" of the independent case). This is an introductory study on the effect of various schemes of lateral connections on network learning. Using inhibitory connections (except the inverse mexican-hat case where we have also excitatory connections) we search for an improvement of selectivity to different patterns in comparison with the non-connected situation. We also try to preserve the topology of the input space in the signal tuning scheme of our network by choosing the strength of lateral connections as a function decreasing with the distance between neurons. The results of simulations are presented in a series of graphics representing the outputs of the neurons after the presentation of the different inputs.

Results and conclusions

The results here reported , although preliminary, suggest that it is possible to influence the behaviour of a BCM network by changing the strength and distribution of lateral connections even in the situation of linear dependent inputs.

It is observed that inhibitory connections favour a competition between neurons: they tend to reach more efficiently different stable states, when compared to non-connected neurons.

As shown in Figure 1, a more refined organization of neuron tuning is observed

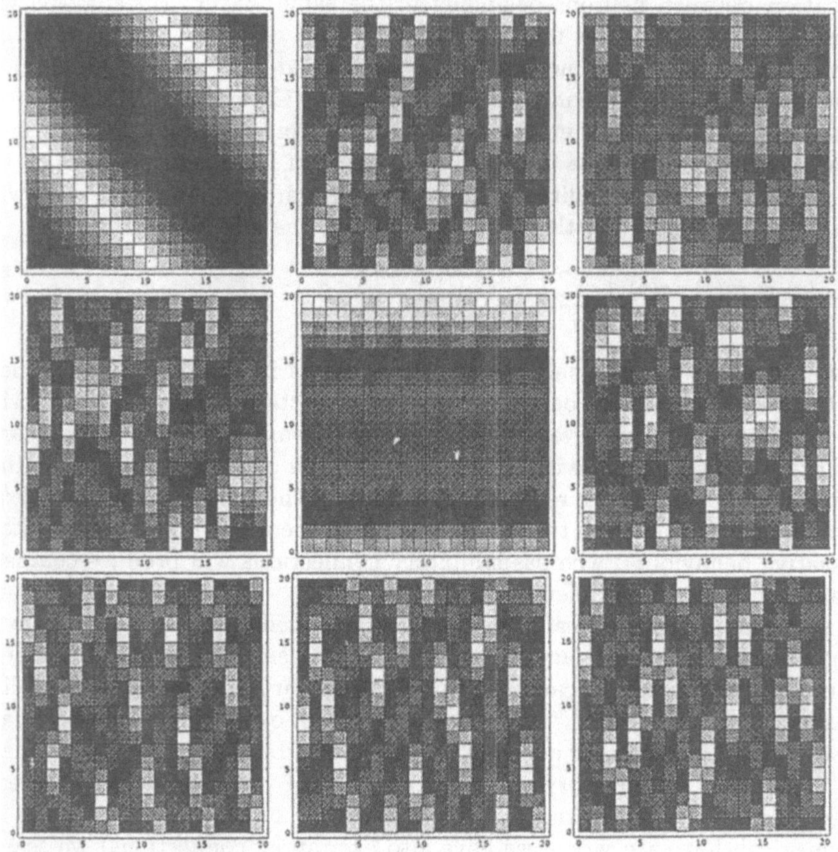

Figure 1: **The effect of lateral connections.** Results of simulations using a 20-neuron network and 20 gaussian-like vectors as inputs. The network is completely connected and the connection matrices are cyclical.
From top-left to bottom-right:
1) Input matrix; each row represents an input.
2-9) Network responses with different connections (each column represents the outputs of a neuron for the input set):
2) not connected neurons ($L_{ij} = 0$);
3)uniformly distributed random inhibitory connections (L_{ij} =Rand[-k,0]);
4) uniform ("all-equal") inhibitory connections ($L_{ij} = k; \quad k < 0$);
5) inverse mex-hat connections (short-range inhibitory, long-range excitatory);
6) linearly decaying inhibitory connections ($L_{ij} = k_1 - k_2 \cdot (i - j)$);
7) inverse-radius decaying inhibitory connections ($L_{ij} = \frac{k_1}{i-j}$) ;
8) exponential decaying inhibitory connections ($L_{ij} = k_1 \cdot e^{k_2(i-j)}$);
9) gaussian decaying inhibitory connections ($L_{ij} = k_1 \cdot e^{k_2(i-j)^2}$).

according to the different schemes of connections. In particular, connections whose strenght depends on the distance between neurons result in patterns of neural activity with a more regular spatial organization having some similarities with cortical maps observed in mammals. On the contrary, random and uniform connections seem to lead to less regular patterns. We can notice, from a general biological point of view, that these patterns of connectivity may reflect simple "design" principles, like a distance dependence in the strenght of lateral connections that could result from a dendritic spreading favouring the connections between closer neurons.
The present work is a preliminary approach to the general problem of BCM network learning with linear dependent inputs that more realistically represent the natural visual environment. Moreover, the analysis of linearly dependent inputs is a key point in pattern recognition and image analysis.

References

Castellani, G. C., Intrator, N., Shouval, H., and Cooper, L. N. (1999). Solutions of the bcm learning rule in a network of lateral interacting non-linear neurons. *Network.*

Cooper, L. N. and Scofield, C. L. (1988). Mean-field theory of a neural network. *Proceedings of the National Academy of Science*, 85:1973–1977.

Intrator, N. and Cooper, L. N. (1992). Objective function formulation of the BCM theory of visual cortical plasticity: Statistical connections, stability conditions. *Neural Networks*, 5:3–17.

Remondini, D. (1997). *Studio dell'effetto di connessioni inibitorie su neuroni BCM per la simulazione di un modulo della corteccia visiva.* Tesi di Laurea, Universitá di Bologna.

An analog on-chip learning architecture based on the weight perturbation algorithm and on current mode translinear circuits

M. Valle, F. Diotalevi, G.M. Bo, E. Biglieri, and D.D. Caviglia

Department of Biophysical and Electronic Engineering
University of Genoa, Via all'Opera Pia 11/A, 16145 Genoa, Italy
ph.: +39 010 3532288, fax: +39 010 3532777, email: francesco@dibe.unige.it

Abstract

The analog VLSI on-chip learning implementation looks attractive for a wide range of applications due to its promising performances. On the other hand, precision requirements seem to be too difficult to be satisfied through the analog circuit approach.

In this paper we present the analog on-chip learning implementation of a gradient descent learning algorithm; to overcome main problems, we adopted a Weight Perturbation learning algorithm and, from the circuit implementation point of view, current mode and translinear operated circuits. The proposed architecture is very efficient in terms of speed, size and power consumption; moreover it exhibits also an augmented scalability and modularity.

1 Introduction

In this paper we present the analog on-chip architecture of a gradient descent based learning algorithm. Among others, two main classes of supervised gradient descent learning algorithms can be considered: Back Propagation (BP) [6] and Weight Perturbation (WP) [7]. BP was developed first but, even if some examples of on-chip implementations have been presented in the literature [4], it is considered not hardware friendly because its requirements in terms of precision seem to be overwhelming. The WP was developed just to overcome such problems and to develop more efficient on-chip learning architectures [9]. In the proposed architecture, we implement on-chip a WP hard learning algorithm.

From the point of view of the analog on-chip implementation our goals are [10]: a) small overall size; b) low power consumption and/or low voltage operation (e.g. battery operated systems); c) high speed through high overall parallelism.

The main applications we envisage are in the following fields. a) smart sensors; b) remote sensed data classification; c) ageing and temperature drift compensation (the learning task can be accomplished: i) at the beginning of operation of the device, ii) periodically with a calibration learning procedure); d) high speed signalling i.e.: on line impedance adaptation (e.g. XSDL signalling systems).

From the architectural point of view, our approach focuses on the following issues (between brackets we indicate also the proposed solution):

a) modularity and scalability (i.e.: i) normalisation of variables with respect to the number of inputs, hidden and output neurons, etc., ii) programmability of the neuron transfer function, iii) usage of current mode signals [11] to make easier the sums);
b) complete on-chip learning implementation (i.e.: the external supervisor only gives the start and target signals);
c) robustness with respect to noise and technological spread of parameters (i.e.: differential coding of information, accurate matching of components, current mode computation);
d) wide signal dynamic range (i.e.: current mode computation).

This paper is organised as follows: the analog architecture will be introduced in Section 2 (feed-forward architecture), while in Section 3 the WP-based algorithm that we adopted will be detailed and discussed together with the circuit implementation issues.

All blocks have been designed and implemented in a CMOS 0.8μm channel length, double poly, double metal CMOS technology; the results of the design will be introduced in Section 5 together with the conclusions and the future work.

2 Feed-forward circuit architecture

The circuit implementation is based on the following guidelines: a) low power / low voltage i.e.: weak inversion region of operation of devices, supply voltage in the range [2.5÷3]V or lower; b) translinear operated circuits; c) implementation of normalised sums of currents: both for the synaptic output currents and the output error (i.e. target - output neuron current) currents; d) programmable neuron circuit gain; e) trade off between power and area i.e.: the maximum transistor bias current has been set to 500nA.

In the following we will refer to the AMS 0.8μm channel length, double poly, double metal CMOS technology; let us note though that the proposed architecture is not dependent on the given technology. Basically all variables are coded by differential electrical variables; moreover the computation is performed as current mode. In the following we will refer to a voltage supply of 2.5V; for voltage mode signals the signal ground value is 1.25V.

In Figure 1, the circuit architecture of the feed forward network, is illustrated: the neural architecture we refer to is the two-layer Multi Layer Perceptron (MLP) [1].

The input signals V_{ini} are coded by differential voltages to make easier the management of the inputs from outside the chip. The input voltage range is [-400mV ÷ +400mV] around signal ground. The blocks G (input transconductors) are linearized transconductors [2]: they translate the differential voltage mode inputs into differential balanced current mode signals varying in the range [-125nA ÷ + 125nA]. The bias current is 125nA.

The feed forward synaptic architecture (blocks S) is composed by two blocks: i) a four quadrant feed-forward multiplier; ii) the weight transconductor. The differential input currents coming from the input transconductors are in input to the synaptic feed-forward multiplier; it is implemented as a translinear four-quadrant multiplier [11]. This circuit implementation approach is very efficient when dealing with low current values; to properly operate, the transistors are biased in the weak inversion region of operation. The synaptic multiplier multiplies the differential input current times the current that codifies the weight value. The output current varies in the range [-250nA ÷ +250nA].

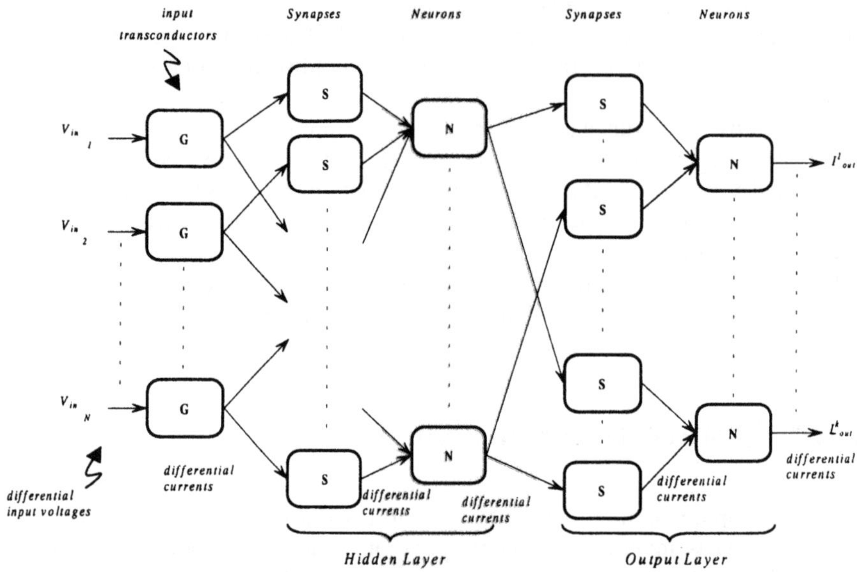

Figure 1. Circuit architecture of the feed forward network

A differential voltage that is dynamically stored on a capacitor controls the weight value; the weight voltage is translated into a single ended current by a two transistors differential pair [3] to reduce the complexity. The weight voltage varies in the range [-500mV÷500mV] around the signal ground; the weight current varies in the range [0÷500nA].

The feed forward neuron architecture (blocks N) is made of two blocks: i) the normalised input current sum block; ii) the neuron transfer function block. The neuron circuit blocks are implemented as translinear circuits.

To increase the modularity of the architecture the sum of the synaptic output currents is normalised with respect to the overall number of synapses connected to a single neuron.

The neuron transfer function block applies a sigmoidal shape transfer function to the normalised synaptic output currents sum. Each current component varies in the range [0÷ 250nA] and the differential neuron output current varies in the range [-250nA ÷ 250nA]. The slope of the neuron transfer function can be programmed through a current mode control signal.

The simulated response time of the cascade of a synapse and a neuron is about 190ns (about 30ns for the synaptic multiplier and 160ns for the neuron circuit); it follows that the feed forward computational power is of about 30 MCPS/synapse.

3 The Weight Perturbation on-chip learning architecture

We focussed our research activity on supervised learning algorithms due to their reliability and wide adoption in real world applications. We can identify two main approaches among gradient descent algorithms: Back Propagation (BP) [1], where the error function partial derivative is "just" computed, and Weight Perturbation (WP) [7], where the error function partial derivative is "measured".

In our previous works [4] we presented an analog on-chip implementation of the BP algorithm. Nevertheless this approach presents some drawbacks: a) high backward signal count for the back propagation of the error terms; b) need of differentiable neuron transfer function; c) fixed (in particular non recursive) feed-forward network topology; d) high sensitivity to circuit offsets ([8]); e) need of high linearity in transfer functions; f) need of circuits for the computation of the neuron function transfer derivative. Following from previous considerations, one can argue that the analog BP implementation may need high precision, then big area (and also high consumption) circuit implementations.

The WP algorithm [7] that we adopted for the on-chip learning implementation was presented in [12]; following that formulation, the weight update learning rule is:

$$\Delta w_{ij} = -\eta' \cdot \Delta\varepsilon \cdot pert_{ij}^{(n)} , \quad pert_{ij}^{(n)} = \begin{cases} +1 \\ -1 \end{cases} \text{ with equal probability} \quad (1)$$

Where η' is the learning rate and $\Delta\varepsilon$ is the variation of the error function caused by the random perturbation $pert_{ij}^{(n)}$ applied to the weight w_{ji} at the nth iteration. Then, to compute the synapse weight update term Δw_{ij}, we only need to compute $\Delta\varepsilon$ and to known $pert_{ij}^{(n)}$. The only term that must be back propagated is then $\Delta\varepsilon$; moreover the signal that codifies $\Delta\varepsilon$ is not further managed by other circuits and then it is not further affected by incremental offset contributions.

We adopted a *by-pattern* learning strategy because in the *by-epoch* approach (usually used when high precision is required) the offsets (due to each pattern presentation) accumulate during each epoch and can compromise the convergence. At the end, one must take into account that even if the on-chip (or chip-in-the-loop) learning implementation can learn around offsets in the feed forward circuits, it cannot do the same for offsets in the learning circuits.

The on-chip learning algorithm, detailed in Table 1., is based on two feed forward phases and some computations which can be divided into two categories: i) computations common to the whole network (see points 4, 7, 8) and ii) computations local to each synapse (see points 5, 10). Then the on chip learning architecture is based on two main parts: i) a block unique for the whole architecture for the *measurement* of the global error function gradient; ii) a block, local to each synapse, which, during the error function gradient *measurement*, applies to each weight a perturbation value (equal in module between all weights but with random sign), and which, during the weight update phase, given the measured value of the error function gradient (in terms of absolute value and sign), computes the single weight update value.

Step	*Comment*
1	all weights are randomly initialised
2	select an input pattern in the learning set
3	feed forward phase
4	output error computation (i.e. $\varepsilon(w) = \Sigma$ |(target neuron output value - current neuron output value)|) and storage
5	weight perturbation
6	feed forward phase
7	perturbed output error computation (i.e. $\varepsilon(w+p^{(n)}) = \Sigma$ |(target output neuron value - current output neuron value)|) and storage
8	measurement of $\Delta\varepsilon=\varepsilon(w+p^{(n)})- \varepsilon(w\varepsilon)$
9	back propagation to all synapses of only two signals: absolute value and sign of $\Delta\varepsilon$
10	weight update (see Eq. (1))
11	if $\varepsilon(w) >$ given threshold goto point 3 elseif end

Table 1. The on-chip learning algorithm

Let us underline again that a unique analog value (i.e. the error function gradient absolute value) is backward propagated during learning; moreover this signal is not further managed by cascade of circuits (as in the back propagation implementations, see for instance [4], [5]), but it goes directly in input to the weight update synaptic circuits. In this way offset contributions don't accumulate and thus the overall architecture is slightly dependent on circuit offsets. Point 11 is managed off chip.

4 Results and Conclusions

The proposed architecture and the circuit blocks have been designed and implemented using the AMS 0.8μm channel length, double poly and double metal CMOS technology. Main physical design guidelines have been taken into account: small silicon area and high precision (i.e. good matching between transistors) circuits.

The complexity increase due to the on-chip learning implementation is due mainly to the learning synaptic circuits: for each synapse the learning circuits occupy about the 40% of the total synaptic area. No learning circuits are needed in the neurons.

The learning of the XOR function has been simulated at circuit level taking into account a 2×2×1 MLP network; the pattern presentation time was 10^{-6}s. After about 250 input patterns presentations, the network learned the XOR function. Please note that the learning task was made more difficult due to the by-pattern operation.

A test chip containing all circuits and a cascade of one synapse and one neuron has been designed and manufactured; preliminary measurements confirm the circuit functionality.

Main figures concerning the synapses are summarised in Table 2. Future work will concern the full test of the experimental chip and the design of an entire network on a single chip.

Power consumption (@ Vdd = 2.5 V)	3×10^{-6} W
Area	0.021 mm^2
Recall speed	30 MCPS/synapse
	1.5 GCPS/mm^2
Learning speed	33 MCUPS/ 10^{-3} W
	4.8 MCUPS/ mm^2

Table 2. Main figures concerning the synapses

References

[1] R. P. Lippmann, "An Introduction to Computing with Neural Nets", *IEEE ASSP Magazine*, Vol. 4, No. 2, pp. 4-22, 1987.

[2] F. Krummenacher, N. Joehl, "A 4 MHz CMOS Continuous-time Filter with On-chip Automatic Tuning", *IEEE JSSC*, Vol. 23, No 3, June 1988.

[3] J. Mulder, and others, "A Reduced-area Low-power Low-voltage Single-ended Differential Pair", *IEEE JSSC*, Vol. 32, No 2, Feb. 1997.

[4] M. Valle, D. D. Caviglia, and G. M. Bisio, "An Analog VLSI Neural Network with On-Chip Back Propagation Learning", *Analog Integrated Circuits and Signal Processing, Kluwer Academic Publisher*, Vol. 9, pp. 231-245, 1996.

[5] T Lehmann. "Hardware Learning in Analog VLSI Neural Networks". Ph.D. Thesis, Electronics Institute, Technical University of Denmark, 1994.

[6] J. Hertz, A. Krogh, and R.G. Palmer, "Introduction to the theory of the Neural Computation", *Addison-Wesley Publishing Company*, 1981.

[7] M. Jabri and B. Flower, "Weight Perturbation: An Optimal Architecture and learning Technique for Analog VLSI Feedforward and Recurrent Multilayer Networks", *IEEE trans. Neural Networks*, vol. 3 (1), pp. 154-157, 1992.

[8] B. K. Dolenko and H. C. Card, "Tolerance to Analog Hardware of On-Chip Learning in Backpropagation Networks", *IEEE Trans. on Neural Networks*, Vol. 6, No. 5, pp. 1045-1052, 1995.

[9] A. J. Montalvo, R. S. Gyurcsik, J.J. Paulos, "Towards a General-Pourpose Analog VLSI Neural Network with On-Chip Learning", *IEEE Trans. on Neural Networks*, Vol. 8, No 2, March 1997.

[10] E. A. Vittoz, "Analog VLSI Signal Processing: Why, Where and How", *Journal of VLSI Signal Processing*, Vol. 8, pp. 27-44, 1994.

[11] C. Toumazou, F.J. Lidgey, D.G. Haigh Eds., "Analog IC Design: the Current-Mode Approach", *IEE Circuits and Systems 2*, 1990.

[12] F. Diotalevi, G.M. Bo, D.D. Caviglia and M. Valle, "Evaluation and Validation of Local and Adaptive Weight Perturbation Learning Algorithms for Optical Characters Recognition Applications", *Accepted for publication in SOCO99*, Genova, Italy, 1999.

Section 7
Applications

Gesture Recognition using Hybrid SOM/DHMM

Andrea Corradini*, Horst-Michael Gross
Technical University of Ilmenau, Germany
e-mail: andreac@informatik.tu-ilmenau.de

Abstract

This paper describes a method for the recognition of dynamic gestures using a combination Neural Network/ discrete Hidden Markov Model. This work deals with four topics. First a reliable and robust person localization task is presented. Then we focus on the view-based recognition of the user's static gestural instructions from a predefined vocabulary based on both a skin color model and statistical normalized moment invariants. The segmentation of the postures occurs by means of the skin color model based on the Mahalanobis metric. From the resulting binary image containing only regions which have been classified as skin candidates we extract translation and scale invariant moments. Further a Kohonen Self Organizing Map (SOM) is used to cluster the feature space. After the self-organizing process we modify the SOM weight vectors using the Learning Vector Quantization (LVQ) method causing the weights to approach the decision boundaries and we quantize each of them into a symbol. Finally, the symbol sequence extracted from time-sequential images is used as input for a system of discrete Hidden Markov Models (DHMMs).

To train and test the system we gathered the data from four people performing five repetitions of each of five movements from our vocabulary (stop, go to left, go to right, hello-waving left and hello-waving right). The system uses input from a color video camera and is user-independent.

1 Introduction

Body language and gestures play an important role in a human-human interaction context. Each person exploits the functions of the gesture already from his early childhood. Long before being able to speak, we convey our desires and needs by means of the gestures. Moreover the more a person becomes adult the more these abilities in gesticulating improve and become natural and intuitive.

In processes acting as intermediary agents between humans and computers, people must be allowed to concentrate their attention and efforts on the content of the interaction. Therefore the optimal interaction does not require any remembrance and should be similar to that they are familiar. The user should know how to interact just by his experience [6].

For this reason gestures with regard to *Human-Computer-Interaction* (HCI) purposes have became an intensive field of research. By now the dominant input devices to capture gestural information have been intrusive dispositives like the keyboard, mouse or glove. These approaches suffer from both not being

*European Commission TMR Marie Curie fellow # ERBFMBICT972613

natural and limiting the freedom of movement of the user. On the contrary non-intrusive camera-based input devices [11] are user-friendly. They do not require the user to wear additional instruments but suffer from both the computational costs for real-time image processing and the difficulty of extracting information from two-dimensional visual images. To sense gestures with a camera also limits the user to face it and requires ***highly constrained environments.*** The analysis and recognition of gestures has to deal with the time and space variability of the hand movements. Even if a person perform the same movement twice, slight changes of position and speed of the hands will occur. The position of the user facing the camera can vary, too. Further in a monocular system the 3D space is projected on a two-dimensional plane, resulting in a loss of information.

Different attempts at gesture recognition have appeared in the literature over the past years. Most of them deal with the two-dimensional set of movement patterns. An early effort by Rubine [9] used mathematical functions as recognizer for the 2D trajectory patterns of mouse gestures. Due to the extraction of some low-level features from the raw data, the choice of the proper feature set is very important. Yamato [12] used DHMMs to recognize image sequence of six tennis strokes with respectable accuracy. Starner [10] and Nam [7] proposed a similar HMM-based recognition technique of 3D hand movements. They modeled the hand gestures as a sequence of movement primes as the unit of recognition.

2 Person Localization: an Overview

In our system a good person localization task is essential for any further gesture recognition process. In the saliency system multiresolution pyramids transform the input images into a multiscale representation. Two cue modules sensitive to *facial structure* and *structure of the head-shoulder contour*, respectively, operate at all levels of a grayscale pyramid. That cue module processing for *skin color* detection uses the original color image and bases on a statistical parametric color model. For the generation of the skin color model we segmented manually a set of images containing skin regions and then we modeled a probability density function given these hand-segmented regions. As parametric model we used the Gaussian distribution. The segmentation of the color module result is transformed into a pyramid representation, too, to obtain an uniform data structure for the different cues.

The utility of the different parallel processing cue modules is to make the saliency system robust and independent of the presence of one certain information source in the images. Hence, we can handle varying environmental circumstances much easier, which, for instance, make the skin color detection difficult or almost impossible.

The output of the cue modules serves as the input for a 3D dynamic neural field. Since we use five fine-to-coarse resolutions we actually can localize persons even in different distances. Therefore, a neural field for selection the most salient region should be three-dimensional. That field can be described as recurrent nonlinear dynamic system with a dynamic behavior which leads to *one* local region of active neurons successfully competing against the others. A more detailed description of the 3D neural field and in general about the localization task can be found in [3].

3 Posture Analysis

3.1 Posture Segmentation

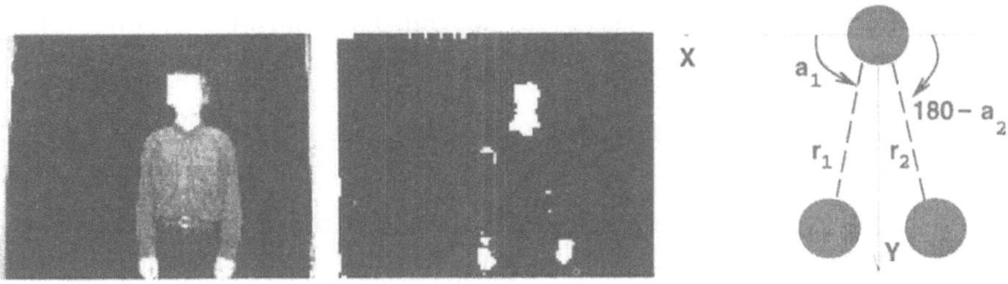

Figure 1: *From left to right: head localization result with head box, thresholded skin classification by means of an adapted color model derived from the pixel distribution inside the head box, modeling of the three greatest regions as circle around their centers of mass. In this last image the new defined coordinate system with origin centered in the COG of the head is depicted.*

The segmentation of face and hands as the gesture relevant parts is exclusively based on skin color processing. After detecting the location of the head as described above, we consider a window subregion around it which we call *head box* (Fig. 1,a). Then we characterize the distribution of the pixel values inside that subregion by a multidimensional Gaussian with centroid location and a covariance matrix describing the local distribution around the centroid. By doing that we adapt the skin color model to fit more specific for the illumination and the skin type at hand. Therefore the detection of skin colored regions can be improved. We handle multiple scales by choosing head boxes of different sizes according to the level in the pyramide.

By using the chromatic projection $r = \frac{R}{R+G+B}$ and $g = \frac{G}{R+G+B}$ of each pixel inside the head box the actual color model is uniquely determined by the multivariate normal density

$$p(\vec{x}) = \frac{e^{-\frac{1}{2}(\vec{x}-\vec{\mu})^T \Sigma^{-1} (\vec{x}-\vec{\mu})}}{(2\pi)|\Sigma|^{1/2}} \tag{1}$$

where the mean $\vec{\mu}$ is a two-dimensional vector, Σ is a 2×2 covariance matrix, and $|\Sigma|$ represent its determinant. Using the quantity appearing in the exponent of equation 1 (also called Mahalanobis distance from $\vec{x}$ to $\vec{\mu}$) each pixel $\vec{x}$ of the image is then classified to be or not a member of the skin class according to a threshold value.

Now we apply to the resulting binary image (Fig. 1,b) a winner-take-all (WTA) algorithm [2] to obtain the regions corresponding to the hands and head (which we assume to be the three greatest regions) and we determine their centers of gravity (COG). Then we model each of these regions as a circle around their COGs with constant radius (Fig. 1,c). That avoids problems deriving by the shape of each region due to the choice of the color threshold.

3.2 Moment-based Posture Description

From that binary image we compute a feature vector $\vec{v}$ containing 13 translation and scale invariant elements characterizing the shape of the segmented scene.

Given a pixel distribution $f(x, y)$ its two-dimensional $(p+q)$th order central moments are defined by

$$\mu_{pq} = \sum_{x,y} x^p y^q f(x - \bar{x}, y - \bar{y}) \tag{2}$$

where $\bar{x}$ and $\bar{y}$ represent, respectively, the x and y coordinate of the image's COG. Applying the theory about algebraic invariants [4], it is straightforward to show that the values

$$\nu_{pq} = \frac{\mu_{pq}}{\mu_{00}^{(1+\frac{p+q}{2})}} \tag{3}$$

known as the *scale normalized moments*, remain unchanged under image translation and size changes. In our work we take them up to the third order yielding the first 9 invariant values of our feature vector. The computation of them for binary image yields theoretically an error-free estimate of the continuous moments which is also independent of illumination as opposed to the value deriving from greyvalue images.

To compute the remaining 4 feature vector elements we operate as follows. To compensate the shift variation of the person gesticulating in front of the camera we choose for each image a suitable coordinate system by fixing its origin point at the current determined head's center of mass. It allows to calculate a feature vector relating to the head position and regardless to the user's position within the image. In this new coordinate system in order to ensure invariance also with respect to image size change, we use the polar coordinates of both hands's COG. If (r_1, a_1) and (r_2, a_2) represent the polar coordinates of the hands's COG (Fig.1,c) as last feature vector elements we take the four-elements vector $(\frac{r_1}{\max\{r_1,r_2\}}, a_1, \frac{r_2}{\max\{r_1,r_2\}}, a_2)$.

4 Gesture Recognition

4.1 Self-organizing Maps for Symbol Production

The goal of the posture analysis is the extraction of local features along the hand trajectory, yielding a sequence of time ordered multi-dimensional feature vectors. The further step is concerned with the quantization of that feature vectors into a sequence of symbols by means of a Self-Organizing Map (SOM) [5].

In the training phase the data vectors are presented to the input of a 40×20 Kohonen feature map. The weight adjustment is carried out using the Euclidean distance between input and connecting weight vectors times a time-dependent learning rate. After the clustering process each neuron of the network correspond to a cluster in the input feature space. Proceeding from the self-organizing process we modify the weight vectors using the LVQ method causing the weights to approach the decision boundaries [5].

In order to utilize the SOM for classification purposes we divide each gesture of our vocabulary in *subgestures* or *posture classes* (see e.g. Fig. 2). Then we map (label) a class to each cluster and therefore to each neuron within the network according to the subgesture symbols by using recorded training samples as input.

We divide the gesture of our vocabulary into altogether 32 subgestures/symbols (9 for each left-,right-waving; 5 for each go left/right; 4 for stop).

4.2 Discrete HMM for Recognition

In the previous section we assigned each feature vector of time-sequential images to a symbol which corresponds to a codeword in the codebook created by LVQ. The feature vectors of the data set for training are vector quantized. Therefore, a sequence of images is converted into a symbol sequence. Because we consider a gesture as a sequence of subgesture the recognition process consists in testing if a given sequence of symbols (each corresponding to one posture class) is part of a prestored movement (Fig. 2).

Mostly applied in the field of speech recognition, HMMs have shown to be able to deal with time-sequential data apart from providing time and scale invariability in recognition [8]. Taking into account the nature of the dynamic gestures we therefore use a recognizer based on discrete Hidden Markov Models.

A DHMM system is a statistical system model able to deal with time series signals. It consists of a finite number of states each with own probability density function, transition and initial probability. To recognize symbol sequences we create one HMM for each gesture to be detected.

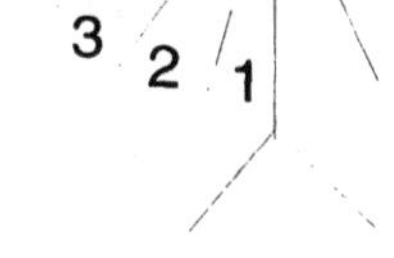

Figure 2: *Waving-right subgestures.*

In the learning phase the model parameters of each HMM are optimized so as to describe the training symbol sequences from the corresponding gesture. The parameter of each model are estimated with symbol sequences of the according gesture samples applying the Baum-Welch training algorithm [1]. For example the symbol sequences $\{1, 1, 2, 3, 4, 4, 5, 6, 7, 7\}$ and $\{1, 1, 2, 2, 2, 3, 4, 5, 6, 6, 7, 8\}$ are both candidates for the gesture in Fig. 2 and can be used to train the HMM modeling the waving-right movement.

In the recognition phase the gesture associated with the model which best matches the observed symbol sequence is chosen as the recognized movement.

The C++ Myers' Hidden Markov model software[1] originally written for automatic speech recognition is used. The code implements a basic left-right DHMM of the first order. It means that a full probabilistic description of the system requires specification of the current state as well as the predecessor one.

[1]http://www.speech.cs.cmu.edu/comp.speech/Section6/Recognition/myers.hmm.html

5 Results and Future Work

To train and test each model we gathered the data from four people performing five repetitions of the gesture to be described. The categories to be recognized were five. The performances were captured by a color camera (25 frames/second) and digitized into 120 × 90 pixel RGB images. We achieved a recognition accuracy up to 76% on average. Up to 8.2% of the pattern could not be classified. We consider an input as not classified if, after feeding it into each HMM, the difference between the highest and the second highest HMM output is not over an heuristically determined threshold.

Up to now we used as well discrete observation symbols from a finite alphabet (and therefore a discrete probability density function within each HMM state) as a limited training data set. We are currently extending the system with the aim to overcome these limitations. The performance of the whole system depends not only on the number of training pattern but also how well that patterns are representative for each class. It means that the training patterns have to cover the maximum test pattern scatter as possible. Moreover due to the continuous inherent character of the feature vectors, the conversion of them into discrete symbols via vector quantization can worsen the recognition task. To date we are developing a continuous HMM with mixture of Gaussian as probability density function in order to model multi-dimensional continuous signal directly [8].

References

[1] L. Baum, T. Petrie, 'Statistical Inference for Probabilistic Functions of Finite State Markov Chains", *Ann. Math. Stat., 37, 1554–1563*, 1966.

[2] A. Corradini et al., "Visual Person Localization with Dynamic Neural Fields: Towards a Gesture Recognition System", "Proc. of the X Italian Workshop on Neural Nets", pp. 201–206, 1998.

[3] A. Corradini et al., "Contour-based person localization by 3d neural fields and steerable filters", *Proc. IAPR Work. on Machine Vision Appl.*, pp. 93–96, 1998.

[4] K. Hu, "Visual pattern recognition by moment invariants", *IRE Trans. on Information Theory*, pp. 179–187, 1962.

[5] T. Kohonen, "Self-Organizing Maps", *2nd Edition, Springer Verlag*, 1997.

[6] A. Mulder, "Hand Gestures for HCI", *Tech.Rep. NSERC Project, Simon Fraser Uni., Burnaby*, 1996.

[7] Y. Nam, K. Wohn, "Recognition of Space-Time Hand-Gestures using Hidden Markov Model", *Proc. of the Workshop on the Integration of Gesture in Language and Speech*, pp. 175–184, 1996.

[8] L. R. Rabiner, B. H. Juang, "An introduction to Hidden Markov Models", *IEEE ASSP Magazine*, pp. 4–16, 1986.

[9] D. Rubine, "Specifying Gesture by Example", *Computer Graphics, (25):4*, pp. 329–337, 1991.

[10] T. Starner, A. Pentland, "Visual Recognition of ASL using HMMs", *Proc. of the Int. Workshop on Autom. Face and Gesture Rec.* , pp. 189–194, 1995.

[11] R. Watson, "A Survey of Gesture Recognition Techniques", *Trinity College, Dublin, T.R. TCDCS9311*, 1993.

[12] J. Yamato, J. Ohya, K. Ishii, "Recognizing Human Action in Time-Sequential Images using HMM", *Proc. Comp. Vis. and Patt. Rec.* , pp. 379–385, 1992.

A Fuzzy Neural Network for Urban Environment Monitoring System: the Villa San Giovanni study case

Domenico Marino (1), Francesco Carlo Morabito (2)
(1) University of Messina, Economics Dept.
(2) University of Reggio Calabria, Faculty of Engineering, DIMET
Consorzio *Innova*Reggio, Reggio Calabria, Italy
Phone: + 39 965 875 224 Fax: +39 965 875 220

Abstract

This paper presents a preliminary study about the design of an intelligent on-line monitoring system to urban environment modeling and prediction. The area of interest is that of the small town of Villa San Giovanni, in Southern Italy, which experiences a very peculiar condition of practical as well as scientific appeal. NNs are used as a basis for modeling interactions among many variables in order to predict well in advance a strong pollution episode may take place. The NN model is supported by a fuzzy knowledge based system which is able to extract the underlying correlation among different types of variables. The fuzzy logic approach is also used to derive statement which may help local politicians to explain their planned actions in order to prevent acute pollution episodes.

1 Introduction

This paper describes the design of a system for predicting the evolution of some quantities directly related to air pollution which strongly affect the urban environment. The case-study regards Villa San Giovanni, a town which is located at the tip of Southern Italy, just in front of Sicily. The urban environment of the small town is strongly affected by its very nature of crossing-city, which is interested in the traffic movement toward and by Sicily. This means in practice that the typical values of some pollution parameters are comparable to those of cities ten times larger. The urban environment of Villa S.G. is thus everyday subject to slow changes due to human activities and natural processes related to urban development, movement of traffic, also as terminal of ferries through Sicily, as well as changing coastlines. The prediction of the evolution of several parameters is consequently important for town planning and global environmental management.

The above described topics are now under study, and an European Community funded project named SMAURN (Monitoring System of Urban Environment through Neural Networks) was started under the auspices and management of Regione Calabria. The project foresees the design and installation of a network of mobile terminals for monitoring some parameters and predicting in real time their evolution. The distributed integrated network of sensors carries out biological, chemical, physical and meteorological investigations of air, and it is able to monitor on-line and continuously the state of the urban environment and air quality. The system manager will issue regular reports on trends in urban environmental aspects (e.g., toxic substances, air pollution and trends in the health of inhabitants).

Extensive observations serve as the basis of the SMAURN activity. Unfortunately, the lot of data gathered by the system is manageable with great difficulty from both the viewpoint of high data throughput and data interpretation. Also Neural Networks (NN's) meet difficulties in solving this problem of time-series prediction. This is due to the high number of inputs of the model, the neglecting of their statistical correlation, the lack of an easy way to embody within the model some evidences derived from expert knowledge. The system could benefit of any prior information based on knowledge of physical as well as chemical aspects of the problem, and of the results of similar studies carried out previously (an important mirror-case is that of Messina, in front of Villa S.G., which experiences the same problems). Fig. 1 shows a remotely sensed image of the area.

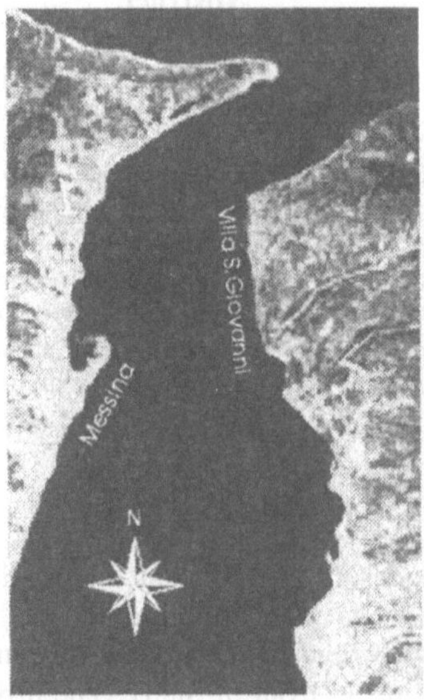

Fig.1: The study area of the Messina Strait. Flag indicates the location of the monitoring station under analysis (Archimede station).

The proposed system integrates this information in terms of fuzzy variables and models. The project also aims to suggest the optimal locations of the sensors in the town (e.g., point of access by highways, ferry-station, ...). In this respect the fuzzy modelling is invaluable, because it gives an initial guess of the distribution based on previous results. Tools derived from fuzzy logic concepts are also used to propose alternative solutions to the problem to decision makers. Fuzzy engineering is thus suitable to meet politicians needs, because it is able to produce explanation of decisions.

2 Monitoring of air quality

Atmospheric pollution arises from the adverse effects on the environment of a variety of substances contaminants emitted into the atmosphere by natural and man-made processes. The physical and chemical processes undertaken during an air pollution episode are, to a large extent, understood. The effects of individual environmental factors and man-made pollutant on human life are also known, but this is mainly through single factor. What is largely unknown is the effect of differing levels of multiple factors, such as humidity, temperature, global radiation, wind speed and directions, light, ground-level ozone-concentration [1]. The ability to predict very early evolution of pollutant levels towards peak episodes is a basic asset for decision makers.

Environmental data are typically very complex to model due to the underlying correlation among several variables of different type which yield an intricate mesh of relationships. It is ordinary to have multivariate dependency with non-linear behaviour. Furthermore, typical variables do not meet the very common Gaussianity assumption. To simplify statistical complexity, many attempts to model these interactions fix some of the micro-climatic conditions or ignore their effects, thereby making any results dependant on a tight set of conditions [2]. These methods always involve expert knowledge and are not suitable for automatic systems. Standard statistical techniques fail to adequately model complex non-linear phenomena. In contrast, NN's are widely used because they showed ability to model non-linear data and their non-reliance on previously assumed equations. Predictions based on NN's depend on extensive collections of data, which are used to learn the model. Unfortunately, the knowledge that a NN gain about a problem is encoded in the connection's weights whose strength is fixed after learning. For political reasons, it is impossible to make decisions based on black box models, particularly for taking unpopular actions, like stopping cars traffic in a town. In addition, in the case of a transit-town, a local decision may have a strong impact on a national level. It is thus essential to derive a set of rules which can support the decisions. On the other hand, the prediction of a pollution episode may benefit of knowledge on the typical meteorological, seasonal and traffic conditions of a particular area. Also, extracting meaning from a trained NN may offer a novel insight on not observed relationships.

3 The SMAURN system

A pollution episode is considered to have taken place if the hourly measurements of some air pollutants have surpassed in duration and in intensity an allowable level. The definition of air pollution criterion is based on the legislation of a particular country. For the EU members a set of pollutant air quality standards has been established.

Urban air pollution is mainly due to transports, but also house-heating has been proved to play an important role in terms of emissions, especially if secondary pollutant must be considered: as a consequence, several different aspects have to be studied to be able of properly modeling air pollution.

The definition of efficient control strategies are based on the analysis of data gathered by an on-line monitoring system. The aim of SMAURN project is to devise a monitoring system of on line sensors for air quality measurements in the Villa S. G. area. The measurements are automatically transferred for the monitoring station to central database. Calibration and raw data controls are performed in order to approve the final storage database, where simple statistics and data visualization systems are used to check the validity of the data. A central data acquisition system collects and stores the data coming from the stations. It is foreseen the installation of three stations in different monitoring sites. Non real time analysis of the data gained by means of the monitoring network is also commonly carried out.

4 The Messina's database (Archimede station)

Since the design of the network of monitoring stations is the output of the SMAURN project, and since the collection of data have been just recently started, the database analyzed in this work refers to the Messina case. The site of the test monitoring station strongly resembles the actual Villa S. G. case (in particular, the station whose location is foreseen in the railway station square). Investigations presented here refers to the period January 1996 - March 1998. The database contain hourly readings for various environmental and pollutant factors. The measured pollutants concentrations are TSP (total suspended particulate), NOx, CO, SO_2, O_3, and the meteorological parameters are wind speed and direction, and temperature. Also, a traffic parameter is considered of help. The system calculates all the required parameters according to a prescribed scheme: e.g., SO_2 episodes are calculated in 24-hour period, NO_2 and O_3 in hourly periods. The data analysed in this paper refers to a station (Archimede) located downtown, along an important way which collects the vehicular traffic from Peninsular Italy. Also, this is a highly populated area. Figs. 2-3 refers to SO_2 concentrations.

The NN models used to predict the time evolution of the pollutants are basically of the feedforward multilayer perceptron type. The inputs of the models are samples of the measured concentrations and the outputs are the predicted concentrations for the next hours and for the following days. In the neuro-fuzzy model we are studying, there is also an explanation of this coefficient in terms of activated rules. Typical errors on the predictions vary over a relevant range. The RMS error is very acceptable for predictions which uses not only the last hours concentrations but also an additional set of inputs representing the measured concentration value at the same hour in recent days and in days with similar meteorological conditions. The errors are increased for one-week prediction but are always within 10%. The model benefits from adding a rule-based knowledge engine with fuzzy inputs. The fuzzification of some parameters is what politicians ask for. Indeed, they express concept like: *a high concentration of ozone, high wind speed, very low car traffic*, and so on. The presentation of too precise numerical results is not well accepted, because they are suspicious about the techniques and the hardware which output data. Thus, also common sense suggests an effort to "fuzzify" concepts. Furthermore, the binary logic of "do nothing" vs. "act quickly" is quite refused. They like to know that "*the day after tomorrow we may have an acute episode thus we should act by tomorrow in this way*".

Also, the dependency of some pollutant concentration parameters from wind speed and direction, temperature, and hourly demand of traffic can be easily predicted. Wind analysis plays a relevant role for the study of natural conditions occurring during the development of an air pollution episode as well as for the extraction of long term meteorological features.

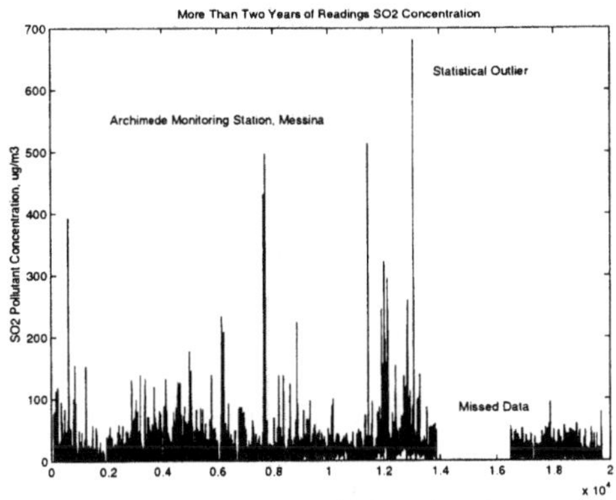

Fig.2: The observed SO_2 concentrations within the inspected time-period (pollutant concentrations in µg/m3 vs. days from 1/1/96 to 31/3/98).

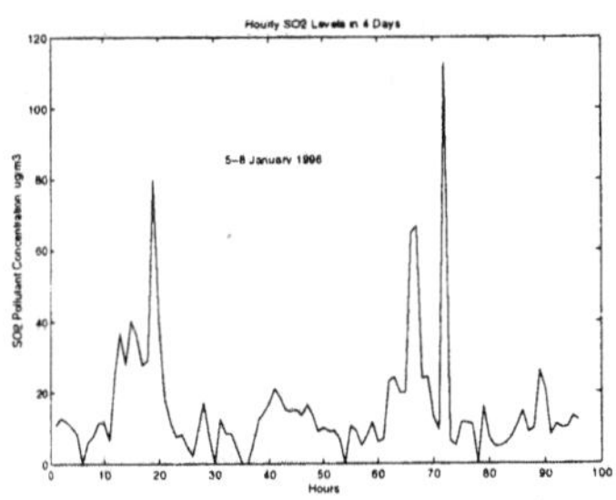

Fig. 3: Evolution of level within four days. It is evident the impact of a holiday (6 January) on the presence of pollutants during peaks (after 4 p.m.).

Fig. 4 reports the distribution of probability of the wind directions measured at the Archimede station. The histogram was further manipulate to have the wind direction chart in the wind rose form. The diagram is easily understood if one refers to Fig.1. From this plot, it is easy to derive an a priori map of the distribution of pollutants in the area. The correlation with other meteorological variables, like temperature and solar radiation, explain the need of fuzzifying the map. The correlation among meteorological condition and evolution of state variables is thus embodied in the model, and the knowledge of the effect of different wind directions is useful in predicting the zones of the town interested by air pollution in different hours of the day. The study of SO_2 levels in connection with the wind speed and directions in the area is carried out. Some wind direction with medium speed may transport pollutants towards selected areas thus creating pollution problems. The wind in this area blows mostly along the NE / SW axis with the prevailing wind directions being similar in summer, fall, winter and spring. Another very simple but essential component of a good prediction is the observation of special behaviour of pollutant concentrations in some conditions. Fig. 5 shows that the typical peaks of

SO_2 experienced each day after 5 p.m. are absent in holidays (namely, January, the 6th). Also this kind of knowledge has been introduced in the neuro-fuzzy model in terms of fuzzy rules. The prediction RMS error improves to about the 5 % by using these evidences.

5 The Villa San Giovanni study area.

Villa S. G. is an important example of town strongly subjected to periodic as well as random acute pollution episodes. Apart from what discussed in the introduction, there are geographical reasons which explain its peculiarity. Villa S. G. is rounded by hills except for the sea side which is distributed along the coast of Messina Channel. The town is subject to very high levels of solar radiation during summer, when temperature often rises up close to 40 (°C), increasing the effect of photochemical aspects in determining air quality over the town. All these unfavorable meteorological conditions tends easily to create very acute episodes of air pollution. The monitoring systems to be designed aims to predict with a high degree of confidence the exceeding of threshold values in NO_2 levels ($\mu g/m^3$) and SO_2 levels ($\mu g/m^3$).

The optimal location of the monitoring stations in the area of interest is also among the most interesting outcomes of our project and is related to the complex geo-morphological characteristics of the area under study. The ultimate definition of the network has to be related according to the standards, which prescribe a preliminary analysis of the area by using mobile stations. In selected areas, the measurements of pollutant related to car traffic is needed, in other ones the measurements of pollutants related to density of inhabitants may be recommendable.

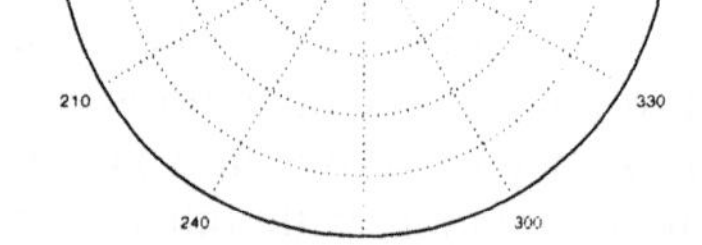

Fig. 4: Probability distribution of the wind direction in a polar diagram (to be compared to both rose wind and Fig.1).

References.

[1] N. Moussiopoulos, H. Power, C. A. Brebbia, Eds., Air Pollution, Vol.3, Urban Pollution, Computational Mechanics Publications, 1995;

[2] C. M. Roadknight, G.R. Balls, et al, Modeling Complex Environmental Data, IEEE Transactions on Neural Networks, Vol. 8, No.4, p. 852, July 1997;

[3] D. Marino, B. Ricca, Neural Networks for Econometric and Environmental Models for Territory Management, Proceedings of the XVII Italian Congress on Regional

Real Time Neural Network Disruption Prediction In Tokamak Reactors

Francesco Carlo Morabito
Mario Versaci

Associazione Create-Enea-Euratom
University of Reggio Calabria, Faculty of Engineering, DIMET
Via Graziella, Loc. Feo di Vito, Reggio Calabria, Italy
e-mail: morabito@unirc.it, versaci@ing.unirc.it

Abstract

This paper proposes the use of Fuzzy Neural Network (FNN) approaches for the early detection of disruption in tokamak plasmas. The fuzzy neural models is able to combine signals from various different plasma diagnostics in order to make an estimation of the expected time of disruption. This is, in turn, useful for having sufficient margin to initiate a disruption avoidance action. The inclusion of many diagnostic measurements results in a much more accurate prediction than that provided by traditional physical approaches. The use of fuzzy logic concept is suggested by the consideration that those previous techniques make use of expert knowledge for deciding about the onset of a disruption. In addition, learning approaches allow to tune the model. The proposed method appears to be a step forward with respect to more conventional NN approach.

1 Introduction

Disruption in a tokamak reactor is a sudden loss of confinement and transfer of plasma energy to the surrounding structure. During a disruption, the plasma current and the thermal energy content of a tokamak plasma discharge collapse in an uncontrollable way, thereby generating mechanical forces and heat loads which threaten the structural integrity of surrounding structures and vacuum vessel components. The machine walls and support structure may be damaged as a result of a disruption. As a consequence, disruption avoidance is important for cost effective operation of tokamak devices. In present day tokamaks, the exploration of high performance regimes frequently implies a disruption event. The objective of this paper is to design an alarm system for detecting the onset of a disruption in tokamak plasma discharges in due time for control actions to be taken. In order to overcome the real time operational requirements and with the aim of using a very large number of potentially predicting variables, Neural Network (NN) models seem to be the most viable tool. NN tool is also more amenable to hardware implementation of the alarm system. The paper will also show that fuzzy inference systems (FIS) can play a relevant role in tokamak disruption prevention.

2 The ASDEX-Upgrade machine

A disruption oriented database of all ASDEX Upgrade discharges has been set up by the ASDEX Team [3]. In this database, there are stored general information about the plasma shots, with special reference to disruptive discharges. A large number of disruptions in flat-top (520 plasmas with 173 disruptions in flat-top) were analyzed with the purpose of finding the technical causes, the precursors and the physical mechanisms of disruptions [4]. Most of the plasma disruptions in ASDEX Upgrade happen in a plasma parameter regime (poor L-mode confinement) which is far away from the desired operational space (H-mode, high beta). In addition, disruptions are usually announced by well identified precursors (detachment, MARFE, growth and locking of resistive tearing modes) which can be detected by the available diagnostics. Fig.1 depicts a general view of the ASDEX-Upgrade machine.

From a physical viewpoint, it is possible to discriminate the majority of the disruptions according to the so called MARFE mechanism. This event can be identified by a number of diagnostics, in particular, by the two bolometer channels (bli) looking horizontally respectively below and above the X-point [4]. The MARFE-indicator is the signal (bli7/bli11) and the other relevant parameters is the ratio between the internal inductance (li) before disruption and the steady state operational region in the li-q diagram. According to those physical evidences, we are able to state two simple conditions:

IF (li disr / li ss) IS (GT 1) AND (bli7/bli11) IS (LT 1)
THEN it IS a MARFE disruption.

This kind of statement should be implemented in a decision tree which allows to categorize the different shots in various physically significant leaves. The bolometer and internal inductance data were also used as input to a NN model, in a vector set of 24 inputs (safety factor, plasma category, input power, li_HL, li_ss, ne/j, gas fluxes, energy confinement time, normalized plasma energy and bolometer channels). The output of the procedure, apart from a categorization of the shots, is the time left before disruption (*ttp*). It is obvious that the importance of a correct *ttd* prediction assumes different importance with respect to the stages of the event. For this reason the cost function of the NN learning procedure is typically weighted by a physically determined figure.

3 The Fuzzy Neural Method

The technical origin of the disruption is quite often identifiable. This means that most of the disruptions could be attributed to some technical external source and could be classified as avoidable. In other words, the onset of a disruption is signaled by means of the detection of plasma instabilities, or *precursors*. Some physical measurable quantities are directly related to the disruption. As a consequence, a group of available diagnostic signals of different physical nature can be used in order to determine if the plasma is approaching the disruption limit. A theoretic limit based on physical approximations can be used for performance comparison. We use firstly a static NN trained on a suitable database of real experimental data based on both non-disruptive and disruptive discharges. A simple dynamic NN model aiming to overcome the limitations of the static models coming from the possibility of facing different types of disruptions not included in the main database was also studied. It was found that the use of 3 past samples of the input measurements is enough to allow the early detection of precursors in real time. In other words, we have used a standard feed-forward NN with a buffering stage that reads three successive measurement samples and eventually carries out some linear transformation on the data. Finally, since some physical knowledge could advantageously be included in the network framework by using the fuzzy system approach, the original model was accordingly modified by obtaining a fuzzy neural system. We have also used the Matlab tool "Fuzzy Logic" with the GENFIS and ANFIS routines in order to design a system entirely based on fuzzy techniques (FIS). One of the most advantageous aspects of this approach is the possibility of selecting among a set of candidates the most important sensors [2].

4 Experimental Results

The proposed NN based model far exceeds the performance of standard techniques. The NN approach has been successfully applied for solving different problems to be faced during work, namely, the detection of redundancy in the input data; the ranking of the input variables, and finally, the automatic extraction of rules from the database to be proposed to experts for further interpretation. The most interesting result achieved was that the NN approach allows to automatically classify the different kinds of possible disruptions. This is obtained by means of an auto-associative model of NN, which learns in an unsupervised way. The NN ats as a bottleneck in which the hidden layer was reduced to just two nonlinear sigmoidal nodes. The NN also has two linear layers. The final model is a 19-10-2-10-19 multilayer perceptron. The use of two nodes in the hidden layer allows the plotting of the database in a two-dimensional diagram. The different classes of disruptions, each related to a different physical mechanism of initiation are separated in the plot. The NN yields a nonlinear boundary separation among classes. Fig.2 shows the results achieved in ranking the sensors by using a technique based on fuzzy curve concept while Fig. 3 reports the qualitative plot used for classifying different disruptive shots. In order to show how the pattern evolves towards the disruption, we plot three different discharges (ASDEX Upgrade shots #10246, #10265, #10285) for a total of 985 patterns. From bottom to top, Fig. 3 reports a gray-scale representation of the three considered discharges. In the plot different values of the various (19) measurements are mapped to different gray-levels. By taking a glance at the plot, it is clear that we are approaching a well identifiable regime. The figure also shows how data visualization could be useful by itself for extracting information from the volume of data. It is evident that visual clues should be properly transformed into viable algorithms: in this sense the fuzzy approach is invaluable, since it yields a natural framework for implementing such qualitative clues. From the reading of the pattern, we can thus simply express a guess about target. In the fuzzy approach, the measurement space should be divided into multiple subspaces, and for each subspace, the relationship between the input pattern and the target category are described by IF-THEN fuzzy rules. A fresh pattern is classified by fuzzy inference. It is clear that in order to be workable, this methods implies the ranking and selection of the input variables. Figs. 4 and 5 reports the results of the FIS procedure in terms of scatter plots (estimated time to disruption vs. target time value). The general conclusion is that a FIS model that uses a reduced number of diagnostics as input is sufficient for interpreting the evolution of the shots, once the kind of shot has been categorized. The use of past input samples and of the buffering stage can be of help in avoiding an excess of categories to be discriminated.

5 Comparison with other approaches

As far as our concern, there are no previous works in the literature which report the use of the combination of fuzzy techniques and NNs. The paper also shows a novel application of unsupervised techniques for generating self organized classification. In this sense, the paper extends the work carried out in USA, in Orincon Corporation, which just uses multilayer perceptron to prediction [1]. In that work, the output of the procedure was the predicted value of β at the time of disruption. The main limit of the present and of past work is the device dependency of the analysis. On the other hand, it seems clear that any applications pointing towards real time is directed to a device implementation. Thus, this cannot be considered a limiting factor for the present work.

6 Conclusions

This paper proposes a real time operating NN-based tool for prediction of incipient disruption in tokamak plasma physics applications. Future works will focus on analyzing the impact of measurement noise on the model and on the data fusion and integration aspects.

6 References

[1] D. Wroblewski, G.L. Jahns, J.A. Leuer, Tokamak disruption alarm based on a Neural Network model of the high- limit, Nuclear Fusion, Vol. 37, No. 6, 1997, pp. 725-740;

[2] M. Versaci, On the Use of Fuzzy Systems (FIS) for Classifying and Identifying Plasma Columns in Tokamak Reactors, Ph.D. Thesis, 1998;

[3] H. Zohm, K. Lackner, C. Ludescher, Statistical Analysis Disruptions in ASDEX, Nuclear Fusion, Vol.33, No.4, pp.655-662, 1993;

[4] G. Pautasso et al, Causes, Precursors and Mechanisms of Disruptions in ASDEX Upgrade, Proc. Intl. Conf. EPS, Wien, 1997.

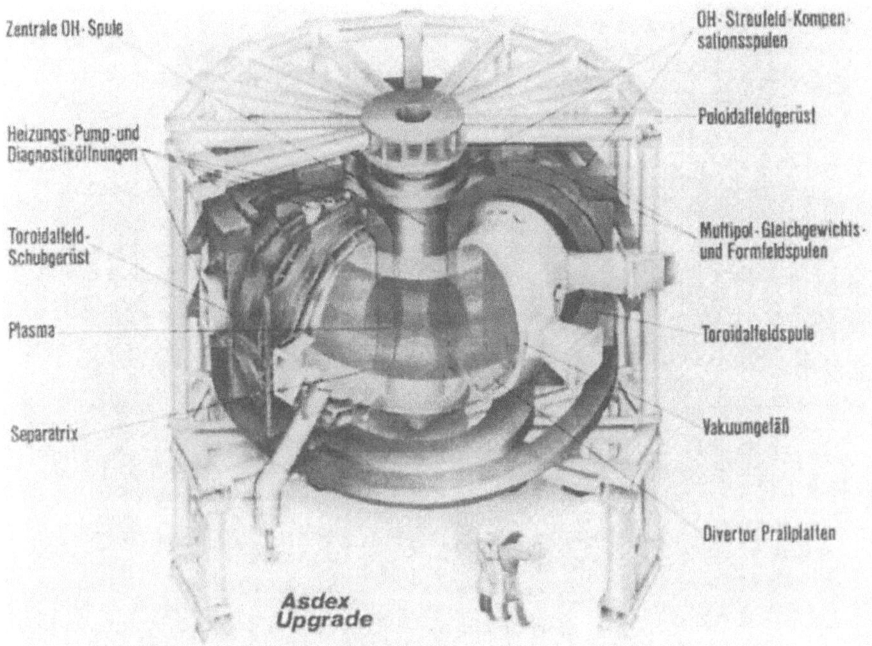

Fig. 1 – A pictorial view of the ASDEX-Upgrade machine (http://www.ipp.mpg.de/aug/aug.eng.html)

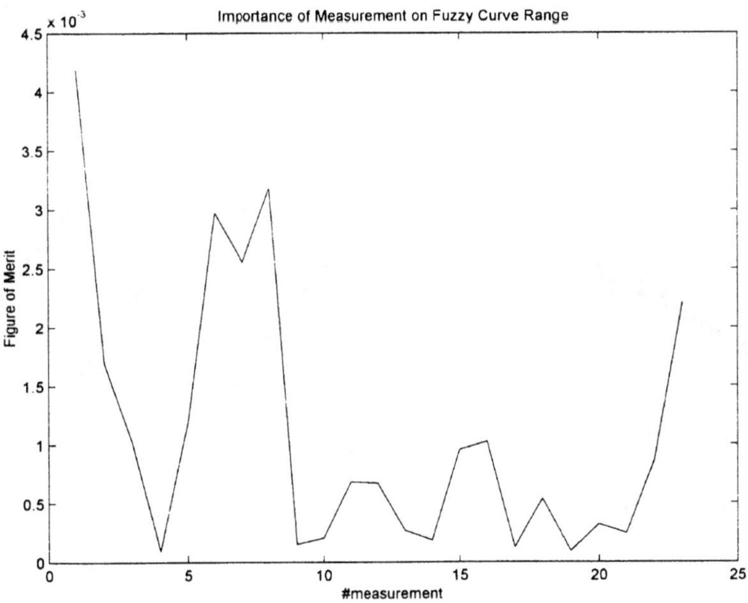

Fig. 2 – Ranking of the input sensors

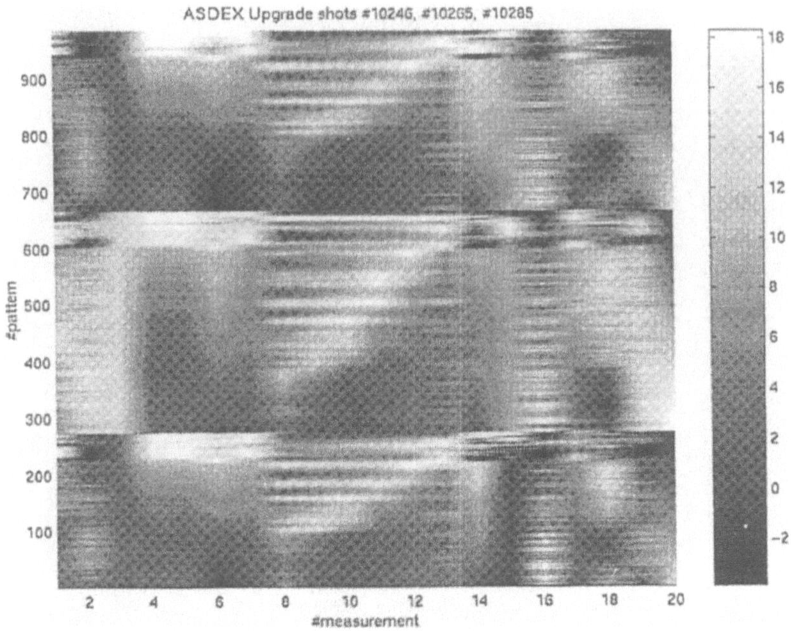

Fig. 3 – Gray-scale representation of the three discharges. It is evident that certain measurements are representative of a dynamic evolution towards the disruption and also the possibility of writing down suitable fuzzy rules

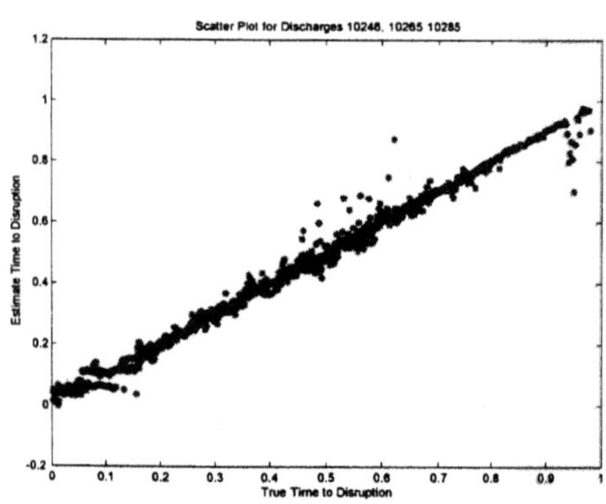

Fig. 4 - Scatter plot for the regressed variable

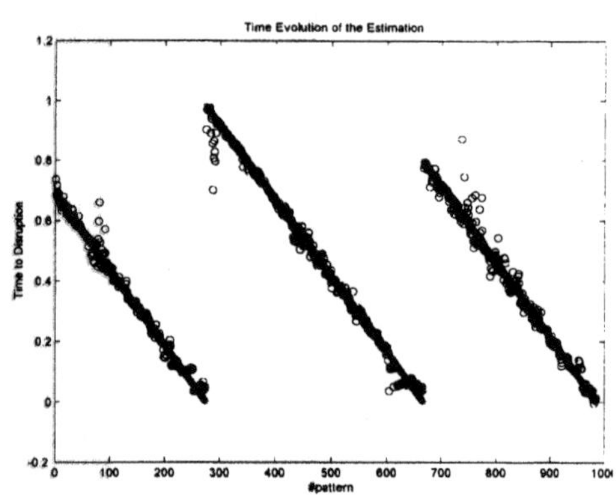

Fig. 5 - Time evolution of the estimation for the three discharges

On-line Quality Control of DC Permanent Magnet Motor Using Neural Networks

Mirko Solazzi(*) and Aurelio Uncini(**)

(*) Department of Electronics and Automatics – University of Ancona
Via Brecce Bianche, 60131 Ancona – Italy.
Fax:+39 (071) 220-4464 e-mail: mrksol@eealab.unian.it

(**) INFOCOM Department – University of Rome "La Sapienza"
Via Eudossiana 18, 00184 Rome – Italy.
Fax +39 (06) 487-3300 e-mail: aurel@infocom.ing.uniroma1.it

Abstract

This paper addresses the use of neural network methods to perform real-time quality control in industrial assembly lines of DC Permanent Magnet Motors (PM). This task can be viewed as a difficult non-linear inverse identification problem. Due to long parameter setting time, noisy environment and in some cases human supervision requirements, these methods are not adequate for real-time applications. Moreover, PM quality control requires the satisfaction of particular specifications rather than simple model parameter identification. So neural networks seem to be a promising paradigm.
In this study we apply fast adaptive spline networks to perform complex model inversion and reduce the effect of noise on the data. Experimental results demonstrate the effectiveness of the proposed method.

Keywords: Neural network quality control, DC motor identification, adaptive spline neural networks.

1. Introduction

On-line electrical machine quality control can be viewed as a difficult non-linear inverse problem parameter estimation in noisy environment. Electrical machine parameter estimations are well-known techniques and largely reported in electrical engineering literature.

Several papers have been published that study techniques used to estimate the parameters [1]-[6] or for fault diagnosis [10] of synchronous machines. In some of these, the estimation method consists of fixing the rotor of the PM machine in a specific position, applying a simple DC voltage source and evaluating the parameters using experimental techniques. This method is called standstill time-domain test. In [1] a method for developing a standstill dq-axis equivalent circuit model with a general number of damper windings and estimating the parameters is derived. These procedures, based on estimation theory, usually require several electrical measures and very long setting time.

These *standard* techniques are not adequate for on-line quality control. In fact, in industrial assembly lines, the diagnosis time should be as short as possible and fast non-intrusive techniques are wanted. Moreover, in motor quality control, steady-state specification estimation rather than parameter estimation is required.

The aim of this study is to develop a neural network method to estimate PM motor specifications, such as relations between torque and speed, torque and mechanical power and torque and efficiency, from a unique test in the no-load situation. A model with a given number of parameters is derived to simulate the PM machine and create synthetic input and output test data to be used during the identification process. Next some features are extracted from the test data to train a feed-forward neural network. The use of the network is essential to perform complex model inversion, reduce the effect of noise on the data and fits the mask of the industrial quality specifications.

2. Permanent Magnet Motor Model

The DC machine model used in this study is shown in fig. 1 [7][8]. Since this type of motor uses a permanent magnet to generate the magnetic field in which the armature rotates, the electrical circuit in the armature alone can model the motor. In this simple model R_a and L_a indicate the equivalent armature coil resistance and inductance respectively and v_a the voltage supplied by the power source. Kirchoff's voltage law leads to the following equation:

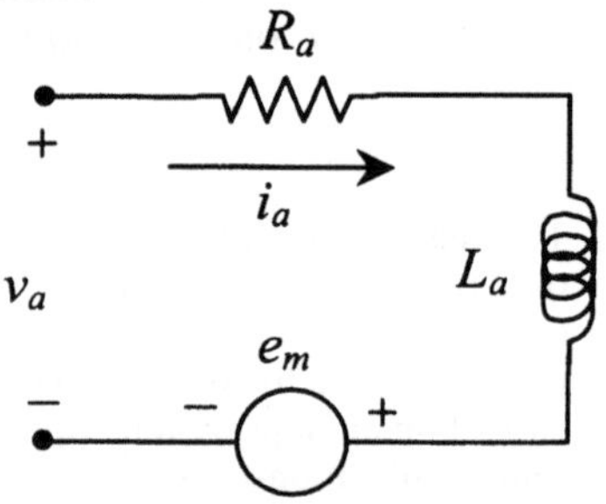

Fig. 1: PM machine armature equivalent circuit.

$$v_a = e_m + R_a i_a + L_a \frac{di_a}{dt} \tag{1}$$

where e_m is the back electromotor force. By examining the effect of the magnetic field in the motor, and realizing that magnetic flux is constant, we can arrive at the following two equations relating the torque t_m and the motor speed output ω to the supplied current and voltage:

$$e_m = K_v \omega \tag{2}$$

$$t_m = K_m i_a \tag{3}$$

The constants K_v and K_m are dependent on the particular motor, but if they are expressed in SI Units, their values are always equal.

In addition to the electrical equations, we should consider the load effect. It can be treated generally as sum of three terms: a speed independent resistant torque (t_0), a torque proportional to the speed according to a frictional coefficient (F) and a torque proportional to the speed derivative according to an inertia moment (J):

$$K_m i_a = t_0 + F_t \omega + J_t \frac{d\omega}{dt} \tag{4}$$

In (4), the symbols F_t and J_t include both load effects and the effects those are provided by the frictional losses within the motor and the rotor inertia respectively. We can rewrite the above relations in compact state space form as:

$$\frac{d}{dt}\begin{bmatrix} i_a \\ \omega \end{bmatrix} = \begin{bmatrix} -\frac{R_a}{L_a} & -\frac{K_v}{L_a} \\ \frac{K_m}{J_t} & -\frac{F_t}{J_t} \end{bmatrix} \cdot \begin{bmatrix} i_a \\ \omega \end{bmatrix} + \begin{bmatrix} \frac{1}{L_a} & 0 \\ 0 & -\frac{1}{J_t} \end{bmatrix} \cdot \begin{bmatrix} v_a \\ t_0 \end{bmatrix} \tag{5}$$

Let Σ, *t*he unknown parameter set to be estimated defined as:

$$\Sigma = [R_a, L_a, K_v, K_m, F_t, J_t\,], \tag{6}$$

we remark that: i) not all the parameters are observable; ii) the identification task requires a specific input signal.

3. Neural Network for Parameters Estimation

3.1 Training Algorithm

The first step in identifying the permanent magnet machine using neural networks is to define the appropriate input and output vectors for the net. Since our aim is to reduce the evaluation time, we consider two phases: an off-line training process and an on-line fast estimation one. Fig. 2 illustrates the block diagram of the training procedure: a DC step voltage signal is applied to the machine for a period of time longs enough so that the system reaches the steady state. During this time, supplied current and rotational speed responses are collected as experimental data. Next, some features are extracted from them and sent to the neural network. Neural outputs, representing the parameter estimations Σ_e, are compared with the known machine parameter values Σ and the error is used to adjust the network weights.

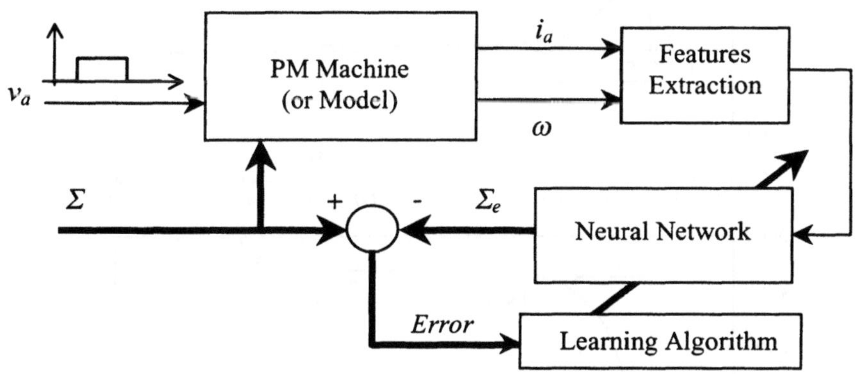

Fig. 2: Off-line neural network weights adjusting algorithm.

Once the network is trained with an appropriate training data set, it can be applied to estimate the PM machine parameters directly from the machine outputs as shown in Fig. 3. There are several advantages in using of this method: first of all the response time of the system is very short, so that it's possible to increment the diagnosis efficiency in industrial application; neural network approach, allows the estimation of the non-observable parameters Σ_e. Moreover, the neural network provides a generalization capability useful to reduce the effect of noise on the estimation process and recognize cases not occurred in the training phase.

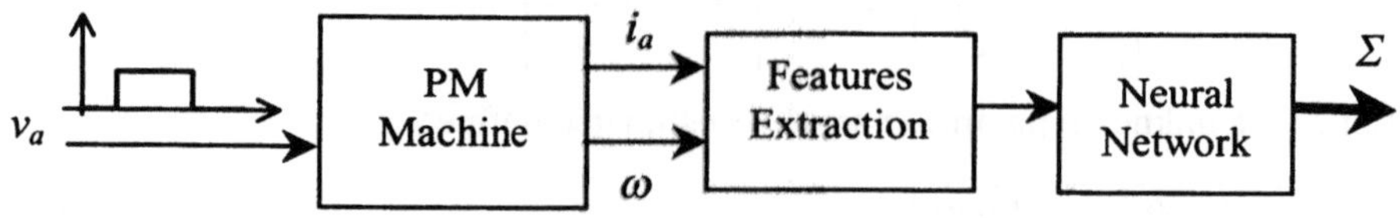

Fig. 3: On-line parameter estimation.

3.2 Training Data Set and Features Extraction

The proposed method raises two questions: how do we choose the training set and which features do we have to extract? We can answer to the first question considering the context in which we operate. Usually parameter estimation is required for diagnosis purposes; in this context we suppose each parameter varying into a restricted range, between a min and a max. We define Σ_o the set of central values of such intervals Σ_{min} and Σ_{max} the sets of minimum and maximum values respectively. Since for each parameter we consider three values, in total, for N parameters we have 3^N training patterns. This large number of patterns can be reduced considering most significant parameters only.

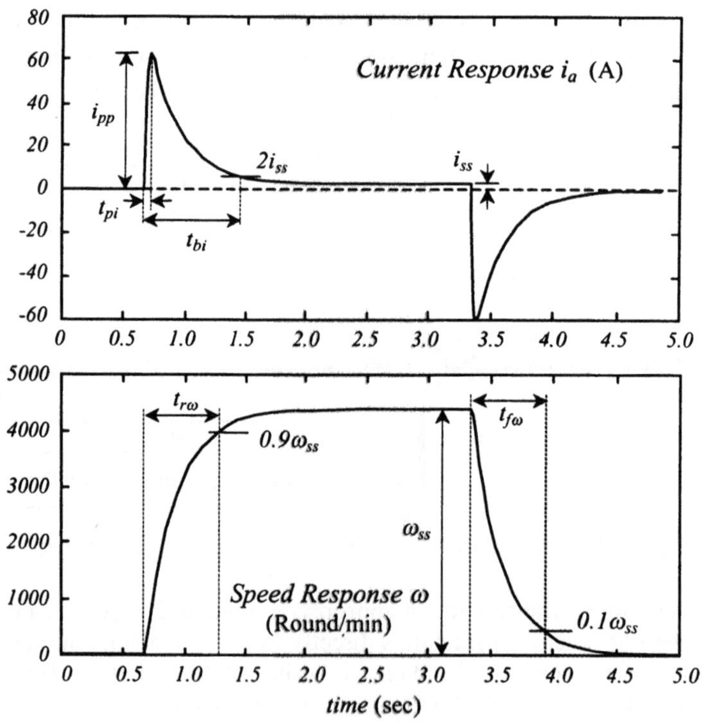

Fig. 4: Features extraction from the current and the speed responses.

As regards feature extraction, some special characteristic points of the PM model output responses are considered in Fig. 4. The number of features should be at least equal to the number of parameter to be estimated.

In our implementation, 7 features were sent to the neural network as inputs: current (i_{ss}) and speed (ω_{ss}) steady state values, current peak value (i_{pp}), speed rise time ($t_{r\omega}$) and fall time ($t_{f\omega}$), current peak time (t_{pi}) and bedding time (t_{bi}).

For an actual PM machine, before performing the feature extraction, the unwanted noise produced on the current response from the brushes (Fig. 5) should be remove using a simple low-pass filter.

As regards network structure, we have adopted a feed-forward structure with two layers. Due to their generalization capability neural network with spline activation functions is used. For more details refer to [9]-[11].

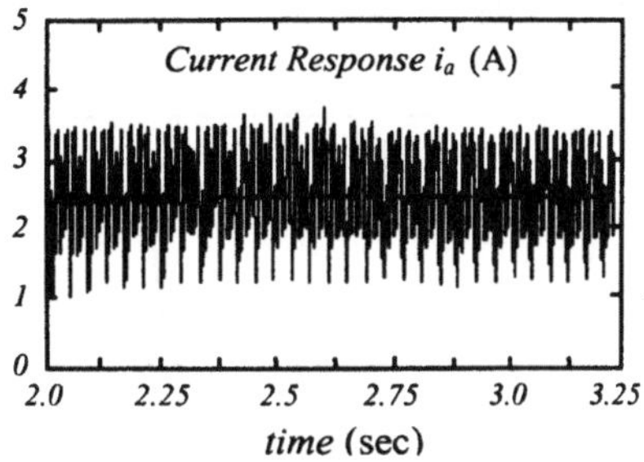

Fig. 5: Noisy current response.

4. Simulation and Results

The described procedure was first validated using simulation studies. A PM machine was simulated to create synthetic data by the model above. For each combination of parameter values, the supplied current and rotational speed responses were derived. Following the defined criteria, features extraction was executed and features were collected to create the neural network training data set.

After network training, a test data set was created in order to test the method proposed. To this purpose, parameter values belonging to the defined ranges were selected and relative responses, obtained from model, were sent to the network. Table 1 shows the estimation results for some fundamental parameters.

R_a (Ω)		K_v=K_m (V·s)		F_t (W·s^2/rad)	
Original	*estimated*	*original*	*estimated*	*original*	*Estimated*
0.120	0.112	0.0225	0.0229	$0.73 \cdot 10^{-4}$	$0.87 \cdot 10^{-4}$
0.224	0.225	0.0225	0.0224	$0.73 \cdot 10^{-4}$	$0.76 \cdot 10^{-4}$
0.120	0.113	0.0275	0.0272	$0.73 \cdot 10^{-4}$	$0.74 \cdot 10^{-4}$
0.224	0.216	0.0275	0.0270	$0.73 \cdot 10^{-4}$	$0.76 \cdot 10^{-4}$
0.120	0.110	0.0225	0.0224	$2.18 \cdot 10^{-4}$	$2.34 \cdot 10^{-4}$
0.224	0.225	0.0225	0.0229	$2.18 \cdot 10^{-4}$	$2.38 \cdot 10^{-4}$
0.120	0.119	0.0275	0.0277	$2.18 \cdot 10^{-4}$	$1.86 \cdot 10^{-4}$
0.224	0.218	0.0275	0.0274	$2.18 \cdot 10^{-4}$	$1.97 \cdot 10^{-4}$

Table. 1: Estimated parameters from simulation.

As it can be seen from the table, estimated values are in some cases quite different respect to the original values; this occurs because the system is not observable, so different parameter sets can provide identical output responses. In fact, comparing model responses from estimated parameters and from original parameters we could check their likeness also in the worst case (Fig.6).

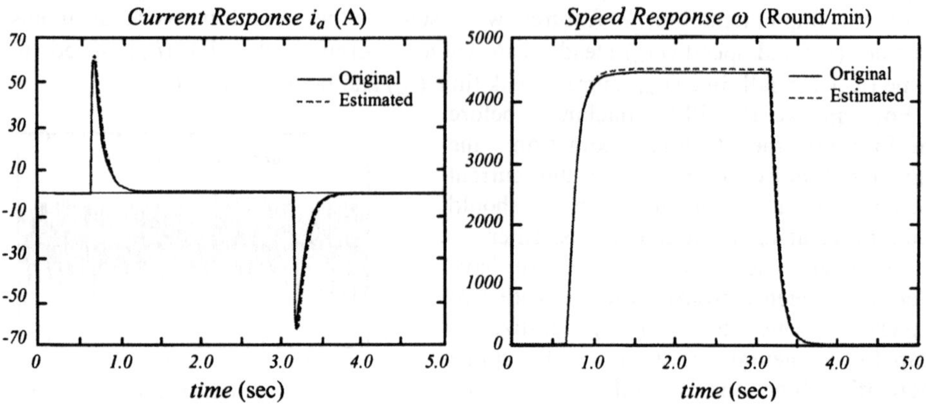

Fig. 6: Original responses and estimated responses comparison.

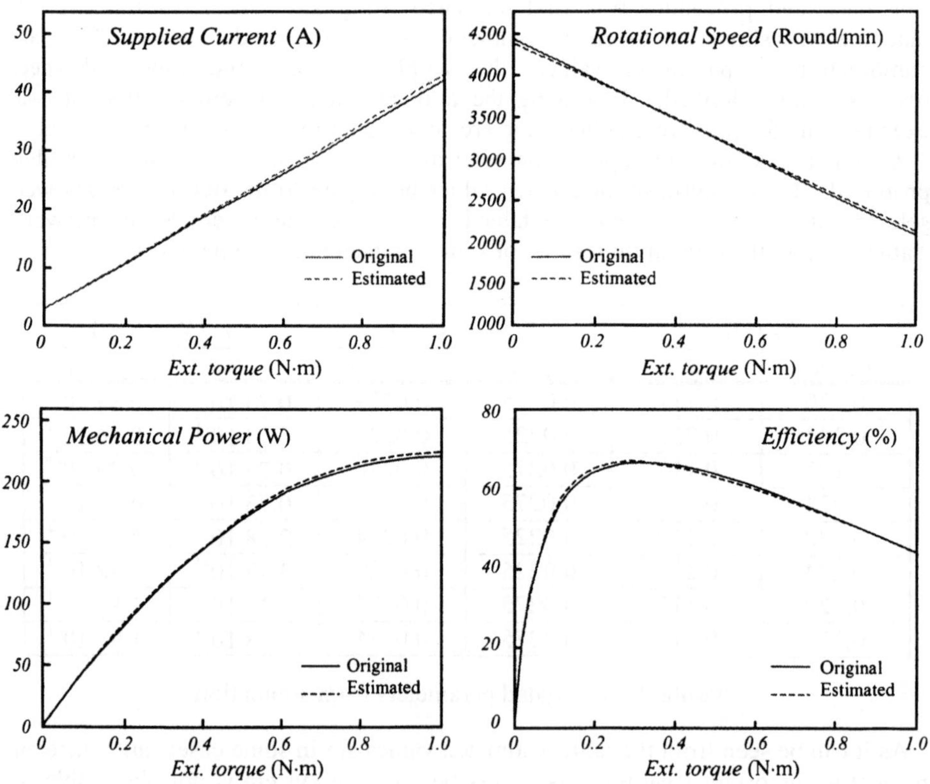

Fig. 7: Steady-state curves comparison: plots of supplied current, rotational speed, mechanical power and efficiency versus external torque.

Non-perfect parameter estimation doesn't compromise the purpose of the method. Let's consider the quality control problem; in this context we are interested basically in the steady-state motor behavior. In particular motor specifications such as relations between current, speed and mechanical power versus external torque are required. Once obtained motor parameters in the no-load situation, we can work on them to simulate other cases. Load presence affects the mechanical equation and could be treated as a contribute in frictional coefficient and in inertia moment according to the following relations:

$$F_t = F_{no-load} + F_{ext} \tag{7}$$

$$J_t = J_{no-load} + J_{ext} \tag{8}$$

Since we are interested in the steady-state condition, only the first contribute can be considered. Fig.7 shows plots of supplied current, rotational speed, mechanical power and efficiency versus external torque. All these curves are obtained from the model varying the term F_{ext}. Solid lines represent values computed by the model from original parameters, while dashed lines are the ones derived from the estimated parameters. As we can see, simulation results validate the method proposed.

5. Conclusions

The present paper focuses on the on-line permanent magnet machine quality control using neural networks. First a neural network with spline activation functions was trained with an appropriate training data set. Next, the neural model was applied for on-line parameter estimation in the no-load situation. Estimated parameters were exploited to calculate steady state values in other cases. Comparing actual specifications with those estimated we could ratify the procedure validity. The proposed method provides better performances both in terms of estimation time and noise reduction respect to standard identification ones.

References

[1] S.Weisgerber, A.Proca and A.Keyhani, "Estimation of Permanent Magnet Motor Parameters". *IEEE Industry Application Society, Annual* Meeting, New Orleans 1997.

[2] A.Keyhani and S.I.Moon, "Maximum Likelihood Estimation of Synchronous Machine Parameters and Study of Noise Effect from Flux Decay", in *IEE Proc.*, vol. 139, no. 1, pp. 76-80, Jan. 1992.

[3] I.Kamwa, P.Viarouge and M.Ferfra, "Modeling and Identification of Permanent Magnet Synchronous Machines from Standstill Time Response Tests Using a Non-Linear Method", IEEE 1993.

[4] A.Keyhani, H.Tsai and T.Leksan, "Maximum Likelihood Estimation of Synchronous Machine Parameters from Standstill Time Response Data", in *IEEE Trans. Energy Conv.-9 (1)*, pp. 98-114, March 1994.

[5] E.S.Boje, J.C.Balda, R.G.Harley and R.C.Beck, "Time-domain Identification of Synchronous Machine Parameters from Simple Standstill Tests", in *IEEE Trans. Energy Conversion,* vol.5, *pp.* 164-170, July 1969.

[6] S.Horning, A.Keyhani and I.Kamwa, "On-line Evaluation of a Round Rotor Synchronous Machine Parameter Set Estimated from Standstill Time-Domain Data", *IEEE Trans. Energy Conversion* 1997.

[7] D.C.Hanselman, "Brushless Permanent-Magnet Motor Design", *Mc Graw Hill,* New York 1994.

[8] P.Vas, "Parameter estimation, condition monitoring diagnosis of electrical machines", Oxford Science Publication, *Oxford*, 1993.

[9] S.Guarnieri, F. Piazza and A.Uncini, "Multilayer Feedforward Networks with Adaptive Spline Activation Function", accepted for publication on *IEEE Transaction on Neural Network.*

[10] F.Filippetti, A.Uncini, F.Piazza, P.Campolucci, C.Tassoni and G.Franceschini, "Neural Network Architectures for Fault Diagnosis and Parameter Recognition in Induction Machines", *IEEE Melecon'96*, Bari, Italy, May 1996.

[11] L.Vecci, F.Piazza and A.Uncini, "Learning and Approximation Capabilities of Adaptive Spline Activation Function Neural Networks", *Neural Networks*, Vol. 11, No. 2, pp.259-270, March 1998.

Section 8
Neural Networks in Economics

Fuzzy local algorithms for time series analysis and forecasting

Silvio Giove
Dept. of Applied Mathematics
University of Venice, Dorsoduro, 3825/E, Venice (ITALY)
E-mail: sgiove@unive.it

1. Introduction

The analysis of non linear dynamic systems is a very important field of research for a lot of real applications. Mainly for economical and financial dynamic systems, sometimes it is very difficult to define a mathematical model, both owing to the lack of knowledge relative to the input-output relationships, and the unpossibility to reproduce artificially the data with repeated experiments. On the other side, a fuzzy logic system can identify a model even from a scarce data base, with other advantages, that are robustness, adaptive, and noise filtering properties, easiness of implementation, maintenance, comprehensibility. For real financial markets we hypothyse that time series shown some characteristic dynamic behaviour that can be observed during the time evolution, the sampled values being the realisation of an unknown, and maybe complex, dynamical system generated by an unknown function f, like the following one:

$$x(t+1) = f\{x(t-k+1),\ldots,x(t-1),x(t)\} + \varepsilon(t) \qquad (1.1)$$

where $\varepsilon(t)$ is the error component. This idea is the main point of all the quantitative analysing methods, like auto-regressive linear model, non linear approach, up to Artificial Intelligence methods. The *efficient market* hypothesis, see [20], [50], contrasts with the presence of a deterministic component, but now this point of view is supported both by theoretical and statistical consideration, partially justified by the psychological behaviour of the investors, see [18]. In this way, some typical *patterns* can be observed and classified. Let us point out again the difficult that we usually encounter when analysing a financial time series, which often shows complex dynamic behaviour like *chaotic* motion [32]. On the other side, let us note that fuzzy logic application showed optimal performances in pattern recognition analysis, see [9], [38], with the aim to discover characteristic patterns in a lot of engineering applications, and it seems to be a very promising approach even for time series analysis.

2. *Local* and *global* fuzzy models

Let us consider a time series $\{x(t)\}$, that is the realisation of an *unknown* and not observable dynamic process. From the sampled values, a model of the dynamical system can be derived, and this model can be used to obtain an estimation of forecasted value of the series k steps ahead , $\hat{x}_{t+k|S}$, where S is the *informative set* available in t, the collection of all the measurable information. In what follows, for notational simplicity, we write $\hat{x}_{t+k}$ instead of $\hat{x}_{t+k|S}$.

In contrast with a *global* approach, we speak of a *local* model if the identification procedure refers only to a suitably defined neighbour of the current point; conversely, if the model extracts knowledge of the dynamic behaviour for all the state space (or a subset of the state space, depending on the samples availability), we refer to a *global* model. Methods based, for instance, on the nearest neighbour are local models, while the inference rule approach, the cluster analysis algorithm and so on, are global ones.

We describe two fuzzy logic approaches to compute $\hat{x}_{t+k}$, a fuzzy version of the *nearest neighbour* approach [14], [15], and a *local* estimation procedures [47], with an application to a filtering pre-processing of a noise corrupted time series [27]. The main concept used in fuzzy modelling is the *fuzzy similarity* measure between k-patterns [52]. The *incomplete* *k*-pattern at time i is the vector $\underline{X}_i^k = \{x(i-k+1),\ \ (i-k+2),...,x(i-1)\}$, and the *complete* *k*-pattern at time t is the vector $\underline{X}_i^k = \{x(i-k+2,...,x(i-1),x(i)\}$. Let φ be a function of the two *k*-patterns $\underline{X}_i^k, \underline{X}_j^k$; φ is a *similarity* measure between $\underline{X}_i^k, \underline{X}_j^k$ if it satisfies the following two properties:

$$\begin{aligned} &\varphi(\underline{X}_i^k, \underline{X}_j^k) = \varphi(\underline{X}_j^k, \underline{X}_i^k) \\ &\varphi(\underline{X}_i^k, \underline{X}_i^k) \geq \varphi(\underline{X}_i^k, \underline{X}_j^k), \quad \forall i, j \end{aligned} \tag{2.1}$$

Often φ is defined by a *triangular* or a *pseudo-gaussian* membership function. In the first case, we have the following analytical form for the function φ:

$$\varphi(z) = \begin{cases} 1-\left|\dfrac{z}{h}\right|, & |z| \leq h \\ 0, & |z| > \mathrm{h} \end{cases} \tag{2.2}$$

which depends on the value $h \in R^+$. The values of h and k, the length of the incomplete pattern, need to be optimised. To this aim, in most of practical applications the *cross-correlation* method is used, see [54], varying the values of $h \in R^+, k \in N$ until the quadratic forecasting error is minimised.

Conversely, a global model tries to describes the input-output relationships over the complete input-output space. A lot of methods exist in the specialised literature for global fuzzy models identification, see among others [10], [31], [36], [44], [51], [55], [56], [58], [59], [60]. Limiting our interest to the so-called *Takagi-Sugeno-Kang* form, *TSK* for brevity, the fuzzy rules representation of a dynamical system appears as follows, the function $f()$ representing the *consequent* of the fuzzy inference rule:

$$\begin{aligned} &IF\ \{x(t)\ is\ A_{i1})\ AND\ (x(t-1)\ is\ A_{i2})\ AND...AND\ (x(t-R+1)\ is\ A_{iR}\} \\ &THEN\ x(t+1) = f\{x(t),...,x(t-k+1)\}\ i = 1,..,L \end{aligned} \quad (2.3)$$

If we suppose to observe the time series $\{x(t)\}$, the input variables are the last sampled R values in the series, that, collected together, form the current pattern $\underline{X}_t^k$. The fuzzy sets $A_{ij} \in V = \{L, M, H, VH,..\}$ are suitable *term-sets* (the *linguistic labels* $L, M, H,...,$ stay for natural language attribute, like *Low*, *Medium*, *High*, and so on), whose membership degree is represented by the following continuous fuzzy number $A_{ij}(x_{t-j})$, see [59]:

$$A_{ij}(x(t-j)) = \exp(\frac{x_{t-j} - \mu_{ij}}{\sqrt{2}\sigma_j})^2, \quad j = 0,..,k-1 \quad (2.4)$$

The conjunction *AND* is implemented using a suitably selected *triangular norm*, see [24], and the normalized antecedent membership degree for the *i-th* rule, say v_i, is computed as follows:

$$v_i = \frac{\tau_i(x_t,..,x_{t-k+1})}{\sum_{l=1}^{L} \tau_l(x_t,..,x_{t-k+1})} \quad (2.5)$$

where:

$$\tau_i(x_t,..,x_{t-k+1}) = \prod_{j=0}^{k-1} A_{ij}(x_{t-j}) \quad (2.6)$$

In this case, the model output $\hat{x}_{t+1}$ is given by:

$$\hat{x}_{t+1} = \sum_{i=1}^{L} v_i(x_t,..,x_{t-k-1}) \cdot f\{x(t),...,x(t-k+1)\} = \sum_{i=1}^{L} v_i \cdot f(\underline{X}_t^k) \quad (2.7)$$

where:

$$\underline{\theta}_i^T = \left[c_{i,0}, b_{i,0}, b_{i,1}, .., b_{i,k-1}\right] \tag{2.8}$$

and the model parameters are $\mu_{ij}, \sigma_{ij}, \underline{\theta}_i, j = 1,...,k;\ i = 1,..,L$. We speak of a *zero-order* model if the consequent is constant for all the rules, and of a *first order* model if $f()$ is a linear function, in this case, the inference rules and the model output become as follows:

$$\begin{aligned} &IF\ \{x(t)\ is\ A_{i1})\ AND\ (x(t-1)\ is\ A_{i2})\ AND...AND\ (x(t-k+1)\ is\ A_{i-k}\} \\ &THEN\ x_{t+1} = c_{i,0} + b_{i,1}x(t) + ... + b_{i,k-1}x(t-k+1),\ i = 1,..,L \end{aligned} \tag{2.9}$$

$$\hat{x}_{t+1} = \sum_{i=1}^{L} v_i(x_t, .., x_{t-k-1}) \cdot \{c_{i,0} + b_{i,0}x_t + .. + b_{i,k-1}x_{t-k-1}\} = \sum_{i=1}^{L} v_i \cdot (\underline{\theta}_i^T \underline{X}_t^k) \tag{2.10}$$

where the dependency of v_i on its arguments is omitted for notational simplicity. It is evident the similarity between the fuzzy *TSK* dynamic model, and the so called *threshold non linear* model, discussed in the specialised econometric literature.

3. The fuzzy nearest neighbour

At the time t, we observe the incomplete pattern $\underline{X}_t^k$, and our aim is to compute $\hat{x}_{t+1}$, the *one step ahead forecaster* of the time series, that is the best estimation of the future value of the series, $x(t+1)$, computed in t. Let us firstly recall the classical *nearest neighbour* method, see [29], that considers the minimum distance between $\underline{X}_t^k$ and all the other observed patterns in the sampled series. Let $\underline{X}_i^k$ be the pattern with lowest distance from $\underline{X}_t^k$; the algorithm assigns $\hat{x}_{t+1}$ to $x(i)$, the value of the series that immediately follows the last component of the last component of $\underline{X}_i^k$:

$$\hat{x}_{t+1} = x(i),\quad i : \arg\min_j \varphi(\underline{X}_t^k, \underline{X}_j^k) \tag{3.1}$$

The performances of this procedure, measured on the basis of the squared error, are sometimes not satisfying at all; the algorithm is in some way quite *rigid.* A possible modification (N-nearest neighbour) considers the N patterns with minimum distance from the current pattern $\underline{X}_t^k$. Let I_t be the set containing the N patterns closest to $\underline{X}_t^k$; the N-nearest neighbour assigns to $\hat{x}_{t+1}$ the average value of the $x(j), j \in I_t$, weighted with the values of the similarity function. In [26] the following similarity function was selected:

$$\varphi(\underline{X}_j^k, \underline{X}_t^k) = e^{-\frac{(\underline{X}_j^k - \underline{X}_t^k)^2}{\sigma_t^2}} \tag{3.2}$$

with:

$$\sigma_t^2 = \frac{1}{N} \sum_{j \in I_t}^{N} \left| \underline{X}_t^k - \underline{X}_i^k \right|^2 \tag{3.3}$$

Then the one step ahead N -nearest neighbour forecaster is given by:

$$\hat{x}(t+1) = \sum_{j \in I_t} x(t_j + 1) e^{-\frac{(\underline{X}_j^k - \underline{X}_t^k)^2}{\sigma_t^2}} \tag{3.4}$$

A disadvantage of this method is the high sensitivity to outlier values; in the next paragraph a way to solve it is presented. Moreover, it seems quite reasonable that $\hat{x}_{t+1}$, even if depending on all the values of a suitable neighbour of $\underline{X}_t^k$, such dependency decreases with the similarity degree between $\underline{X}_t^k$ and the other ones. This is the idea that leads to the *fuzzy* nearest neighbour algorithm: $\hat{x}_{t+1}$ should be as close as possible to $x(i)$ if $\underline{X}_t^k$ is close to $\underline{X}_i^k$. This can be done, see [14] for all the mathematical details, using the following weighted sum:

$$\hat{x}_{t+1} = \sum_{i=k}^{t-1} \frac{\varphi(\underline{X}_t^k, \underline{X}_i^k)}{\sum_{j=k}^{t-1} \varphi(\underline{X}_t^k, \underline{X}_j^k)} x(i) \tag{3.5}$$

Note that this formulation is very close to what the statistician call the non-parametric Nadaraya-Watson estimator with kernel function φ, see [42]. If $\varphi(\underline{X}_t^k, \underline{X}_i^k) = 0,\ \forall i$, then it is impossible to compute $\hat{x}_{t+1}$, because there is no similarities between patterns. The performances depend on the pattern length k and on the choice of the function φ whose parameters need to be optimised using some statistical criterion. This method is very simple but efficient, and easy to be implemented. In [14] a variant is proposed, starting from the consideration that for financial time series (for instance, when dealing with returns) the *punctual* value $\hat{x}_{t+1}$ it is not so much important, as the concordance or discordance between the signs of $\hat{x}_{t+1}$, the forecaster computed in t, and the real value, unknown in t. In this case, the values of h, k are optimised trying to minimise the signs error over a pre-defined *learning* set. A first benchmark series was performed using two computer simulated *chaotic* maps, the Hènon map and the Mackey-Glass map. The learning set was formed by 508 and the test set from the other 73 ones. The results were very good in both the cases, as it can be seen from Fig. 1 that reports the results for the Mackey-Glass map [39], the continuous and the dotted lines representing the real series and the forecasted one respectively.

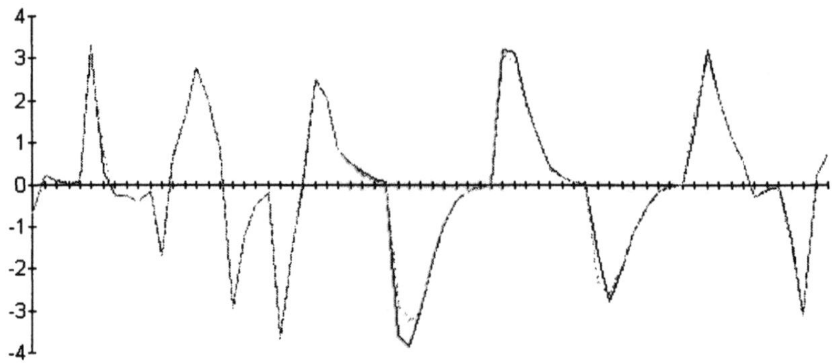

Fig. 1. Mackey-Glass map: real and forecasted values

Subsequently, the algorithm was applied to the one step ahead forecasting of the time series of the Italian COMIT return index, using 508 data as the learning set, and 73 as the test set, with not satisfactory results, measured on the basis of the square forecasting error. Better results were obtained trying to forecast the *sign* of the series. In fact, more than 50% values of the signs are predictable, using the last 7 values as regressors, $k = 7$, and with $h = 1$, being h the membership *window length*, see formula (2.2). Moreover, the value of 38.9 was obtained for the *profitability*, that is a common index used to measure the efficiency of a forecasting strategy. It consists of a measure of what an investor can gain if he uses the forecasted values, against another theoretical investor that has a perfect knowledge of the future values of the series, without taking into account expenditure, taxation, economic frictions such as transaction costs, and so on. In this case, the obtained profitability is considered a good value if compared with other ones obtained using different strategies, such as neural nets, see [1], [2], [3], [11], [18], [28], [33], [34].

4. A joint cluster-nearest neighbour for fuzzy dynamical models

With the aim to introduce a pre-filter analysis of the N nearest neighbour, in [26] a suitable algorithm introduce a preliminary *cluster analysis* is previously applied. This propose, that differs from other fuzzy cluster approaches, see for instance [16], [22], consists into clustering the sampled point not on the basis of the distances between the uncomplete patterns, but, conversely, on the basis of the output values, thus we speak of a *output-cluster* pre-processing. In so doing, the idea present in [56] for zero order models, is extended to any first order *TSK* dynamic models. To perform the cluster analysis, the following new distance function is introduced:

$$E = \sum_{\tau=k}^{t-1} \left\{ x(\tau+1) - \sum_{i=1}^{L} \left[c_{i,0} + \sum_{j=0}^{k-1} b_{i,j} x(\tau - k + j) \right] \right\} \tag{4.1}$$

where L is the number of the clusters $C_1, C_2, ..., C_L$. The value of L can be optimised using the AIC information criterion. A significant computational advantage with respect to other fuzzy cluster algorithms is that, paying attention only to the consequent of each rule, we deal with only one variable. When the cluster procedure stops, we obtain L clusters represented by the parameters of L hyperplanes, that best approach the unknown function over all the space plane. Then, each output sampled point is modified, since it is substituted with the closest hyperplane value. In this way, possible outliers are filtered away, and its noise corrupted value is substituted by the *reconstructed* one. After this initial pre-processing, the forecaster value is then computed by the previously introduced N nearest neighbour method.

The algorithm assigns each k-pattern $\underline{X}_\tau, \tau = k,..,t$ to the hyperplane C_i with minimum distance the relative output value, $x(\tau + 1)$, that is:

$$\underline{X}_\tau \in C_i \Leftrightarrow i = \arg\min_l \left\{ x(\tau+1) - \left[c_{l,0} + \sum_{j=0}^{k} b_{,lj} x(\tau - k + j) \right] \right\}^2 \tag{4.2}$$

and $x(t)$ is then substituted by the following locally *filtered* value, $\hat{x}(t+1)$:

$$\hat{x}(\tau + 1) = \underline{\theta}^T \underline{X}_t^k, \forall \tau \le t - 1 \tag{4.3}$$

with $\underline{6}$ is given by (2.8); note the difference between $\hat{x}(t+1)$ and $\hat{x}_{t+1}$: the first one is the filtered value obtained from (4.3), while the other one is the value of the forecaster. After the output-cluster phase, σ_t^2 and $\hat{x}_{t+1}$ are compute from (3.3), (3.4). The following Fig. 2 shows the output-clustering pre-processing, with two clusters (hyperplanes); note how the sampled point are moved to one of the two hyperplanes, and the effect is more sensitive for the grey bolded point in the left part, a suspected anomalous value.

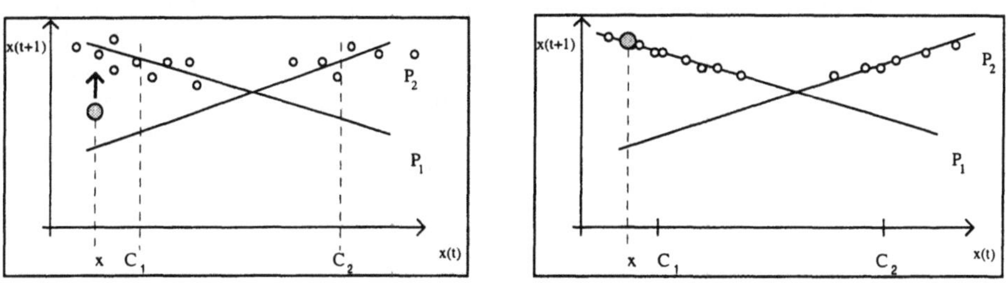

Fig. 2 original sample points (left side) and moved points, with two output clusters.

Some experimental tests showed good robustness properties in the presence of artificially introduced outliers. Moreove , applying the procedure to real market time series, better results were obtained with respect to the mere use of the *N-nearest neighbour* by alone. The first numerical test uses the artificially reproduced MacKey-Glass time series, added with an uniformly distributed noise and a *shock* noise in the time instants $t_1 = 70, t_2 = 170, t_3 = 270$. Using the first 500 data as the learning set and the remaining ones as the test set, after the optimisation of the parameters, we found a slightly better result using the pre-proceesin algorithm; in particular the average quadratic error was 0.23 against 0.25, obtained using the *N*-nearest neighbour with no output-cluster. Subsequently the procedure was applied to forecast the time series of the returns of the Italian COMIT index sampled from march 1995 up to july 1997, obtaining interesting results for what concerns the average profitability.

5. The algorithm *FLLA*

An other local algorithm is the so called *FLLA* (Fuzzy Linear Local Approximation), that is a refinement of the nearest neighbour method; in fact, instead of locally approximation of the unknown function with piecewise constant values, this algorithm computes an approximation using a set of locally linear hyperplanes. To this aim, let be $(\underline{X}_i^k,1)^T = \underline{X}_k^t \cup \{1\} = \{x(i-k+1), x(i-k+1),..$ $.., x(i),1\}^T$ the *augmented* uncomple *k*-pattern at time i, and let $\underline{\beta}$ be the *regressors* vector obtained solving the following weighted problem:

$$\underline{\beta}_t := \arg \min_{\underline{\beta} \in R^{k+1}} \sum_{i=k+1, i \neq t}^{t-1} \left[X_i - \underline{\beta}_t^T \cdot (\underline{X}_i^k,1)^T\right]^2 \cdot \varphi(\underline{X}_t^k, \underline{X}_i^k) \tag{5.1}$$

Then, $\hat{x}_{t+1}$ can be computed as follows, see [47]:

$$\hat{x}_{t+1} = \underline{\beta}^T{}_t \cdot (\underline{X}_t^k,1)^T \tag{5.2}$$

In [27] the *FLLA* approach was used to *filter* and *reconstruct* a noise corrupted time series, implementing an *outlier* detection algorithm. The algorithm consists of a two-phase procedure; in the first one a fuzzy filtering local method uses the fuzzy similarities degrees between patterns to select the suspicious anomalous values (outliers), while in the second phase the identified outliers are substituted with values computed by the *FLLA* algorithm. To this aim, a particular *similarity ratio* index is computed for the current value $x(t)$ in the following way:

$$\zeta_t \equiv \begin{cases} \dfrac{\widetilde{W}_t}{W_t}, \text{if } W_t \neq 0 \\ 0, \quad \text{if } W_t = 0 \end{cases} \tag{5.3}$$

where:

$$\widetilde{W}_t \equiv \sum_{i=k+1}^{t-1} \varphi(\underline{X}_t^{k+1}, \underline{X}_i^{k+1}),\ W_t \equiv \sum_{i=k+1}^{t-1} \varphi(\underline{X}_t^{k}, \underline{X}_i^{k}) \tag{5.4}$$

The index ζ_t measures the ratios between the similarity degrees of complete patterns versus the incomplete ones. Moreover, ζ_t satisfies some interesting properties, in particular $0 \leq \zeta_t \leq 1$. It seems quite rational that the more ζ_t is small, the more $x(t)$ will be classified an outlier; conversely, $x(t)$ is not an outlier if $\zeta_t \cong 1$. Of course, ζ_t depends on h and k, whose optimal values empirically determined, see the quoted paper for the technical details. The method was applied to a set of data created by using the Mackey-Glass time series artificially corrupted by *observational* and *dynamical* noise; from empirical simulation it was able to identify the outliers artificially introduced, otherwise impossible to detect, even from visual inspection of the time series. Subsequently the algorithm was applied to the Italian Stock Index COMIT, with no any other data pre-processing, showing good robustness properties. It is theoretically verified that, under some general hypotheses, the same method can be used to compute a non parametric estimation of the conditional probability density of the time series, with no other hypothesis about the unknown stochastic process; this approach can be used to obtain the empirical distribution of the so called *scenario* analysis, to be subsequently used for a dynamic portfolio optimisation, see [12], [30], [41], [49], [57].

6. Conclusions

In this paper some fuzzy logic local method for dynamic systems analysis and time series forecasting are briefly revised and discussed. In particular, some methods based on modifications of the *nearest neighbour* method and the *linear local approximation* algorithm were proposed. The procedures were tested both on artificially created chaotic time series and on real market index series, showing good performances, in particular for the high robustness and noise filtering properties. Some problems still remain open, we limit to quote the definition of efficiency procedures for the parameters optimisations, in fact the most used approach, the cross-correlation algorithm does not guarantee the global optima. Moreover, it was not here discussed the problem of selection of the *best* regressor set, that is formed by the input variables that really affect the output value, some suggestions for this topic can be found, for instance in [37], [40], [43], [56].

References

[1] Apostolos Refenes P., ed., *Neural networks in the capital markets*, John Wiley & Sons, Chichester, England, 1995.

[2] Azoff E., *Neural networks time series forecasting of financial markets,* Wiley & Sons, Chichester, 1994.

[3] Baestaens D.E., et al., *Neural networks solutions for trading in financial markets*, Pitman Publishing, London, 1994.

[4] Barnett V., Lewis T., *Outliers in statistical data*, Wiley & Sons, Chichester, 1994.

[5] Bezdek J.C., et al., Generalized k-nearest neighbor rule, *Fuzzy Sets and Systems*, n. 18, 1986.

[6] Bezdek J.C., *Pattern recognition with fuzzy objective function algorithms*, Plenum Press, New York, 1981.

[7] Birge J.R., Decomposition and partitioning methods for multistage stochastic linear programs, *Operational Res*., vol. 33, 989-1007,1985.

[8] Birge J.R., Mulvey J.M., Stochastic programming, in: *Mathematical programming for industrial engineers*, Avriel M, Golany eds., M. Dekker, N.Y., 543-5741996, 1996.

[9] Bishop C.M., *Neural networks for pattern recognition*, Clarendon Press, Oxford, 1995.

[10] Bo Cho K., Wang B. H., Radial basis function based adaptive fuzzy systems and their applications to systems identification and prediction, *Fuzzy Sets and Systems*, 83, 1996.

[11] Buckley J.J., Hayashi Y., Fuzzy neural networks: a survey, *Fuzzy Sets and Systems*, 66, 1994.

[12] Bunn D.W., Salo A.A: Forecasting with scenarios, *European Journal of Operational Research*, 68, 291-302, 1991.

[13] Canestrelli E., Giove S., Pellizzari P., Financial Time Series Analysis by Fuzzy Rules and Clustering Methods, *Internal Report n. 18*, Dept. of Applied Mathem. and Computer Sc., 1994.

[14] Canestrelli E., Giove S., Sogliani A., Financial time series fuzzy forecasting, *Badania Operacyjne i Decyzyjne*, n.3, 1996.

[15] Casdagli M., et al., State space reconstruction in the presence of noise, *Physica D*, **51**, pp. 52-98, 1991.

[16] Chiu S. L., Fuzzy model identification based on cluster estimation, *Journal of Intelligent and Fuzzy Systems*, 2, John Wiley & Sons, 1994.

[17] Cootner P.H., *The random character of stock market prices*, Mit Press, Cambridge, 1964.

[18] Deboek G. J., *Trading on the edge*, John Wiley & Sons, Chichester, England, 1994.

[19] Dupachová J., Multistage stochastic programs: the state of the art and selected bibliography, *Kybernetika*, vol. 31, n. 2, 151-174, 1995.

[20] Fama E. F., Efficient capital markets a review of theory and empirical work, *Journal of Finance*, 25, 1970.

[21] Fan J., Gijbels I., *Local polynomial modelling and its applications*, Chapman and Hall, Lonadon, 1996.
[22] ForgyE.W., *Cluster analysis of multivariate data*, Biometrics, 21, 1965.
[23] Fuller R., *Fuzzy reasoning and fuzzy optimization*, Turku Centre for Computer Science, n. 9, 1998.
[24] Fuller R., *Neural fuzzy systems,* Abo Akademi, Ser. A:443, 1995.
[25] Genther H., Glesner M., Advanced data preprocessing using fuzzy clustering techniques, *Fuzzy Sets and Systems*, 85,1997.
[26] Giove S., Mendo M., Un metodo congiunto di cluster analysis e del metodo nearest neighbour per la previsione di serie storiche, Atti del Convegno A.I.R.O. '98, Treviso, 1998.
[27] Giove S., Pellizzari P., Time series filtering and reconstruction using fuzzy weighted local regression, in: *Soft Computing in Financial Engineering*, Physica-Verlag, Heidelber, 1999.
[28] Goonatilake S., Treleaven P., *Intelligence systems for finance and business,* Wiley & Sons, Chichester, 1995.
[29] Hattoti K., Torii, Effective algorithms for the nearest neighbour method in the clustering problem, *Pattern Recognition*, n.26, 1993, pp. 741-746.
[30] Holmer M.R., Mckendall R., Vassiadou-Zenoiu C., Zenios S.A., Dynamic model for fixed-income portfolio management under uncertainty, *Department of Public and Business Administration*, University of Cyprus, march 21, 1996.
[31] Hong T.-P., Lee C.-Y., Induction of fuzzy rules and membership functions from training examples, *Fuzzy Sets and Systems*, 84, 1996.
[32] Hsieh D.A., Chaos and nonlinear dynamics: application to financial markets, *Journal of Finance*, 46, 1991.
[33] Ishikawa M., Learning of modular structured networks, *Artificial Intelligence*, 75, 1995.
[34] Ishikawa M., Moriyama T., Prediction of time series by a structural learning of neural networks, *Fuzzy Sets and Systems*, 82, 1996.
[35] Ivanknenko A. G., et al., The review of problem solvable by algorithms of the group method of data handling (GMDH), Pattern recognition and image analysis, 5, n.4, 1995.
[36] Lee M., et al., A new neuro-fuzzy identification model of nonlinear dynamic systems, *Int. Jour. of Approximate Reasoning*, n. 44, 1994.
[37] Lin Y., Cunninghma G.A., Coggeshall S.V., Input variable identification - fuzzy curves and fuzzy surfaces, *Fuzzy Sets and Systems*, 82, 1996.
[38] Looney C.G., *Pattern recognition using neural networks*, Oxford University press, Oxford, 1997.
[39] Mackey M., Glass L., Oscillation and chaos in physiological control systems, *Science*, **197**, pp. 287-289, 1977.
[40] Miller A.J., *Subset selection in regression*, Chapman and Hall, London, 1990.
[41] Mulvey J.M., Ziemba W.T., Asset and liability allocation in a global environment, *Handbooks in Op. Res.*, 9, Elsevier Science, 1995.
[42] Nadaraya E., On estimating regression, *Th. Prob. Appl.*,**9**, pp. 141-142, 1964.
[43] Nelles O., Structure optimization of Takagi-Sugeno fuzzy models, *Int. Journal of Uncertainty, Fuzziness and Knowledge-Based Systems*, 6, n. 2, 1998.
[44] Nozaki K., Ishibuchi H., Tanaka H., A simple but powerful heuristic method for generating fuzzy rules from numerical data, *Fuzzy Sets and Systems*, 86, 1997.

[45] Parzen E., On estimation of a probability density function and mode, *Ann. Math. Statist.*, vol. 33, 1962.
[46] Pellizzari P., Pizzi C., Fuzzy-like conditional density estimation in time series outliers detection, submitted to *Technometrics*, 1996.
[47] Pellizzari P., Pizzi C., Linear local approximation for financial time series forecasting, *Rendiconti per gli Studi Economici Quantitativi,* University of Venice, 1997, available at URL:http:\\pluto.dma.unive.it\~paolop \rendi.ps.Z.
[48] Pellizzari P., Pizzi C., Fuzzy-like conditional density estimation in time series outliers detection, *Technical Report* 1996-1, Dipartimento di Statistica, Venezia.
[49] Rockafellar R.T., Wets R.J., Scenario and policy aggregation in optimization under uncertainty, *Mathematics of Operations Research*, vol. 16, n. 1, 119-147, 1991.
[50] Scheinkman J.A., Le Baron B., Nonlinear dynamics and stock returns, *Journal of Business*, 62, 311-337, 1989.
[51] Simpson P.K., Fuzzy min-max neural networks - part 2: clustering, *IEEE Trans. fuzzy systems*, 1, n.1, 1993.
[52] Sneath P., Sokal R., *Numerical Taxonomy*, Freeman, San Francisco, 1973.
[53] Specht D.F., A general regression neural network, *IEEE Trans. On Neural Networks*, vol. 2, n. 6, 1991.
[54] Stone M., Cross-validatory choice and asessement of statistical predictions, *J. Royal Statis. Soc.*, B36, 1977.
[55] Sugeno M., Kang G.T., Structure identification of fuzzy model, *Fuzzy Sets and Systems*, 28, 1988.
[56] Sugeno M., Yasukawa T., A fuzzy logic based approach to qualitative modeling, *IEEE Trans. On Fuzzy Systems*, 1, n.1, 7-31, 1993.
[57] Wallace S.W., Solving stochastic programs with network recourse, *Networks*, 16, 1986.
[58] Wang B.H., Mendel J.M., Fuzzy basis functions, universal approximation, and orthogonal least squares learning, *IEEE Trans. on Neural Networks,* 3, 1992.
[59] Wang B.H., Mendel J.M., Generating fuzzy rules by learning from examples, *IEEE Trans. on System Man and Cybernet.*, SMC-22, 1992.
[60] Wang L., Langari R., Sugeno Models, fuzzy discretization, and the EM algorithm, *Fuzzy Sets and Systems*, n.82, 1996..

Neural Networks Applications in Economics: a Statistical Point of View[1]

N. Carlo Lauro, Cristina Davino, Domenico Vistocco
Dipartimento di Matematica e Statistica
Università degli Studi di Napoli - Federico II
Napoli, Italy

Abstract

Nowdays neural networks (NN) are applied in the most various fields and are actually receiving a lot of attention among the researcher's community. In this paper we will provide a review of some NN applications in economics. We distinguish the applications according to the main objectives achieved by NN in this field: *prediction*, *classification* and *modeling economic theory*. It is a matter of fact that NN share with statistics a lot of methodological and computational aspects [9] as well as many fields of application. In this framework we introduce a *general strategy for a statistical approach to NN* which allows to use NN in a statistical context taking into account typical statistical applications problems, such as the selection and coding of the variables, the sample representativeness but more the interpretation, visualization and stability of the results.

1 Introduction

Research activities in Neural Networks (NN) have seen an explosion of interest over the last few years. Inspired by biological systems, NN can be defined as computational models represented by a collection of simple units, the neurons, interlinked by a system of connections, the weights [9]. A training set, i.e. a sample of observed units, is used as input information of the neurons so that the network updates the weights (learning phase) to satisfy an optimum criterium. Once the final weights are identified, the neural network can be used for prediction of unseen cases.

The wide success of NN can be attributed to some key features which make these models valuable and attractive for many tasks. First, NN are able to generalize the results to unknown situations without imposing distribution hypotheses and are data-driven methods as they learn from experience even if the underlying relationships are unknown. Second, by emulating human brain through the parallel transmission of the information, NN can be implemented in parallel computer architectures. Third, NN are able to analyze complex problems characterized by non-linear relationships among data. Finally, NN

[1]Finalcial support: MURST60% "Nuovi strumenti e metodi interpretativi per l'impiego delle reti neuronali in statistica" (Resp. Prof. Carlo Lauro).

are universal function approximators, as they can approximate any continuous function to any desired accuracy.

It is a matter of fact that the literature on NN applications in economics is growing quickly. Some typical applications of NN can be found in the following fields: *finance* (e.g portfolio selection, stock market and exchange rate prediction), *economics* (e.g. game theory, modelling economical laws), *banking* (e.g. loan scoring and bankruption prediction) and *marketing* (e.g. market segmentation).

The aim of this paper is, first of all, to provide a *review of some NN applications in economics.* These applications are distinguished according to the main objectives achieved by NN in economics: *prediction*, *classification* and *modeling economic theory.* In this paper we consider only feed-forward multilayer networks, which are those ones used to solve the previous problems.

Many of NN applications in economics can be easily recognized as fields where traditional statistical methods are applied. At this regard, the second aim of this paper is to view *NN from a statistical perspective* introducing a *Statistical Neural Network* i.e. a statistical approach to NN which allows to use NN in a statistical context.

2 Neural Networks Applications in Economics

2.1 Prediction

The prediction of the financial markets is a topic of great interest: the recent literature shows that NN are one of the most used tools in this framework. The objective of the prediction methods is to use a window of n past values to predict a window of m future values (windowing technique), without any prior specification of the model.

The *exchange rate prediction* is one of the example of applications of NN. The traditional economical theories are able to explain the past history of the exchange rates but are not able to make predictions while NN show good results also in case of non-linear and complex problems.

Many companies use NN to face financial problems, for example the Banca San Paolo of Brescia has developed an integrated previsional system based on NN (ISTRIA) [17]. One of the possible use of ISTRIA is the prediction of the exchange rates. The application presented gives a lot of attention to the preprocessing phase and to the best model selection, through a genetic-evolutive approach.

Peroni [40] compares traditional methods such as ARIMA models with NN to predict the DM/$ exchange rate. The NN architecture is defined using three criteria (SIC, AIC and PSC) and splitting the sample in training and test set. The results show that recurrent networks are most suitable to analyze temporal data.

Belcaro, Canestrelli e Corazza [5] use NN to predict the daily values of the COMIT index on the italian market. The trend of these prices is typically non-stationary, characterized by many outliers and with observations time dependent. A logaritmic transformation has been applied on the original data. As regard to the architecture selection, the authors have analyzed different possibilities. The number of input neurons has been fixed in 15 days (three market weeks) or 22 days (one month week), while the number of output neurons, that is the number of daily values predicted, in 1, 2, 3, 5 days. Moreover, the dimension of the hidden layers has been defined through three approaches: Kolgomorov, DARPA, Bottleneck networks. During the learning phase a linear reduction of the learning rate and a stopping rule with a particular gaining function are used. The predictions are evaluated through some indicators in order to choose the best network. Even if the prediction results are satisfactory and they try to interpret the final weights, no attention is devoted to the comparison with alternative methods.

The *prediction of the industrial production index* has been realised by Centrale dei Bilanci [8] using monetary and economic indicators. Each phase of the application seems to be faced empirically, the results are evaluated through the R^2 coefficient, without any comparison with other methods.

Various methods of predicting a stock's future systematic risk (regression techniques, aritmetic means and NN) are discussed in the paper of Wittkemper and Steiner [57]. They use the financial statements of 67 German corporations from 1967 to 1986 and show that the most precise forecasting are given by NN whose architecture is determined by a genetic algorithm.

2.2 Classification

As regard to the classification objectives the *detection of failures* is one of the most usual applications. This kind of analysis can be distinguished in *business failures* and in *credit scoring.*

The detection of business failures is based on ecomical and financial indicators of the enterprises; the objective is to identify financial and economical lack of balance that can cause the business insolvency.

Lacher et al. [32] propose a neural network to estimate the future financial health of businesses. Their approach, based on the cascade correlation algorithm, allows to overcome the restrictive assumptions of the Multiple Discriminant Analysis, the architecture selection problem of NN and to obtain good percentages of correct classification.

The detection of company failures made by Bardos and Zhu [4] is also based on a comparison of Discriminant Analysis and NN. The authors give particular attention to the initial selection of the variables (descriptive analysis, sensitivity analysis, etc.) and to the stability of the results (cross-validation) while the architecture selection is realized through a "trial and error" procedure.

Credit and banking institutions are very interested in the credit scoring problem that is in the research of models which allow to identify bad and good loans. Recently the role of artificial intelligence techniques such as NN in helping bankers make loans and detect frauds has been highlighted by many papers.

Usually the authors compare the ability of NN and traditional techniques. Desai et al. [16], for example, consider linear discriminant analysis, logistic regression and NN in building credit scoring model. They use the large database of three credit unions in the Southeastern United States for the period 1988 through 1991. The techniques utilized are compared not only through the percentages of correct classification but also using a paired t-test on these percentages.

Yobas, Crook and Ross [52] compare the predictive performance of a traditional technique such as linear discriminant analysis with evolutionary techniques such as NN, genetic algorithms and decision trees in order to distinguish good and slow payers of bank credit card accounts. The best NN is choosen empirically trying different network topologies, learning and momentum rates, activation functions and epoch numbers. The comparison among different techniques is only based on the percentage of correct classification.

2.3 Modelling of Economic Theory

In the framework of economical theory, different problems are faced: *artificial agent simulation*, *bounded rationality models*, *game theory*, *equilibrium selection*.

Terna [49] develops an ad-hoc heuristic for the application of NN to the construction of adaptive artificial agents. This kind of agents exploits NN capability to learn and to manage complex and non-linear situations, as agents have to modify their behaviour according to the external environment and to actions of other agents. Terna proposes to look inside the black-box through the study of the derivatives of the output values with respect to the input values, so it is possible to extract some rules from the networks in order to identify the role of each element of the network. An original tool to define the NN architecture and to interpret the parameters is proposed with the *Cross Target* tecnique. This is a way to use NN in a particular context exploiting their advantages and fitting them to the specific aspects of the economic theory.

The same requirement to take into account an environment continuously updating is expressed in Baldassarre [2]. The bounded rationality models ignore the cognitive processes of the economic agents as they are based on a typology of perfectly rational agents. NN and other artificial intelligence techniques such as genetic algorithms, can be used when the agents behaviour seem not to be due to optimization criteria but to cognitive processes such as learning, perception, memory.

3 Neural Networks and Statistics

The use of NN in economic applications gave rise to an unvoidably overlap of these computational models with some statistical models so that now NN share with statistics a lot of methodological and computational aspects as well as some fields of applications.

At this regard the statistical community is divided into *two opposite streams.* On one side some people are crazy about NN as they are only worried about the best perfomance. On the other side some statisticians hardly criticize these models for the lack both of originality and of theoretical support:

"*There is nothing new or magical about neural networks. Artificial neural networks are sometimes used as "black-box models" with lots of parameters [...] Neural network research seems to have contributed almost nothing at all save for jargon and confusion*" (E. Robert Tisdale, user group comp.ai.neural-nets, november the 6th 1994).

With regard to this point, many NN can be viewed as reproductions of statistical methods but using a *different terminology*; in the following list some statistical terms are associated to the corresponding NN terms [43]:

- model $\longleftrightarrow$ architecture;
- statistical unit $\longleftrightarrow$ pattern;
- independent variable $\longleftrightarrow$ input neuron;
- dependent variable $\longleftrightarrow$ output neuron;
- estimation $\longleftrightarrow$ training;
- parameters $\longleftrightarrow$ weights;
- Regression Model $\longleftrightarrow$ supervised network;
- Binary Segmentation $\longleftrightarrow$ supervised network;
- Discriminant Analysis $\longleftrightarrow$ supervised network;
- Principal Component Analysis $\longleftrightarrow$ supervised and unsupervised network;
- Correspondance Analysis $\longleftrightarrow$ supervised and unsupervised network
- Cluster Analysis $\longleftrightarrow$ unsupervised network;

The structure of a NN can be associated to the statistical model: input and output information are related respectively to explanatory and response variables, whereas the weights are the parameters to be estimated on the basis of the individual units which are the sample of observations; statistical criteria,previously defined, are optimized by the learning algorithm.

In the last few years, many NN have been constructed to reproduce statistical methods. The simple *perceptron* with linear transfer function can be used to estimate the parameters of a *Multiple Regression Model* [55]; using a logistic transfer function the final weights of the neural network are equal to the coefficient of the *Logistic Regression Model* [45]. The *Kohonen maps* [30] can be compared to the *Cluster Analysis* while a *Multi-Layer Perceptron* with one hidden layer can reproduce the *Correspondance Analysis* [34] and the *Principal*

Component Analysis [3], and a *Multi-Layer Perceptron* with two hidden layers can reproduce a *binary tree* [46].

In this framework we consider very important to investigate on the *complementary use of NN and statistics* so to exploit the advantages of each approach and supply the weaknesses:

"*Neural Networks and Statistics are not competing methodologies for Data Analysis. There is considerable overlap between the two fields......Better communication between the fields of Statistics and Neural Networks would benefit both*" [43].

Moreover, using *NN in a statistical context* requires to solve both typical NN problems, such as the architecture definition, parameters tuning, the weights initialization, and typical statistical applications problems, such as the selection and coding of the variables, the sample representativeness and the interpretation, visualization and stability of the results.

At this aim the construction of the Statistical Neural Network (SNN) can be structured in the following phases:

- *pre-processing data*: a descriptive analysis of the data, a possible data reduction via a factorial analysis or a graphical model, and the coding of the variables;
- *architecture selection*: the selection of the number of hidden layers and the number of neurons in each of them;
- *neural model definition:* the functional relationship among variables;
- *learning*: the estimation of the weights through suitable training algorithms;
- *stability*: the evaluation of the network performance and the validation of the results through resampling techniques;
- *post-processing:* the de-coding of data for the interpretation of the results;
- *visualization and interpretation of the results*: the construction of typical statistical charts such as residual plots, distributions of the estimates, etc.

In the following sections we will concentrate only on the phases which highlight how statistics can contribute to neurocomputing.

3.1 Pre-processing and Post-processing of the data

NN are often considered "computational machines" that take numerical input and produce numerical output. In other words they are able to analyze information without any prior analysis of the data. In fact, NN requires, as many statistical techniques, a *pre-processing* phase which consists of a descriptive analysis of the data, a possible data reduction, the coding and transformation of the variables. This phase is very important as it influences the generalization ability of the network, the computational costs and the time required.

- *Descriptive analysis of the variables.* Univariate statistics, frequency distributions, hystograms and other descriptive tools can be useful to have a first view of the data and to decide possible data transformations.

- *Reduction of the variables.* Many multivariate statistical methods allow to reduce the number of variables to the best informative ones. At this aim the Principal Component Analysis [36] reduces a data set with p variables to $q < p$ variables that synthetize the structural part of the variability. Binary Segmentation [6] procedures automatically select the most discriminant variables ignoring redundant or irrelevant information.

- *Coding of the variables.* The variables used in a neural network are coded in different mode according to their nature (continuous or nominal). The formers are represented in the neural network by a single neuron, while the latters by as many neurons as the number of modalities they have. In particular, the qualitative variables need to be transformed from a reduced coding to a complete disjoint coding by generating m binary variables, each associated to a modality. As a matter of fact, it is sufficient to represent such a qualitative variable with *m-1* neurons, as the last modality is univocally determined.

- *Transformation of the variables.* The standardization of the variables is advisable when the variables are expressed in different units of measure or to avoid computational problems. Moreover, in other cases depending on the transfer function used, a sort of normalization should be realised in the following way:

$$ValNorm = ValMin + \left(\frac{Val - Min}{Max - Min}(ValMax - ValMin)\right)$$

  where *ValMin* is the lower limit of the normalization range, *ValMax* is the upper limit of the normalization range, *Min* and *Max* are, respectively, the minimum and the maximum of the variable to be normalized, *Val* is the value of the variable to be normalized.

  When a logistic transfer function is used, for example, the quantitative input variables and the binary output variables have to be normalized. In that case, if the value of the parameter α which defines the slope of the logistic function is equal to 1, the parameters *ValMin* and *ValMax* have to be set respectively to -5 and 5 for the input neurons and 0.1 and 0.9 for the output neurons.

  Once the analysis is complete and the results are obtained, a ***post-processing*** phase is necessary, to bring back the results to the original scale of the data. It should be noted that the accuracy obtained by the NN should be based on the rescaled data set that is the performance measures should be calculated on the post-processed output values.

3.2 Architecture Selection

NN are widely used for their capability to "learn from experience" and to generalize the results to unknown situations that is they are able to analyze data belonging to the same sample space of the training set but with different characteristics. It is very important to control the training phase trying to balance the fitting of the observed sample with the generalization of the network. This ***generalization ability*** is not only related to the training time but also to a ***proper architecture selection***: a neural network with a complex architecture, even if it is able to fit perfectly the training set, will be not able to generalize the results to new data (this phenomenon is well known as ***overfitting***). This means that from a certain time during the training period the network does not improve its ability to solve problems any more, but starts to learn some random regularities contained in the set of training patterns [44].

The empirical solution used to decide a proper selection of the neural network architecture consists in training several NN with different architectures in order to choose the one which gives the lowest error rate combined to the simplest architecture.

The *Cascade Correlation* algorithm [20] solves the problem of the architecture selection during the training phase by training, in the first step, a simple network (***single layer network***), then adding neurons to the network until the desired goal is reached.

In this framework we believe that some statistical methods can help to determine a neural architecture which guarantees a good generalization. At this aim, in the present paper we will describe the pruning methods, the binary segmentation method, some algebraic estimates and some recent contributions.

3.2.1 Pruning Methods

The pruning methods reduce the complexity of the network through the "pruning" of neurons and/or connections not used [41]. These methods can be distinguished in two categories according to the elements pruned (neurons or connections) and the way these elements are pruned: the ***sensitivity algorithms*** and the ***weight decay algorithms.***

- ***The sensitivity algorithms.*** The network is trained until a fixed error goal is reached and then a ***saliency measure*** is computed to measure the importance of each neuron and each connection. The element with the lowest value of the saliency is removed and the new network is trained again. This process stops when the error obtained after the elimination of an element is higher than the error obtained before this elimination. The sensitivity algorithms differ one from another for the saliency measure computed [50].

  The simplest algorithm, called ***magnitude based pruning***, removes after each training the connection with the lowest absolute value.

Le Cun et al. [35] proposed the *optimal brain damage algorithm* which is based on the effect produced on the error function by the elimination of a connection.

The *Skeletonization method* proposed by Mozer and Smolensky [38] is again based on the effect that the elimination of an element produces on the error function. The difference with the previous method is that Mozer and Smolensky introduce a *gating* term (α) in the net input of each neuron so that the output of the neuron j is:

$$o_j = f\left(\sum_i w_{ij}\alpha_i o_i\right)$$

If $\alpha = 0$ the neuron has no importance in the network while if $\alpha = 1$ the neuron is not removed. The saliency measure of the *Skeletonization method* is equal to the difference between the value of the error function in case the neuron is removed and in case it is not removed:

$$\rho_i = E_{\alpha_i=0} - E_{\alpha_i=1} \approx \left.\frac{\varrho E}{\varrho\alpha_i}\right|_{\alpha_i=1}$$

If a connection has to be removed the saliency measure is:

$$S_{ij} = E\left(w_{ij}=0\right) - E\left(w_{ij}=w_{ij}^f\right)$$

where the first element is the value of the error function when w_{ij} is removed and the second element is the value of the error function if the connection is not removed and after the training is equal to w_{ij}^f.

The saliency measure proposed by Karnin [29] is:

$$S_{ij} = -\frac{E\left(w_{ij}\right) - E\left(0\right)}{w_{ij} - 0} w_{ij}$$

where w_{ij} is the value of the connection after the training, 0 the value of the connection if it is removed and $E(0)$ the corresponding error.

- *Weight decay algorithm.* This class of methods adds a cost term to the error function which penalizes complex network solutions: during the training phase a decay term is introduced so that the weights weaker are removed. The Weight decay algorithms differ for the decay term they add to the error function.

  Hanson e Pratt [26] proposed the following cost function:

$$C_\lambda\left(w\right) = \lambda\sum_i w_{ij}^2$$

where the parameter λ ($\lambda \in [0, \infty)$) control the strenght of the decay term.

Weigend et al. [56] proposed:

$$C_\lambda(w) = \lambda \sum_i \frac{(w_{ij}/w_0)^2}{1 + (w_{ij}/w_0)^2}$$

Using the Weight decay algorithms the training and the pruning phases go on together but the setting of the parameter λ is arbitrary.

3.2.2 Binary Segmentation

A binary tree obtained from a binary segmentation procedure can be used to decide the number of hidden layers, the number of neurons in each layer so as to solve the so called "credit assignment problem" that is the definition of the desired output of the hidden neurons during the learning phase [46].

A binary segmentation procedure [21] classifies N units, on which some explicative variables and a dependent variable are observed, in homogeneous groups through binary partitions of the sample. The result of a binary segmentation procedure is a binary tree where each path from the root node to the terminal nodes can be considered as a sequence of AND logic operations plus an OR logic operation in the terminal nodes.

Let $\mathbf{x}_1, \mathbf{x}_2, ..., \mathbf{x}_I$ be the explicative variables observed on N cases and J the classes of the response variable.

Sethi [46] proposed to represent a binary tree in a multilayer network with two hidden layers. In particular a binary tree T with t non terminal nodes and $t+1$ terminal nodes can be represented in a neural network R with I input neurons, t neurons in the first hidden layer, $t+1$ in the second hidden layer, J in the output layer and with threshold transfer functions.

According to this approach the architecture of the network depends on the structure of the tree: the number of neurons in the input layer is equal to the number of explicative variables, in the first hidden layer to the number of non terminal node, in the second hidden layer to the terminal nodes and in the output layer to the classes of the response variable.

The first hidden layer is the *partitioning layer* that divides the features space into several regions, the second hidden layer is the *ANDing layer* that performs ANDing of partitioned regions and the output layer is the *ORing layer* that logically combines the results of the previous layer to produce disjoint decision regions.

Once the neural network architecture is defined, it is possible to train the network modifying the transfer functions with soft functions as the logistic one. In this way it is possible to improve the results obtained with the binary segmentation procedure.

3.2.3 Algebraic Estimates of the Prediction Risk

The algebraic estimations of the prediction risk can be used for estimating the quality of model predictions and for model selections [37]. Many useful algebraic estimates have been proposed:

- *Mean Squared Error* (MSE):

$$MSE = \frac{1}{N}\sum_{n}\sum_{j}(t_{nj} - o_{nj})^2$$

where t_{nj} is the target output and o_{nj} is the observed output for each pattern n and for each output neuron j. The MSE may not be appropriate for NN model building since it ignores the important information about the number of parameters (weights) the model has to estimate. From the statistical point of view, as the number of weights goes up, the degrees of freedom of the model go down, thus favouring the possibility of overfitting in the training sample. At this regard, the following estimates of the prediction risk take into account this number of parameters of the model.

- *Generalized Cross Validation* (GCV) [23]:

$$GCV = MSE\frac{1}{(1-\frac{k}{N})^2}$$

where k is the number of parameters of the model and N the number of patterns.

- *Final Prediction Error* (FPE) [1]:

$$FPE = MSE(\frac{1+\frac{k}{N}}{1-\frac{k}{N}})$$

Note that although GCV and FPE are slightly different for small sample, they are asymptotically equivalent for large N.

- *Predicted Squared Error* (PSE) [27]:

$$MSE + 2\hat{\delta}^2\frac{k}{N}$$

where $\hat{\delta}^2$ is the estimation of the residual variance.

- *Generalized Prediction Error* (GPE) [31]:

$$GPE = MSE + 2\hat{\delta}^2{}_{eff}\frac{\hat{k}_{eff}}{N}$$

where $\hat{k}_{eff}$ is the estimation of the number of parameters of the model and $\hat{\delta}^2{}_{eff}$ is the estimation of the residual variance.

- *Schwarz Information Criterion* (SIC) [31]:

$$SIC = \log \frac{SSE}{N} + \frac{k \log N}{N}$$

where N is the number of patterns, SSE is the sum of squared errors and k is the number of parameters.

- *Akaike Information Criterion* (AIC) [31]:

$$\log \frac{SSE}{N} + \frac{2k}{N}$$

- *Predictive Stochastic Complexity* (PSC) [31]:

$$PSC = \frac{1}{N-k} \sum_{n=k+1}^{N} (t_n - o_n)^2$$

3.2.4 Some Recent Contributions

The problem of the selection of the NN architecture has been indirectly faced by some authors who reproduced some multidimensional data analysis tecniques in a neural network.

A supervised neural network [13] has been proposed for *Latent Budget Analysis* [15]. This method analyses a contingency table where the row variable is the explanatory variable (with I modalities) and the column variable is the response variable (with J modalities) in order to approximate the conditional frequencies distributions (observed busgets) by theoretical budgets obtained as linear combination of K latent budgets with mixing parameters. A neural network with I inputs, K hidden neurons and J outputs, can be used to estimate the parameters of a latent budget model. The latent budget network proposed provides some advantages both to NN and to statistics. From one side, it allows to previously define the number of input, hidden and output neurons and to interpret the weights as probability (parameters of the latent budget model); from the other side, this approach can be used for complex models parameters estimation and for a different interpretation of the latent budget model as it is possible the prediction of the values of the response variable for a new case.

A neural network has been proposed [12] to reproduce the *Principal Component Analysis onto a Reference Subspace* (PCAR) [10]. This method analyses the dependence between a set of dependent variables **J** and a set of explicative variables **I** through the principal components. The architecture of the Multilayer Perceptron used to reproduce the PCAR is thus previously defined and it is composed by I explicative input neurons, J dependent output neurons and C hidden neurons that correspond to the number of principal component associated. The neural network proposed could be easily generalized to reproduce non linear PCAR by changing the transfer functions between layers from linear to non linear functions thus solving the problem of complex models parameters estimation. More, the typical black box effect of NN is removed and it is possible to interpret each hidden neuron and each weight of the network as they correspond respectively to the principal components and to the eigenvectors of the PCAR.

3.3 The Stability of the Results

Once a network has been trained, it is very important to evaluate the accuracy of the predictions and, in particular, the ***stability of the results.*** A standard procedure for evaluating the performance of a model is the so called ***stopped training rule*** which consists in splitting the data into a ***training set*** (used to modify the weights), a ***validation set*** (used to determine the stopping point of the training phase) and a ***test set*** (used to test the final values of the weights). The neural network is trained for a certain number of epochs using the training set; the network performance is evaluated using the validation set and the network is trained again. The procedure stops when the error on the validation set does not improve anymore. Performing the training procedure after such break-point would not reduce the misclassification error rate and would cause overfitting. This approach can be very sensitive to the specific sample splitting and can be not advisable when the sample is not very large. An alternative way to evaluate the stability of the neural network results uses the ***resampling techniques*** [39] [33] [11].

The concepts of internal and external stability [24], which refer respectively to the influence of anomalous observations in the sample and to the representativeness of the sample, can be used in a neural network framework to evaluate the capability of the NN to hold out to noisy affected data or anomalous data (***internal stability***) and the capability to generalize the results (***external stability***).

The first type of stability (internal) can be evaluated through the jackknife and cross-validation methods. The *jackknife* [18] consists in eliminating at each step a row of the data matrix to test the importance of this row.

The *cross-validation* [48] can be considered a generalization of the jackknife as the sample £ with N observations is divided in V subsets $£_1, £_2, ..., £_V$ with $d = N/V$ elements so that $£_1 \cup £_2, ..., \cup £_V$ e $£_i \cap £_j = 0 \; \forall \; i \neq j$.

The neural network is trained using each of the V samples $£ - £_v$ ($v=1,..., V$) as training sets and the $£_v$ ($v=1,..., V$) samples as test sets.

The estimation of the error of prediction (E_{cv}) is a weighted mean of the errors of prediction obtained after each training (E_V):

$$E_V = \frac{1}{d} \sum_{y_v, x_v \in £v} [y_v - \hat{y}_v]^2$$

$$E_{CV} = \frac{1}{V} \sum_{v=1}^{V} E_v$$

The *cross-validation* is very useful when the number of observations in the sample is not very large and it takes into account all the elements of the sample in the learning phase and in the test phase.

The external stability of the neural network, that is the generalization capability, can be evaluated using the bootstrap method [19]. Let $\mathbf{X}(N\times P)$ be the data matrix and $\mathbf{y}$ the corresponding output vector $(N\times 1)$, the *bootstrap* consists in resampling with replacement B matrices and the corresponding output vectors $\mathbf{y}^{*b}$ from the $\mathbf{X}$ matrix (b=1,...,B).

Both the internal and the external stability can be evaluated using the empirical distributions of the mean squared errors obtained training the jackknife, cross-validation and bootstrap matrices. The value of the MSE, whose stability has to be evaluated, is thus compared with the interval of the empirical distribution centered on the mean and with a width of two times the standard deviation of this empirical distribution ($\mu \pm 2\sigma$). This interval is a confidence interval with a significance level of 5%.

3.4 The Visualization of the Results

A *visual representation* of statistical data is a picture on a two-dimensional surface using symbols, lines, areas, and text to display possible relations between the variables. A NN analysis is rarely completed with graphical displays which could provide one or more of the following *purposes*: recognize patterns in the data, explore relationship among variables, assess the adequacy of the model fitted to the data, suggest remedial actions, etc. [25]. After the neural network has been trained, it can be advisable to evaluate the fit of the model through plots of residuals in order to check if the functional relationships used are right, if outliers and influential observations are present.

4 Concluding remarks

A view of NN economical applications from a statistical point of view show that the leading criterion is very often the search of a good performance in spite of methodologies traditionally used to solve the same problems. In some cases this can be considered satisfactory and sufficient as the specific problem could not be solved in a different way (e.g. short term prediction) or it requires a timely solution even if not entirely transparent (e.g buy-sell decisions). In other cases a good performance is not sufficient as it is necessary to justify the way the performance is obtained. In these situations it is important to have a theoretical support which guides the model selection phase, the interpretation and validation of the results. These features should be implemented also in *NN software*. Actually, NN users can easily find many packages, both freeware and commercial, characterized by an easiness of learning and of use, but also by a lack of other *requirements* that should be taken into account [53]. For example, an intelligent checking of user actions could assist the user in each phase of the analysis, from the pre-processing to the post-processing phase, in order to avoid incoherent choices. The possibility to interface with main statistical packages and to add user defined methods could provide some tools for a proper use of NN in a statistical context.

Thus, the proposed statistical approach to NN should provide a *methodological and practical support* so to exploit the possible synergy that can derive combining statistical methods and NN:
"*Neural Networks and Statistics are not competing methodologies for Data Analysis. There is considerable overlapping between the two fields [...] Better communication between the fields of Statistics and Neural Networks would benefit both*" [43].

References

[1] Akaike H., Statistical Predictor Identification, **Ann. Inst. Statist. Math.**, 22, 203-217, 1970.

[2] Baldassarre G., Reti Neurali ed Algoritmi Genetici per i Modelli Simulativi di Teoria della Razionalità Limitata. Un'Applicazione ai Mercati Oligopolistici, **Rivista di Politica Economica**, 115-157, 1997.

[3] Baldi P., K. Hornik, Neural Networks and Principal Component Analysis: Learning from Examples without Local Minima, **Neural Networks**, vol.2, n.1, 53-58, 1989.

[4] Bardos M., W. Zhu, Comparison of Discriminant Analysis and Neural Networks Applications for the Detection of Company Failures, **Bio-mimetic approaches in management science**, 1997.

[5] Belcaro P., E. Canestrelli, M. Corazza, Modelli Previsivi Neurali: un'Applicazione al Mercato Finanziario Italiano, **Atti del Ventesimo Convegno Annuale A.M.A.S.E.S.**, Università degli Studi di Urbino, 1996.

[6] Breiman L., J.H. Friedman, R.A. Olshen and C.J. Stone, **Classification and Regression Trees**, Wadsworth, 1984.

[7] E. Cavalli, Reti Neurali e Previsione di InFsolvenza, **Atti del ventesimo convegno annuale A.M.A.S.E.S.**, Università degli Studi di Urbino, 1996.

[8] Centrale dei Bilanci (a cura di), Reti Neurali Artificiali: le Applicazioni in Campo Economico e Bancario, **Bancaria**, n. 2, 1996.

[9] Cheng B., D.M. Titterington, Neural Networks: a Review from a Statistical Perspective, **Statistical Science**, vol. 9, n. 1, 1-54, 1994.

[10] D'Ambra L., N.C. Lauro, Analisi in Componenti Principali in Rapporto ad un Sottospazio di Riferimento, **Statistica applicata**, 15, 1982.

[11] Davino C., M. Gherghi, D. Vistocco, Sulla stabilità delle Regole di Classificazione nelle Reti Neuronali, In: A. Bellacicco e C. Lauro (a cura di), **Reti Neurali e Statistica**, 137-147, FrancoAngeli, 1997.

[12] Davino C., G. Giordano, N.C. Lauro, Analisi dei Dati e Reti Neurali per la Conjoint Analysis, In: **Atti del Convegno SIS "La Statistica per le Imprese"**, 299-306, 1997.

[13] Davino C., F. Mola, R. Siciliano, Un Modello Neuronale Simultaneo per l'Analisi dei Bilanci Latenti a Tre Vie, In: **Atti del Convegno SIS "La Statistica per le Imprese"**, 291-298, 1997.

[14] Davino C., F. Mola, R. Siciliano, D. Vistocco, A Statistical Approach to Neural Networks, In: K. Fernandez-Aguirre, A. Morineau (eds.) **Analyses Multidimensionnelle des Données - IV Congrès International NGUS'97**, 37-51, CISIA Ceresta, Paris, 1997.

[15] de Leeuw J., P.G.M. van der Heijden, The Analysis of Time-Budget with a Latent-Time-Budget Model, In Diday et al. (eds.), **Data Analysis and Informatics**, 159-166, North-Holland, 1988.

[16] Desai V.S., J.N. Crook, A Comparison of Neural Networks and Linear Scoring Models in the Credit Union Environment, **European Journal of Operational Research**, 95, 24-37, 1996.

[17] Di Pasquale S., G. Maifredi, M. Temporin, L'Analisi dei Mercati Finanziari con Reti Neurali. Il Modello ISTRIA della Banca San Paolo di Brescia.

[18] Efron B., Bootstrap Methods: Another Look at Jackknife, **Annals of Statistics**, 7, 1979.

[19] Efron B., R. J. Tibshirani, **An Introduction to the Bootstrap**, Chapman & All, 1993.

[20] Fahlman S.E., C. Lebiere, **The Cascade-Correlation Learning Architecture**, Technical Report CMU-CS-90-100, School of Computer Science, Carnegie Mellon University, Pittsburgh, 1990.

[21] Fielding A., Binary Segmentation: the Automatic Interaction Detector and Related Techniques for Exploring Data Structure, In C.A. O'Muircheartaigh and C. Payne (eds.), **The analysis of survey data**, 221-257, 1975.

[22] Giannini C., Reti Neurali Artificiali in Economia: un Commento, **Rivista Italiana di Economia**, 107-110, n. 0, 1995.

[23] Golub G., H. Heath, G. Wahba, Generalized Cross Validation as a Method for Choosing a Good Ridge Parameter, **Technometrics**, 21, 215-224, 1979.

[24] Greenacre M.J., **Theory and Applications of Correspondence Analysis**, Academic Press, 1984.

[25] Hadi A.S., Graphical Methods for Linear Models, In: Rao C.R. (ed.) **Handbook of Statistics**, vol. 9, 775-802, 1993.

[26] Hanson S. J. e L.Y. Pratt, Comparing Biases for Minimal Network Construction with Back-Propagation, In: D. S. Touretzky (eds.), **Advances in Neural Information Processing (1)**, 177-185, 1989.

[27] Hastie T.J., R.J. Tibshirani, **Generalized Additive Models**, Chapman & Hall, 1990.

[28] Hill T., O'Connor M., Remus W., Neural Networks Models for Time Series Forecasts, **Management Science**, 1082-1092, vol. 42, n.7, 1996.

[29] Karnin E.D., A Simple Procedure for Pruning Back-Propagation Trained Neural Networks, **IEEE Transactions on Neural Networks**, vol. 1, n. 2, 239-242, 1990.

[30] Kohonen T., Self-Organized Formation of Topologically Correct Feature Maps, **Biological Cybernetics**, 43, 59-69, 1982.

[31] Kuan C.H., T. Liu, Forecasting Exchange Rates Using Feedforward and Recurrent Neural Networks, **Journal of Applied Econometrics**, 10, 347-364, 1994.

[32] Lacher R.C., P.K. Coats, S.C. Sharma, L.F. Fant, A Neural Network for Classifying the Financial Health of a Firm, **European Journal of Operational Research**, 85, 53-65, 1995.

[33] Le Baron B., A. Weigend, Evaluating Neural Network Prediction by Bootstrapping, **Proceedings of International Conference**, 2, 1207-1212, 1994.

[34] Lebart L., Correspondence Analysis, Discrimination and Neural Networks, **Proceedings of the International Federation of Classification Societies**, 1996

[35] Le Cun Y., J.S. Denker e S.A. Solla, Optimal Brain damage, In: D.S. Touretzky (ed.), **Advances in Neural Information processing (2)**, 598-605, 1990.

[36] Lebart L., A. Morineau, M. Piron, **Statistique Exploratoire Multidimensionnelle**, Dunod, 1995.

[37] Moody J., Prediction Risk and Architecture Selection for Neural Networks, In: V. Cherkassky, J.H. Friedman, H. Wechsler (eds.) **From Statistics to Neural Networks: theory and pattern recognition applications**, 146-165, 1994.

[38] Mozer M.C. e P. Smolensky, Skeletonization: A Technique for Trimming the Fat from a Network Via Relevance Assessment, In: D.S. Touretzky (eds.), **Advances in Neural Information Processing (1)**, 107-115, 1989.

[39] Paass G., Assessing and Improving Neural Networks Prediction by the Bootstrap Algorithm, In: S. Hanson, J. Cowan, C. Giles (eds), **Advances in Neural Information Processing Systems 5 (NIPS5)**, 196-203, Morgan Kaufmann, San Mateo, CA, http://borneo.gmd.de/AS/nn/publi/paass93.nips9301.ps.Z, 1993.

[40] C. Peroni, **Modelli di Previsione a Breve Termine dei Tassi di Cambio**, Quaderni di ricerca n. 103, Università degli Studi di Ancona, 1998.

[41] Reed R, Pruning Algorithm - a Survey, **IEEE Transactions on Neural Networks**, vol. 4, n. 5, 740-747, 1993.

[42] Salzano M., Control of the Complex Economy through Fiscal Variables, **Economics & Complexity**, vol. 1, n.2, 45-68, 1998.

[43] Sarle W. S., Neural Networks and Statistical Methods, **Proceedings of the 9th Annual SAS Users Group International Conference**, 1994.

[44] Schittenkopf C., G. Deco, W. Brauer, Two Strategies to Avoid Overfitting in Feedforward Networks, **Neural Networks**, vol. 10, n. 3, 505-516, 1997.

[45] Scumacher M., R. Robner, W. Vach, Neural Networks and Logistic Regression, **Computational Statistics and Data Analysis**, 21, 1996.

[46] Sethi I.K., Entropy Nets: from Decision Trees to Neural Netwoks, **Proceedings of the IEEE**, vol. 10, n.78, 1605-1613, 1990.

[47] Stern H. S., Neural Networks in Applied Statistics, **American Statistical Association**, vol.38, n.3, 205-214, 1996.

[48] Stone M. , An Asymtotic Equivalence of Choice of Model by Cross-Validation and Akaike's Criterion, **Journal of the Royal Statistical Society**, B39, 44-47, 1977.

[49] Terna P., Reti Neurali Artificiali e Modelli con Agenti Adattivi, Rivista Italiana di Economia, 71-106, n. 0, 1995.

[50] Thimm G., Fiesle E., Neural Networks Pruning and Pruning Parameters, **1st Online Workshop on Soft Computing**, http://www.bioele.nuee.nagoya-u.ac.jp, 1996.

[51] Tibshirani R., **A Comparison of Some Error Estimates for Neural Networks Models**, Technical Reports, ftp://utstat.toronto.edu/pub/tibs/nnerr.ps, 1996.

[52] Yobas M.B., Crook J.N., Ross P., **Credit Scoring Using Neural and Evolutionary Techniques**, Working paper series n.97/2, Credit Research Centre, The University of Edinburgh, 1997.

[53] Vistocco D., C. Davino, Towards a Statistical Approach to Neural Networks: an Integrated Strategy in a New Software Tool, In: **Pre-Proceedings of the International Seminar on New Techniques and Technologies for Statistics**, 521-526, Studio Idea Editrice, Napoli, 1998.

[54] Zhang G., B.E. Patuwo, M.Y. Hu, Forecasting with Artificial Neural Networks: the State of the Art, **International Journal of Forecasting**, 14, 35-62, 1998.

[55] Warner B., M. Misra, Understanding Neural Networks as Statistical Tools, **The American Statistician**, vol. 50, n.4, 284-293, 1996.

[56] Weigend A. S., D. E. Rumelhart, B. A. Huberman, Generalization by Weight-Elimination with Application to Forecasting, In: R. Lippmann, J. Moody e D. Touretzky (eds.), **Advances in Neural Information processing (3)**, 875-882, 1991.

[57] Wittkemper H.-G., M. Steiner, Using Neural Networks to Forecast the Systematic Risk of Stocks, **European Journal of Operational Research**, 90, 577-588, 1996.

Regional Economic Policy and Computational Economics

Domenico Marino[1]
Messina and Reggio Cal. University
Italy
dmarino@unirc.it

Abstract

The main aim of this paper is to design the connection between economic models and new planning tools (neural networks) in a complex economy. Very interesting features of these systems are: self-organisation, learning and self-renforcing mechanisms. These properties are discussed under different points of view in the paper.

1. Introduction

The debate in economics between those who maintain that complexity and its causes plays a decisive role in the construction of models with high levels of realism and those who think that a complete and exhaustive description of economic phenomena can be achieved using linear and equilibrium type models regardless of the complexity of the behaviour of agents and markets is relatively recent.

In this work we would like to further analyse the relationship between complexity and economic policies from the point of view of regional and territorial economics.

The economy as a complex evolving system (Arthur, 1988) therefore implies that:

- individuals are bound to each other by strong relationships;
- dynamic characteristics cannot be represented by means of individual approaches but only by collective properties;
- evolution manifests itself by means of multiple equilibria;
- each equilibrium depends on previous historical interrelations through possible inefficiencies and/or lock-in.

From a conceptual point of view, the main characteristics of the effects that emerge in the dynamic evolution of a system with complex behaviour can be explained by:

- the difficulty prescriptive type regional and territorial policies have in promoting and sustaining economic development;
- the loss of importance of the national dimension: the local dimension clashes with the global dimension;
- the faltering view of economic policy and its propagation mechanism as being based on principles of command and control;

[1 1] Domenico Marino Via Lia, 35 - 89100 Reggio Cal - Italy
Tel +39 348 3815608 Fax +39 965 651042

- the inability of a central planner to govern all the underlying relationships between economic agents at any given time according to linear type response procedures.

Economic, territorial and urban systems are, in this light, considered to be interconnected dynamic systems which present interesting learning, self-organizing and self-renforcing properties (Arthur 1988, Wolfram 1986, Marino 1992b, Gatto-Marino 1994, Diappi et al. 1993).

Governing these dynamic systems and simulating their behaviour becomes extremely complicated if only traditional mathematical instruments are used.

A new branch of economic science called computational economics, which tries to build realistic economic models using cellular automata, neural networks, genetic algorithms and fuzzy variables, has therefore entered the arena.

In this essay we will describe these new mathematical instruments in detail and try to show how it is possible to describe economic-territorial phenomena using these instruments by integrating them with the concept of the Territorial Economic System which can constitute the analytical context of reference which makes these mathematical algorithms applicable to reality.

2. The Territorial Economic System as a Descriptive Category.

The observations and instruments previously introduced would remain mathematical abstractions if they were not ordered in a coherent manner so as to allow their application to reality. It is therefore necessary to define the *scenari* in which these instruments can be applied. In this way the properties of the mathematical instruments are closely linked to the dimensions and characteristics of reality and to the microeconomic behaviour of agents and institutions.

The concept of Territorial Economic System (TES) seems a good way of representing the context in which these instruments find coherence.

A first consideration, which would allow us to use such an approach, is the observation that the relationship between territory and economy has, in recent years, constituted an interesting area of debate from which the need to build an economic space in which the various components, economic agents and environmental factors can interact and synthesize has emerged.

It is therefore from an awareness of the centrality of territory that the need to assign to it an interpretative capacity springs, since a unitary proposition of the elements behind the evolution of an economic system can conceivably be arrived at through it.

As already pointed out, within the TES, the two dimensions that can be considered typical are *proximity* and *resilience.*

These terms are simply another way of expressing the more general concepts of local interaction and self-organization.

The TES therefore constitutes the physical space within which economic agents interact and the equilibrium properties of the system depend on the type of structure it has and, if this space is complex, on the particular attraction basin the system is located in. The concepts of increasing returns, multiple equilibria and history dependence can find justification in the complex space (P. Krugman, 1994).

Spatially extended complex systems, a class to which the TES also belongs, can be reasonably represented by the Ising model (Arthur, 1998; Krugman, 1994; Marino, 1994). The fundamental characteristics of this model are local interaction of a nearest neighbours type (proximity) and the complexity of behaviour (resilience). If we imagine an economic system made up of a series of individuals (be they agents or institutions) interacting with each other and we build for them a sort of local dimension for interactions, then the overall equilibrium of the system is achieved with these local interactions as a starting point.

The widespread "global think, local act" idea behind regional policy finds justification and foundation in such a model. If we think of a territorial economic system as a spatially extended dynamic system, then we can model it according to Ising and consequently consider its dynamic evolution as the motion generated by the possible states arising from certain behaviours of a microeconomic nature. This kind of model also presents an interesting property, the coexistence of the dual concept of competition-cooperation, which regulates the attainment of equilibrium. From the microeconomic point of view, that is, the system is characterized by agents interacting locally in a co-operative manner. Once a sort of local order has been achieved, this system simultaneously enters into competition with the other local orders achieved, so overall equilibrium is a result of competition between these local orders.

A variant on the Ising model is the "spin glass" model. The difference between the two models resides essentially in the nature of the interaction between agents. In the Ising model the single agent is limited to transmitting information, while in a spin glass model each agent can choose whether to transmit or not to transmit (to inhibit) information to neighbouring agents. This makes it possible, for example, to model this kind of system by the use of neural networks, associating a neural network to each agent which will obviously be able to transmit or inhibit information. This last association can constitute a definition, albeit qualitative, of artificial agents.

3. Economic Policies in Spatial Extended Systems: New Paradigms

Description of the evolution of spatialized economies emphasizes the role of new paradigms rather than of classical ones. New factors seem to have replaced the earth, work and physical capital. Natural and environmental resources, human resources and technology are beginning to get the upper hand following the so-called "technological revolution".

Co-operation within businesses and between businesses and business systems takes place on a vertical and a horizontal scale in which the local dimension and the territorial variables constitute the catalyst for processes of development. Technological expertise and social capabilities (Latella - Marino, 1996) are the basic elements capable of explaining the different levels of development that can be observed in different territorial contexts. Territorial variables, in other words, are decisive factors in explaining development differentials, especially when associated with the idea of market conceived as a social construction requiring rules that will guarantee its smooth running given that access rights, exchange mechanisms and opportunities for distribution of the wealth generated not only do not reassemble themselves uniformly and autonomously in time and space (Sen, 1984 and 1985), but almost always require outside intervention to achieve the objectives set for development policies. Re-equilibrium policies thus appear necessary to guarantee a more equitable development process.

Within the market it is necessary to define collective rules which ensure that positive dynamics (increasing return) can develop through the interaction of the agents operating in it. The territorial dimension and the systemic nature of the production process are elements that are fundamental to the understanding and governing of development processes.

Public intervention in such a scenario cannot simply be thought of as a mechanism for allocating resources within the economy but must assume the role of guide and director of processes taking the shape, on the one hand, of a set of actions aimed at defining and guaranteeing individual access rights and, on the other hand, of interventions aimed at developing the exchange capacities of markets and business systems (Bianchi, 1995).

An explanation may be sought in the fact that local communities increasingly interact with the rest of the world in a continuous process of integration and globalization without necessarily responding to stimuli from the central state. This obliges us to re-examine the composition of the economic policy maker's "tool box" and, at the same time, forces us to radically rethink the very meaning of policies government given that the central public authority is no longer able to guarantee the development of the local community in the presence of particular actions enforced by the central authorities (Bianchi, 1995).

Traditional economic policies, when enforced in the context of an open market or of a market characterized by strong interrelations between agents, lose their capacity to produce the expected results because the mechanism of response to the policy maker's input has to deal with a system characterized by high levels of interrelations between individual decisions and which therefore displays collective response characteristics which are different from individual response mechanisms. The consolidated logic of public intervention in economics assumes that the government authority will identify objectives for which the instruments most likely to achieve results which can be verified and therefore simulated are chosen.

Traditional macroeconomic policies only work if acting on a closed system for which it is possible to order objectives and priorities with certainty. In this case the policy maker can govern the system of underlying relations by assuming linear type response mechanisms. If these assumptions are not verified, the complexity of the system makes traditional policies pointless, therefore, to govern a complex system we must equip ourselves with a set of objective instruments and programming actions able to cope with non-linearity and the consequences of complexity.

4. Planning Actions in the TES: Old and New Approach

From the idea that an economy is a "complex evolving system" in which single individuals are linked to each other by strong relationships, it follows that dynamic characteristics cannot be represented by individual approaches but rather by collective properties subjected to subsequent non-reversible scansions (Arthur, 1988). It is thus conceivable that each economic system, in its evolution, might manifest both a multiplicity of equilibria, each dependent on previous historical interrelations, and the presence of inefficiencies and lock in which can be selected during the evolutionary course of the system to the detriment of possible efficient solutions.

Government of an economy thought of as a complex evolving system therefore excludes the possibility that commands might be expressed with a prescriptive type mechanism in mind, as would happen if the system being analysed were essentially closed and characterized by low levels of interactions between agents.

To this we must add the considerable incidence of variables of a territorial nature. Territory cannot be thought of simply as a physical support for business activities but must itself become an active factor conditioning the exploitation of local resources and the capacities of single businesses to cope with international competition. Therefore, the general objective of regional policy becomes that of structural adjustment with a view to greater economic and social territorial integration.

So new regional policy must firstly contemplate a "transactive" rather than a "prescriptive" type of approach and the basis for any action must consider not just "what must be done" but "in what manner, by what procedures and with whom". This means making systematic and widespread use at all levels of the principle of subsidiarity which implies that decisions should be taken as near as possible to the problem and be appropriate to its solution, and individual responsibilities should also be identified using the same criterion.

Thus the main task of decision-makers in each TES is to aim at reassembling the rules and re-establishing the access rights which are the basis of any subsequent action designed to re-appropriate local culture and raise the threshold of contextual knowledge. On these premises it is possible to imagine the transfer of outside knowledge and the creation of networks which build up the basis for the realization of a self-sustained model of development.

To achieve these aims the TES needs to equip itself with instruments capable of identifying moments of participation and complementarity among all the actors that make up the local system. To do this opportunities must be created to allow the human resources to increase the know-how and acquired cognition that will qualify them to introduce innovative codes and routines within the productive system.

If such cognitive improvement occurs, there will be an increase in flexibility and specialization and a greater capacity to understand and govern change and innovation and ultimately an improvement in the overall efficiency of the productive system.

The government of a local system which is complex because of the continuous, strong interrelations between the individuals operating within it cannot be of a deterministic kind unless part of it is isolated from the rest of the relations.
The government of a complex system demands a series of deliberations over interventions, which by their intrinsic nature are irreversible, i.e. they produce permanent changes in the state of the system.
To return to the now extensively examined concept of TES, multiplicity of equilibria, co-operation, proximity, resilience and freedom of access can be pointed to as some important categories in the description and government of a complex system. The conceptual field within which the local system has to move is, in fact, of a bottom up kind and provides the archetype for programming actions capable of leading the evolutionary paths of the TES towards states of greater growth.
Bianchi's (1995) taxonomy of interventions identifies the following three procedures:
1- programming according to exogenous concepts;
2- programming according to critical situations;
3- programming according to integration contexts.
Programming according to exogenous concepts is nothing more than the traditional concept of programming, achieved by means of the exogenous definition of objectives by the policy maker in conjunction with the identification of the instruments necessary to achieve the pre-established goals. If complexity and environmental turbulence are low, this methoth of programming is effective. This type of programming enters a crisis when the system enters those critical areas characterized by high levels of turbulence or uncertainty. In such circumstances it is necessary to programme according to critical situations, i.e. to devise programming capable of self-regulation in the presence of criticality and of varying parameters in order to overcome any lock-in or bottle-neck situations. As long as the critical areas are small in size, this approach is sufficient. If, however, levels of turbulence and complexity are so high that criticality can occur at any moment, then it is necessary to programme according to integration contexts, i.e. considering the system as a whole as an organism capable of adapting continuously to the outside environment.
In this case policies have to take into account the changes they induce in the system itself, i.e. they way the system metabolizes them.
The need for programming according to integration contexts therefore justifies, as fundamental elements for regional policy, forms of structural adjustment whose objective is to lower the costs of transaction and which concern:

- the social dimension, linked to the quality of life and culture;
- the ecological aspect, closely connected to the urban habitat, the landscape and the ecosystem;
- public institutions and productive sectors, with special reference to the organizational aspect and the quest for efficiency.

Public-private co-operation, improved social standards, the construction of R&S networks and appropriate territorial policies designed to provide the basis for integration are irreplaceable instruments for governing the economy and for leading it to the highest levels of development.

5. An Outline of the Transmission Mechanism of Economic Policy in the Presence of Complexity.

The collective properties of a territorial economic system in relation to the link existing between productivity growth and information could be represented in terms of response function.

We would like, at this point, to generalize the previous relationship by constructing an interpretative model which describes the propagation mechanism of economic policy in a situation of complexity. The description of the transmission mechanism logically completes the previous observations regarding objectives and instruments.

Single economic policy decisions, aimed at achieving the j-th objective through the use of the i-th instrument, can be represented as an outside stimulus which superimposes itself on interactions between agents.

Agents in this approach are thought of as being spatially distributed and linked to each other by local mutual interactions (of a nearest neighbours type). We use H to indicate the effect of the economic policy. We can thus define an effective Heff stimulus which includes both outside stimulus and agent interaction.

Heff therefore assumes the form:

$$Heff = H + \int dr' c(r-r') \delta\gamma(r') \qquad (1)$$

where c(r-r') is a function of correlation between agents which can constitute an acceptable means of modelling the concept of proximity, $\delta\gamma(r')$ is a variation in the behaviour of agents induced by the policy applied, the integral can be linked to the concept of resilience.

This type of behaviour arises in the area of a linear response model for systems with collective properties. The effect of an economic policy on a complex system made up of many agents interacting with each other can therefore be described in this way and modelled, as seen in the previous chapter, by means of the response properties of the system itself.

Therefore, in the area of linear response theory we have a cause-effect relationship of the type:

$$E(X)=G(X)\otimes H(X) \qquad (2)$$

where E(X) represents the generalized effect, G(X) the response function, H(X) the generalized cause. It is therefore possible to study the generalized transmission mechanism of the economic policy describing the response function as a sort of susceptivity which comes to depend on the distribution of agents within the market. Obviously the type of response depends not only on distribution, but also on the type of interaction between agents.

6. Models with Interacting Artificial Agents.

Feedforward type neural networks are basically generators of non-linear functions. They "map" a set of inputs and a set of "outputs". This relationship can be represented by a function of the type:

$x'=f(x, u, t)$ (3)

where x is a vector of state and u is a vector of input which is mapped in a vector of output x' which represents the temporal derivative of vector x. A feedforward type neural network is therefore capable of identifying an appropriate temporal space situation, which means that it is capable of learning. In one sense a neural network can represent an agent acting inductively. The importance of this property is linked to the capacity to adapt to the outside environment, which means that the network has self-tuning capacities.

On the basis of the above, the construction of a model for interacting artificial agents can be formalized.

Let us consider, therefore, a model made up of N interacting agents, each of which is represented by a neural network in the sense that the behaviour of the i-th agent is completely described by the functioning of the network. Let I be the state of each agent described by a vector $x^I(t)$. If each agent is subjected to an input $K^I(t)$ the state of the system evolves according to the following law:

$x^I(t+1)= S^I(A^I x^I(t)-K^I(t))$ (4)

where A is a real, symmetrical square matrix and S an appropriate function.

Spin glass models and the concept of an artificial agent, positioned within the TES enable us to model a large number of economic phenomena taking different types of agent rationality into account.

7. The Irreversibility of Economic Policies.

As we have already seen, single economic policy decisions aimed at achieving the j-th objective by the use of the i-th instrument can be represented as an outside stimulus which superimposes itself on interactions between agents.

We consider now the model proposed in paragraph 5 and synthesised by equation (1). We will model this relation by using the neural networks.

To do this we consider representative agents; agents, that is, that do not react to changes in the outside environment according to certain rules (dictated by economic rationality or conventional wisdom or, more simply, by adaptive type mechanisms).

In this representation the effect of a particular economic policy can be seen as a shock which upsets a situation of equilibrium. The disturbance will spread to the whole system in an unpredictable manner, giving rise to new equilibria.

Obviously the type of response depends not only on the type of interaction between agents but also on their distribution and previous configuration.

If we associate each agent with a node in the neural network, our agent will respond to a stimulus in a pre-defined manner. We will, that is, have assigned a model of behaviour to our artificial agent. Since agents are n and rules of behaviour can be m with $n>m$, the output of the network will depend on the interaction of the models of behaviour of the n agents which are nevertheless interrelated.

This means that the effect of economic policy on the i-th agent depends on its situation at the previous moment and on its interaction with neighbouring agents.
In this way a dissipative system is defined that is capable of describing complex behaviours macroscopically. Non-ergodic systems belong to this class. The possibility of introducing and considering increasing returns thus becomes feasible. In particular, a system with increasing returns and/or network externality presents itself as a dissipative structure (Prigogine, 1967). The underlying processes are irreversible in the sense that in them the arrow of time plays an uneliminable role and they can display self-organization, dissipative chaos and other complex phenomena. This is a very important property because in such circumstances the dynamics of the system lose all their characteristics of measurability and Birkoff's Theorem is no longer valid.
One property of dissipative systems is that they present temporal asymmetry and therefore basic irreversibility. This means that the effects generated by single economic policies irreversibly upset the economic system and therefore it simply isn't possible to cancel out the effect produced by a given policy by opposite sign interventions.
To return to equation (3) we can define the collective dynamic system as being described by a vector of state

$$x(u,t)=(q_1,q_2,..q_n) \qquad (5)$$

where q_i represents the i-th agent.
The solution of the dynamic evolution of this vector of state according to eq (3) leads to solutions of the form (Haken 1983):

$$x(u,t) =x_0+\Sigma h_0{}^t v_0+ \Sigma h_s{}^t v_s \qquad (6)$$

where $h_0{}^t$ represents the strength of the parameters of order, $h^t{}_s$ the strength of the stable states and v_0 and v_s are eigenvectors which describe the system independently of time. The equation is made up of three parts: the first connected with the initial state, the second connected with jumps between attractors and the third connected with the stable states. The system, although complex, can be governed on the basis of a few parameters of order, presents multiple equilibria and the choice of these depends on the parameters of order. If the eigenvalue of v_0 is zero, the system possesses a single basin of attraction and the parameters of order will identify the exact position of the system. If, conversely, the eigenvalues are greater than zero, then the system presents many basins of attraction. In this case it is possible to move the system into several basins of attraction by means of the parameters of order.
In other words, a new means of interpreting both economic policy and its effects seems to emerge from this scheme. A mono-directional approach to economic policy is replaced by an "interactive" approach in which the economic system actively reacts to policies amplifying or frustrating their expected effects in unpredictable ways.

References

1. Arrow K. J. (1994), "Informazione e organizzazione dell'industria", *Rivista Internazionale di Scienze Sociali, Occasional Paper*
2. Arrow K.J. (1962), "The Economic Implication of Learning by Doing", *Review of Economic Studies*, Vol. 29.
3. Arthur W. B. (1988), "Self-Reinforcing Mechanism in Economics", in Anderson - Arrow - Pines (eds.), *The Economy as an Evolving Complex System,* Santa Fe Institute Studies in the Science Complexity, Vol. V, Redwood City, Addison -Wesley Pu.,
4. Arthur W. B., (1994), "Inductive Reasoning and Bounded Rationality", *American Economic Review,* Vol 84, n.2.
5. Aydalot Ph., Keeble D., (1988), *High technology industry and innovative environments*, London, Routledge.
6. Aydalot Ph. (1986), ed. , *Milieux innovauteur en Europe*, Gremi, Paris.
7. Batty M., 1976, " *Urban Modelling*" Cambridge University Press, Cambridge
8. Baumol W. J. - Wolff E.N. (1992) "Feedback between R&D and Productivity Growth: a Chaos Model", Benhabib J. (cd.), *Cycles and Chaos in Economic Equilibrium*, Princeton, Princeton University Press.
9. Becattini G. - Rullani E., (1993), " Sistema locale mercato globale ", in *Economia e Politica Industriale*, n. 80.
10. Becattini G., (1987), (a cura di), *Mercato e forze locali: il distretto industriale*, Bologna, Il Mulino.
11. Becattini G., (1989), (a cura di), *Modelli locali di sviluppo*, Bologna, Il Mulino.
12. Bellandi M., (1994), "Struttura e cambiamento economico nei distretti industriali", in Garofoli _ Mazzoni (a cura di), *Sistemi produttivi locali: struttura e trasformazione*, Milano, Angeli.
13. Benhabib J., Day R. H., (1982), "Rational Choice and Erratic Behaviour", *Review Economic Study*.
14. Bianchi P. (1992), " Sentieri di industrializzazione e sviluppo regionale negli anni '90: strategie politiche di intervento pubblico nella nuova fase competitiva post-fordista ", in Belussi F. (a cura di), *Nuovi modelli di impresa, gerarchie organizzative e imprese a rete*, Milano, Angeli.
15. Bianchi P. (1995), *Le politiche industriali dell'Unione Europea,* Bologna, Il Mulino.
16. Bielli M., Reggiani A. (a Cura di), (1991), " *Sistemi spaziali approcci e metodologie*" Franco Angeli , Milano
17. Boldrin M. e L. Montrucchio, (1986), "Cyclic and Chaotic Behaviour in Interporal Optimitazion Models", *Mathematical Modelling.*
18. Bramanti A. - Folloni (1992), "Economie e diseconomie ambientali: un nodo problematico per la crescita dell'impresa meridionale", in Camagni _ Hoffman - Latella (a cura di)„ *Mezzogiorno e Scienze Regionali: l'analisi e la programmazione,* Milano, Angeli.
19. Bramanti A., Miglierina C., (1994), "Alle radici della crescita regionale: fattori, fenomeni, agenti", in *Modelli di Sviluppo per le aree arretrate e*

regional competition, Quaderno n. 2. Ricerca di Base, Università Bocconi, Milano..

20. Bramanti A. (1991), "Il modello di sviluppo endogeno interrelato", *Rivista Economica del Mezzogiorno*, n. 2.
21. Brock W.A., (1988), "Nonlinearity and Complex Dynamics in Economics and Finance", in Anderson P.W., Arrow K.J., Pines D. (a cura di) "*Economy as an Evolving Complex System*", Addison Wesley.
22. Brock W.A., Dechert W.D., Skeinkman J.A., (1987), "A Test for Indipendence Based upon the Correlation Dimension", *W. P. University Wisconsin*
23. Bronzini F., Bedini M. A., (1991), " Sistemi spaziali urbani come reti neurali: un approccio connessionistico ", in: atti della *XII conferenza di scienze regionali,* pag. 239-251
24. Camagni R. P., (1991)," Cooperation networks and networks of cities: Toward a taxonomy and a theory", in: atti della *XII conferenza di scienze regionali,* pag 39-62
25. Camagni R. P., (1992), " *Economia Urbana*" La Nuova Italia Scientifica, Roma
26. Cellier F. E.,1991, "*Continuous System modelling*" Springer Verlag New York
27. Ciciotti E. (1993), *Competitività e territorio*, Roma, NIS.
28. Coffey W. J. - Polese M. (1984), " The concept of local development: a stages model of endogenous regional growth ", *Papers of The Regional Science Association*, Vol. 54.
29. Courlet C. (1993), *Industrie e territoire. Les systemes productifs localises*, Actes du Colloque IREDP, Grenoble.
30. Craig S. G., Kohlhase J. E., Papell D.H., (1991), "Chaos Theory and Microeconomics: Applications to Model Specification and Hedonic Estimation", *The Review of Economics and Statistics*.
31. Dei Ottati G. (1995), *Tra mercato e comunità: aspetti concettuali e ricerche empiriche sul distretto industriale*, Milano, Angeli.
32. Deneckere R., S. Pelikan, (1986), "Competitive Chaos ", *Journal of economic theory*.
33. Diappi L., Ottanà M., (1993), " Sistemi reticolari di città: l'approccio neuronale" in: "*Modelli d'analisi e di intervento per un nuovo regionalismo*" Pasquini F., Pompili T., Secondini P. (a cura di) Franco Angeli pp.27-50.
34. Domar E. R., (1946), " Capital Expansion, Rate of Growth and Employment" *Econometrica*, pagg. 137-147
35. Dosi G., (1985), *Institutions and Markets in a Dynamic World*, Dattiloscritto presentato al "The World Congress of Political Science", Paris.
36. Dupuy C. - Gilly J. P. (1993), "Dynamique industrielle, dynamique territorial et strategies des groupes", in Corlet C. (Ed.), op. cit..
37. Eckmann J.P., Ruelle D., (1985), "Ergodic Theory of Chaos and strange attractor" *Review of Modern Phisics..*
38. Gaffard J. L. (1990), *Economie industrielle et de l'innovation*, Paris, Dalloz.
39. Garofoli G, (1991), *Modelli locali di sviluppo*, Milano Angeli

40. Gatto A. - Marino D., (1993), "L'approccio hamiltoniano alla dinamica economica: stabilità e biforcazioni di un sistema hamiltoniano in assenza di shock esogeni", *Atti dell'Accademia Peloritana dei Pericolanti.*
41. Gatto A., Marino D., (1992b), "Sistemi urbani e reti neurali: un approccio attraverso l'uso degli automi cellulari" in: atti della *XIII conferenza di scienze regionali*, pag.17-33
42. Gatto A., Marino D., (1994), "Gli automi cellulari: lo stato dell' arte e le possibili applicazioni" in: "*Modelli d'analisi e di intervento per un nuovo regionalismo*" Pasquini F., Pompili T., Secondini P. (a cura di) Franco Angeli pp.51-62
43. Gillepsie A. (1991), "Advanced communications networks, territorial integration and local development", in Camagni R. (ed.), *Innovation networks. Spatial Perspectives*, London, Belhaven Press.
44. Giunta G.,(1995), *The Informatic Revolution and Social Organization: a quantitative Model*, Forthcoming.
45. Grabher G. (1993), (Ed.), *The embedded firm. On the socioeconomics of industrial networks,* London, Routledge.
46. Grandmont J.M., (1985), "On Endogenous Competitive Business Cycles", *Econometrica.*
47. Grandmont J.M., (1986), "Stabilizing Competitve Business Cycles" *Journal of Economic Theory.*
48. Hahn F., Solow R., (1995), " *A Critical Essay on Modern Macroeconomic Theory* " Mit Press Cambridge.
49. Haken, H,(1983), "*Sinergetics: an Introduction*", Springer-Verlag, Berlino
50. Harrod R.F., (1939, "An Essay in Dynamic Theory", *Economic Journal*, pagg. 14-33
51. Hopfield J. J., (1982), "Neural Networks and Phisical Systems with Emergent Collective Computational Abilities" *Proc. Natl. Acad Sci. USA,* n. 79, pag. 2554
52. Jaynes E. T., (1957), " Information Theory and Statistical Mechanics", Physical Review, vol. 106, n. 4, pag. 620-630
53. Jurgens H., Peitgen H. O., Saupe D., (1990), " Il linguaggio dei frattali ", *Le Scienze,* n.266, pag .42-51
54. Kaneko K., (1986), " Attractors, Basin Structures and Information Processing in Cellular Automata", in: S. Wolfram (a cura di) " *Theory and Applications of Cellular Automata*", World Scientific, Singapore, pag. 367-399.
55. Kirkpatrick S., Gelatt C. D., (1983), "Optimization by Simulated Annealing", , n.220, pag. 671-680
56. Kirkpatrick S., Toulouse G., (1985), "Configuration Space Analysis of Travelling Salesman Problem" *Journal de Phisique,* n.46 pag 1277-1292
57. Kondratief N. O. (1960), Die langen Wellen der Konjunktur, *Archiv fur Sozialwissenschaft und Sozialpolitik.*
58. Krugman P. (1994), Complex Landscapes in Economic Geography, *American Economic Rewiev*, 2.
59. Kurz M., (1994) "On Rational Belief Equilibria" *Economic Theory* 4.

60. Kuznets S. (1930), *Secular movements in production and prices*, Houghton Mifflin.
61. Latella F. (1994), *Mercati e istituzioni nel Mezzogiorno*, Milano, Angeli.
62. Latella F., Marino D., "Diffusione della conoscenza ed innovazione territoriale: verso un approccio sistemico", *Quaderni della ricerca di base "Modelli di sviluppo per le aree arretrate e regional competition"*, 2, Università Bocconi, Milano.
63. Liapunov A., (1907), " Problème general de la stabilité du mouvement" *Annales de la Faculté de Sciences de Toulouse*
64. Lucas R.E. (1988), "On Mechanism of Economic Development", *Journal of Monetary Ecobnomics*, Vol. 22.
65. Lucas R.E. (1990), "Why doesn't Capital Flow from Rich to Poor Countries", *American Economic Review, Pa. & Pro.*, Vol 80, n. 2.
66. Lorentz E., (1963), " Deterministic nonperiodic flow", *Journal of the Atmosferic Science.*
67. Maggioni M.A. (1994), "Modelli ecologici per l'analisi della dinamica industriale regionale", in Pasquini F., Pompili T., Secondini P., (a cura di), *Modelli d'analisi e d'intervento per un nuovo regionalismo*, Franco Angeli, Milano.
68. Malecki E. J. (1980), "Corporate organisation of R & D and the location of technological activities", *Regional Studies*, Vol. 14.
69. Malecki E. J. (1981), "Public and private sector interrelationships, technologicalchange and regioonal development", *Paper of the Regional Science Association*, Vol. 47.
70. Marino D, Giunta G., (1995), "Informazione, diffusione dell'innovazione e crescita della produttività: un modello complesso." paper presentato *al Convegno: "Economia e politica dell'innovazione. Le politiche di diffusione delle tecnologie.*"Piacenza, 23-24 giugno 1995
71. Marino D., (1991), "Disordine, incertezza, informazioni, entropia. Un'analisi in ambito fisico ed economico", seminario presso l'Istituto di Fisica Teorica dell'Università di Messina tenuto il 18/ 10/ 1991. Lavoro dattiloscritto
72. Marino D., (1994), "Una rilettura dello sviluppo endogeno in termini di self renforcing mechanisms", *Atti della XV conferenza di Scienze Regionali* - Matera 3-5 ottobre 1994.
73. Marino D., (,1998), "Policies and Complex Dynamics: an Hamiltonian Approach", forthcoming su: *Economics and Complexity* Vol. 1, n.2, Spring
74. Marino D., (1996), "Sistema Economici Territoriali, Reti Neurali e Automi Cellulari: Modelli Teorici e un'Applicazione Empirica" *Rassegna Economica* n. 3.
75. Marino D., (1992), "Gli automi cellulari : un nuovo strumento per la descrizione dei sistemi dinamici non lineari. Un' applicazione al sistema dei trasporti e ai problemi di congestione del traffico". Abstract presentato a: *AIRO 92, XIII giornate di lavoro dell'Associazione Italiana di Ricerca Operativa,* Acireale 23 - 25 settembre.
76. Marino D. Gatto A., (1998), " *Dinamica economica, complessità e caos: teorie metodi e modelli*" Liguori editori, Napoli

77. Marino D, Timpano F., (1997), "Economia legale ed economia illegale: schemi interpretativi della coesistenza" *Rivista Internazionale di Scienze Sociali,* n.1
78. Mariotti S. (1989), " Efficienza dinamica e sistemi di imprese ", *Economia e Politica Industriale*, n. 64.
79. Martin O.,Odlyzko A., Wolfram S., (1984), "Algebraic Properties of Cellular Automata", *Communications in Mathematical Phisics*, n. 93, pag 219-258
80. May R.B., (1976), "Simple Mathematical Models with Very Complicated Dynamics" *Nature*, 261, pagg. 459-467
81. McCulloch W., Pitts W., (1943), " A Logical Calculus of Idea Immanent in Nervous Activity " *Bullettin of Mathematical Biophisics,* n. 5, pag. 115-133
82. Mignolet M. (1983), *Le processus de diffusion des innovations dans l'espace et le redeploiment economique regional*, Dattiloscritto.
83. Molle W. T. M. (1984), "Potenziali regionali di innovazione nella Comunità Europea", in Camagni et. al. (a cura di), *Cambiamento tecnologico e diffusione territoriale*, Milano, Angeli.
84. Nelson R.R. - Winter S.G. (1977), "In search of a useful theory of innovation", in Stroetmann K.A. (ed.) *Innovation, Economic Change and Technology Policies,* Basel, Birkhauyeser Verlag.
85. Nijkamp P., Reggiani A., (1989), " Comportamenti caotici nell'analisi dinamica delle scelte fra alternative diverse" in: atti della *X conferenza di scienze regionali*, pag 137-164
86. Nijkamp P., Reggiani A., (1991), "Competizione spaziale ,dinamiche spazio-temporali e teoria del caos " in: atti della *XII conferenza di scienze regionali*, pag 211-238
87. Parisi D., (1989), "*Intervista sulle reti neurali*", Il Mulino, Bologna.
88. Paternello S., (1991), "*Le reti neurali*" Franco Angeli Milano.
89. Piore M. G. - Sabel C. F. (1984), *The second industrial divide*, New York, Basic Boooks.
90. Poincarè H., (1982), " *Les methodes nouvelles de la mecanique celeste*", Gautier Villars, Parigi.
91. Porter M. (1990), *The Competitive Advantge of Nations,* London, MacMillan.
92. Prigogine I., (1967), "*Thermodynamics of Irreversible Processes*", J. Wiley, New York.
93. Ptesimer P., (1989), " *Le reti urbane tra decentramento e centralità* ", Franco Angeli, Milano.
94. Rabino G. A., (1991), " Una famiglia generalizzata di modelli di interazione spaziale" in: Bielli M. e Reggiani A. (a cura di) " *Sistemi spaziali, approcci e metodologie* " , Franco Angeli, Milano, pag. 15-32.
95. Rallet A., (1993), "Geographical prroximity and thechnological process: a few questions towards a research agenda", in Courlet C. (Ed.), Op. Cit..
96. Rand. D., (1978), "Exotic phenomena in games and duopoly Models*", Journal of mathematical economics.*
97. Romer P.M., (1986), "Increasing Returns and Long Run Growth", *Journal of Political Economy*, Vol. 94, n. 5.
98. Romer P.M. (1990), "Endogenous Technological Change", *Journal of Political Economy*, Vol. 98, n. 5.

99. Samuelson P.A., (1961), "*Foundation of Economic Analysis*", Harward University Press
100. Schoreder M.., (1991),"*Fractals, Chaos, Power Laws*" w.h. Freeman & Co, New York
101. Shannon C.E , (1949)," *Mathemathical Theory of Communication* " University of Illinois Press , Urbana.
102. Simon H.A, ":*Administrative Behaviour: a Study of Decision Making Processes in Administrative Organization*", Macmillan New York 1957.
103. Solow R. M., (1956), " A Contribution to the Theory of Economic Growth", *Quaterly Journal of Economics*, pagg. 65-94
104. Stohr W. B. (1981), " Development from below : the bottom - up and periphery inward development paradigm ", in Sthor W. B. and Taylor d> R. F. (eds.), *Development from above or below ? The dialectics of regional planning in developing countries* , Chichester, Wiley.
105. Storper M., (1993), " The Limits to Globalization: Techonology Districts and International Trade ", *Economic Geography*, n. 1.
106. Storper M. (1994), " The Resurgence of regional economies. Ten Years Later: The as a Nexus of Untraded Interdipendencies ", *European Urban an Regional Studies*, forthcoming.
107. Stutzer M., (1980), " Chaotic Dynamic and Bifircation in a Macromodel " *Journal of Economic Dynamical Control*, n.1, pag. 377-393
108. Sugden R., (1991), " Rational Choice: a Survey of Contributions from Economics and Philosophy." *The Economic Journal*, n. 101, pag. 751-785
109. Topol R., (1991), "Bubbles and volatility of stock prices: effect of mimetic contagion " *The Economic Journal.*
110. Utterback J. M. - Albernaty W. J. (1974), "A test of conceptual model linking states in firms' process and product innovation", *Working Paper HBS 74 - 34, Harward Business School*, Cambridge (Mass.).
111. Veltz P., (1991), "New models of production organisation and trends in spatial development", in Benko - Dunford (Eds.), *Industrial change and regional development*, London, Belhaven Press.
112. Veltz P., (1993), "D'une geographie des couts a une geographie de l'organisation. Quelques theses sur l'evolution des rapports entreprise / territoire", *Revue Economique*, n. 4.
113. Von Neumann J., (1966), "*Theory of Self Reproducing Automata*", University of Illinois Press, Urbana.
114. Williamson O., (1985), "*The Economic Institutions of Capitalism*", Free Press New York
115. Wilson A.G., (1970), "*Entropy in Urban and Regional Modelling*" Pion Londra
116. Wilson A.G., (1981), "*Catastrophe Theory and Bifurcations. Applications in Urban and Regional Geography*" Croom Helm Londra
117. Wolfram S., (1986), "Statistical Mechanics of Cellular Automata" *Review of Modern Phisics,* vol. 55 n. 3 pag 601-643
118. Wolfram. S., (1984), "Computation Theory of Cellular Automata" *Communications in Mathematical Phisics,* n. 96, pag. 15-57.

Neural Networks for Economic Forecasting

Massimo Salzano
University of Salerno – Dip. Scienze Economiche
Via Ponte Don Melillo, 1
Fisciano, Salerno
Tel. 0039 081 5445899
Email: salzamas@netlab.it

Abstract:

After a survey of literature, we analyse the main experiences of economic forecasting using NN. The review aims at two purposes: it provides a general summary of the work in ANN forecasting done to date and it furnishes guidelines for neural network modelling. Particular attention is given to the peculiarity of economic data (high noise, non stationary, and small sample size signals) and to the contrast between model and data driven approaches. We discuss fundamental limitations and inherent difficulties when using neural networks for the economic forecasting. Recently, the traditional analogy with biological nervous systems has been considered insufficient. A deeper understanding of theoretical foundations of these models is required. Some hints for assessing the correctness of NN implementation and their contribution to a better forecast are reported.

Introduction

In the last period, NNs are increasingly used for economic forecasting[1]. This is both a consequences of the recognition that economic data is often non-linear and of the fact that now this tool is more accepted by statisticians[2]. While NNs provide a great deal of promise, they also embody much uncertainty. To date, researchers are still not certain about what are the key factors that modify the forecasting performance of NNs.

Economic forecasting is challenging. In fact, one of the more stringent economic characterisations is about data. Very often, the economic data is noisy and incomplete. NNs seem to be the principal tool for dealing with this sort of problem. Is there any theoretical hint that can be gained from the usage that economists have made of NN? Could the economic use of NNs be ameliorated, taking into better account the development of NN theory for the case in which data is noisy and incomplete?

For a better understanding of methodological questions, we separate the financial from the micro-macro-approach. In fact, even if the two sectors are similar, they present different problems and have a slightly different tradition in the use of NNs.

During the last period, there have been many important developments in the theoretical basis for neural computing. These have been accompanied by a shift of

---

[1] See, for example, the recent and very good survey by: Zhang, Guoqiang; Patuwo, B. Eddy; Hu, Michael Y. (1998)

[2] C. M. Bishop. (1995).

emphasis away from the biological metaphor of the 1980's towards a more principled view of neural networks during the 1990's.

One of the most important benefits of such a shift is that it allows NNs to be applied to more complex and subtle problems like the solution of inverse problems. In this case, a naive approach (application of the standard least-squares approach) leads to poor results, while the adoption of a more principled viewpoint allows better results. Some authors consider that NN provide modest improvements, if any, in forecast accuracy relative to well-specified regression models[3]. Therefore, we will try to highlight the main question we have to consider for reaching goods forecast by NNs, and relative benchmarks.

Neural networks are non-linear mapping structures. They have been shown to be universal and highly flexible function approximators to any data-generating process[4]. Therefore, they are powerful models for forecasting purposes, especially when the underlying data-generating processes are unknown. However, the appropriate design of the network's architecture and learning rules is crucial for obtaining satisfactory results. Obviously, statistical theory can offer some suggestions for designing optimal network architecture. Some authors[5] try to give a balanced assessment of the potential of artificial neural networks for forecasting and decision making models, comparing artificial neural networks and statistical models. Some work[6] presents surveys of many applications in finance published during those last years and generally grouped in many domains as: portfolio management, business forecasting, bankruptcy prediction, credit approval and so on[7].

Given the brief nature of this survey, we have not attempted to provide an exhaustive list of the many relevant references in the literature.

Theoretical Foundations of Neural Networks[8]

Some Authors[9] shows that neural networks (at least some kinds of them) can be seen as a natural extension of regression models, already well known by the financial community. The majority of current applications of neural networks fall into the category of pattern recognition. They can be broadly divided into regression problems (modelling continuous functions of input variables), classification problems (assigning inputs to one of a number of distinct categories) and density estimation problems (modelling the probability distribution of the input variables) [10].

There are many applications for which the generally used network training (involving minimisation of a sum-of-squares error function) can fail badly. An important class of such applications corresponds to inverse problems. These are characterised by a "forward" problem which is governed by a well-defined mapping and which is often associated with causality in a physical system. The corresponding inverse problem, however, can be multi valued.

---

[3] McMenamin, J. Stuart; Monforte, Frank A. (1998)
[4] Kohzadi, and al. (1995)
[5] Hill, T. and al.(1994)
[6] Zhang, and al. (1998) and: Gorr, W. L. (1994)
[7] de Bodt, E. and al (1995)
[8] Bishop, C. M. (1996)
[9] de Bodt, E. and al (1995)
[10] Bishop, C. M. (1996)

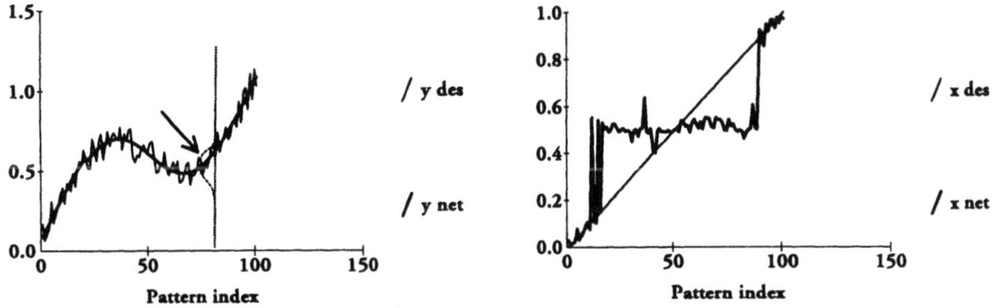

Figure 1: (a) A plot of the data from a simple `forward' problem together with the result of fitting a neural network using leastsquares (solid curve). (b) The results obtained using the same network when the input and output variables are interchanged, corresponding to an `inverse' problem, showing the very poor fit to the data.

A direct application of neural networks to this kind of problem can lead to extremely poor results. This is illustrated with a simple example in Figure 1[11].
This occurs for the statistical principles on which the neural networks are based:
- the goal in network training is not to learn the training data set but rather to model the underlying process which generated the data. The model is based on the "conditional average" of the data. Only in this way will the network have good generalisation properties whereby it can make accurate predictions when presented with new input vectors. This is strongly connected with forecasting questions.
- For the forward problem shown in Figure 1(a), the conditional average gives a good representation of the underlying function from which the data was generated.
- However, for the inverse problem of Figure 1(b) the conditional average provides a very poor model. This problem cannot be resolved by additional training, by increasing the number of data points, or by adding extra hidden units to the network, since it stems from a fundamental property of leastsquares. Although the failure of the simple neural network approach is obvious in the example in Figure 1(b), it may be far from clear in real applications involving spaces of many dimensions.
To resolve this problem some Authors propose a model in which the conditional distribution $p(t|\mathbf{x})$ is given by a mixture model consisting of a linear combination of L component densities[12] (a mixture density network[13]).

Issues in ANN modelling for forecasting

In this process the most critical decisions that influence the performance of an ANN are: the choice of the architecture, the selection of activation functions of the hidden and output nodes, the training algorithm, the use of data transformation, the choice of training and test sets, and that of performance measures.

---

[11] The data set has been generated by drawing values of the input variable x and computing the corresponding values of $y = x + 0{:}3 \sin(2nx)$ and then adding uniformly distributed random noise in the range.

[12] Full details can be found in Bishop. (1995); in Bishop.(1994) and in Bishop and Legleye (1995)

[13] Bishop (1994).

Finding a parsimonious model

With NNs it is more important then usual to use a parsimonious model for at least two main reasons.

1) Minimising the network size generally reduces the data requirement for training and increases the generalisation capability[14]. This is the most important characteristic for forecasting.

2) As NNs are general descriptors, the problem of overfitting could easily occur.

The network architecture

- *The overall appropriate size and structure of NN models for forecasting:* The selection of these parameters (the number of input nodes; the number of hidden layers and hidden nodes; and the number of output nodes) is basically problem-dependent[15]. Several researchers address the issue of finding networks with appropriate size for forecasting. There exist many different approaches for finding the optimal architecture of an ANN. The most used are: a) the pruning algorithm (Smith, 1993; Gorr et al., 1994; Sietsma and Dow, 1988; Karnin, 1990; Weigend et al. 1990, 1991, 1992; Reed, 1993; Cottrell et al., 1995); b) the polynomial time algorithm (Roy et al., 1993); c) the canonical decomposition technique (Wang et al., 1994); d) the network information criterion (Murata et al., 1994); e) statistical stepwise (Cottrell et al.; 1995). f) normalised Akaike information criterion (AIC); (De Groot and Wurtz, 1991; Akaike, 1974) g) hybrid approach combining Box-Jenkins and NNs (Lachtermacher and Fuller; 1995) h) two-step procedure (Kuan and Liu, 1995).

However, as these methods are usually quite complex in nature, difficult to implement, and they are not able to guarantee the optimal solution for all real forecasting problems, they must be considered as heuristic or general guidelines.

Recently, for the optimal design of a neural network genetic algorithms have received considerable attention (Miller and al., 1989; Guo and Uhrig, 1992; Jones, 1993; Schiffmann et al., 1993).

- *The number of input nodes:* The forecasting can be casual or time series forecasting. While, in the first case the number of inputs is usually relatively easy to choose for the second case it is probably the most critical decision variable (Zhang, 1998). In fact, in this case it contains the important information about the complex autocorrelation structure in the data (the time lag structure).

It seems that their number has positive effects on the results when the step-ahead of forecasting increases (Zhang, 1998; Lachtermacher and Fuller, 1995)[16].

- *The number of hidden layers and nodes*: The issue of determining the number of hidden nodes play very important roles for the success of NNs forecasting, while it is a complicated one. In fact, there is a trade off between: a) better generalisation

---

[14] Baum and Haussler, 1989 discuss the general relationship between the generalizability of a network and the size of the training sample

[15] Amirikian and Nishimura (1994) find that the appropriate network size depends on the specific tasks of learning.

[16] For the use of maximum entropy principles to identify the time series lag structure, see: Cheung et al. (1996).

ability - less overfitting problems; and b) power to model and learn the data[17]. More, the total number of hidden nodes is strictly correlated with the number of hidden layers.

Often, for forecasting purposes, the choice is between one or two hidden layer. If one hidden layer is used the network may require a very large number of hidden nodes. The use of two hidden layers may give better results (Zhang, 1998; Barron, 1994; Srinivasan et al., 1994; Zhang, 1994; Vishwakarma, 1994; Grudnitski and Osburn, 1993; Lee and Jhee, 1994; Chester, 1990).

For all this difficulties, the number of hidden nodes is determined by trial-and-error or on rules of thumb, often based on a heuristic constraint related to the number of input nodes (Lachtermacher and Fuller, 1995; Lippmann, 1987; Hecht-Nielsen, 1990; Wong, 1991; Tang and Fishwick, 1993; Kang, 1991).

- *The number of output nodes*: for a time series forecasting problem, the number of output nodes (one or more) generally corresponds to the forecasting horizon: one-step-ahead or multi-step-ahead forecasting (either iterative or direct).

- *The interconnection of the nodes*: The connections between nodes in a network fundamentally determine the behaviour of the network. Generally, the network is "fully connected[18]". It is possible to have sparsely connected networks (Chen et al., 1991) and direct connections from input to output nodes (Duliba, 1991). These may be advantageous to the recognition and forecast power since they can be used to model the linear structure of the data (Tang and Fishwick, 1993).

- *Activation function*: Different activation functions could be used for different nodes. Even if there are some heuristic rules for the selection of the activation function, yet almost the same activation functions (the logistic one), is generally used for all node but the output nodes.

For these, even if there is no consensus[19], generally it is considered that for a forecasting problem that involves continuous target values, it is reasonable to use a linear activation function (Rumelhart et al., 1995; Lapedes and Farber, 1987, 1988; Weigend et al., 1990, 1991, 1992; Wong, 1991; Ginzburg and Horn, 1992, 1994; Gorr et al., 1994; Srinivasan et al., 1994; Vishwakanna, 1994; Cottrell et al., 1995; Kuan and Liu, 1995, etc.).

---

[17] Without hidden nodes, simple perceptrons with linear output nodes are equivalent to linear statistical forecasting models.

[18] All nodes in one layer are only fully connected to all nodes in the next higher layer, except for the output layer.

[19] But see: Klimasauskas (1991).

Training algorithm

The neural network training is an unconstrained non-linear minimisation problem based (generally) on a gradient steepest descent method (backpropagation algorithm). A step size must be specified (the learning rate). For this, there is a trade-off between the velocity of the learning process and network oscillation in the weight space. The trade-off can be reduced connecting the "next" weight change "more or less" with the previous one (momentum parameter) (Rumelhart et al., 1986b). The conclusions with regard to the best teaming parameters are inconsistent even if Tang et al. (1991) report that high learning rate is good for less complex data, and low learning rate with high momentum should be used for more complex data series[20].

Other methods have been proposed. Among them the most efficient are the second-order methods - such as BFGS and Levenberg-Marquardt methods - (Parker, 1987; Battiti, 1992; Cottrell et al., 1995) and a general-purpose non-linear optimiser - GRG2 (Lasdon and Waren, 1986; Hung and Denton, 1993; Subramanian and Hung, 1993).

Objective function

The objective function is typically defined in terms of errors (SSE and MSE), even if for some problems like financial forecasting other objective functions such as maximising the return, profit or utility may be more appropriate.

Data normalisation

Data normalisation (restricting the possible range of data) is often performed before the training process begins Azoff (1994). In this case, the outputs must be rescaled to the original range.

Training, test and validation samples - Performance measures

The literature offers little guidance in selecting the training (used for model development), the test sample (adopted for evaluating its forecasting ability), and validation sample (generally the same data used for the test, utilised to avoid the overfitting problem) (Weigend et al., 1992; Granger, 1993). The only consence is about the fact that for time series forecasting, both the training and test sets must be representative of the population or underlying mechanism.

The most important measure of performance is the forecasting ability for which a suitable measure of accuracy is not universally accepted (see Zahda, 1998).

Theory and comparative studies

By far the most popular type of neural network has been back-propagation. However, there are advantages in using other learning techniques, such as the swift response of the probabilistic neural network (PNN). Some authors[21] suggest the desirability of adapting other models to the predictive function (an array of elementary PNNs with bipolar output).

The economy is evolving (rather slowly) over time. This feature cannot easily be captured by fixed specification linear models, and manifest itself in the form of

---

[20] But see: Zhang (1998)

[21] Kim, Steven H.; Hak Chun, Se (1998)

evolving coefficient estimates[22]. Some articles focus on the construction of prediction intervals[23]. Previous statistical theory for constructing confidence intervals for the parameters (or the weights in an ANN) is inappropriate, because the parameters are unidentifiable. It seems that the problem disappears in prediction, and it is possible to obtain asymptotically valid prediction intervals and to use such intervals to choose the number of nodes in the network.
Time orison: the confrontation of results obtained by NN and traditional structural time series or unobservable components model (UCM) do not show clear cut conclusions as far as one step ahead forecasts are concerned; but, for longer time horizons the ANN has a better forecasting performance[24].
This confirmed the results by a comparative study in economic forecast[25] considering traditional time series methods, such as ARIMA and unobservable components models (UCM) and artificial neural networks (ANN) showed the relative performance of the different forecasting methods. ANN forecast more accurately than ARIMA models, but the comparison with UCM is not quite straightforward. The UCM is found to produce better one step ahead forecasts than the ANN does, but the performance of the ANN for larger forecast horizons is better.
Sometimes, the neural networks are able to find additional model structure over the regression models[26].

Time series prediction[27]

Let a given one-variable time series be represented by the N values x(1), x(2), ..., x(N). Prediction then consists in finding the future values x(N+1), x(N+2), ... Takens[28] has shown that if the series is obtained from a deterministic dynamical system, there exists a scalar d (which is called the *embedding dimension*), a scalar τ (which is an arbitrary delay) and a function f such that for every $t > d\tau$: $x(t) = f(x(t-\tau), x(t-2\tau), \ldots, x(t-d\tau))$
Then the prediction problem consists, given the first N values of a time series, in finding the appropriate d, τ and f. Of course, one usually cannot be sure that a given series is deterministic. Actually, statistical methods do exist to verify if a series is deterministic and to find d as well as τ but they require the size of the series to be on the order of 10**d which is rarely the case in economic practical problems. For the moment, let us assume that we know d and τ and that we want to find f. This is where *forecasting* come in.
Forecasting has long been the domain of traditional linear statistical models (Box-Jenkins or ARIMA). They can be easily understood, analysed, explained and implemented. However, real world systems are often non-linear (Granger and Terasvirta, 1993).

---

22 Swanson, White (1997).
23 Hwang, and Al. (1997)
24 Fernandes; Navaux; Portugal, (1996)
25 Portugal (1995)
26 Gorr; Nagin; Szczypula, (1994).
27 Bengio, Fessant, and Collobert. (1995a,b,c)
28 Takens, (1981)

The problem of predicting non-linear time series: features of NNs for forecasting.

Sometimes, the non-linearity in time evolutions exhibits distinctive features that can be well described by existing parameterised classes of non-linear models such as bilinear or threshold models. More often, the non-linearity derives from highly irregular behaviour with complex time dependencies, and the *prescription of a particular non-linear model can be difficult*[29]. This is where *neural networks* come in: it is a well-known fact that they can be used as universal function approximators[30]. However, these results are valid if and only if there is a sufficiently high number of training data in the series. In the other case, one could overtrain and prediction accuracy could deteriorate. This problem is known as the generalisation problem and is well addressed by the works of Vapnik[31].

Forecasting is one major application area of NNs (Sharda, 1994). NNs provide an attractive alternative tool for both forecasting researchers and practitioners.

- Model-Based Non-Linear Methods (MBNLM): During the last decade, several non-linear time series models such as the bilinear model, the threshold autoregressive (TAR) model, and the autoregressive conditional heteroscedastic (ARCH) model have been developed[32]. However, for them an explicit relationship for the data series at hand has to be hypothesised. They pertain to model-based non-linear methods.

Several distinguishing features of NNs make them valuable and attractive for a forecasting task. They are:

- Data-Driven Non-Linear Approaches (DDNLA): Vice versa, NNs, which are non-linear data-driven approaches, are capable of performing non-linear modelling without a priori knowledge about the relationships between input and output variables. Thus, they represent a more general and flexible modelling tool for forecasting.

As opposed to the traditional model-based methods, NNs are data-driven self-adaptive methods in that there are few a priori assumptions about the models for problems under study. They learn from examples and capture subtle functional relationships among the data even if the underlying relationships are unknown or hard to describe. Thus, NNs are well suited for problems whose solutions require knowledge that is difficult to specify but for which there is enough data or observations. In this sense they can be treated as one of the multivariate non-linear, non-parametric statistical methods (White, 1989; Ripley, 1993; Cheng and Titterington, 1994

The problem with the data-driven modelling approach is that the underlying rules are not always evident and observations are often masked by noise.

- Can generalise for noisy information;
- Are general approximators: Frequently, traditional statistical forecasting models have limitations in estimating the underlying function due to the complexity of the real system. NNs can approximate any continuous function to any desired accuracy

---

[29] Poli, Jones, (1994)
[30] Cybenko, (1989)
[31] Vapnik. (1982)
[32] See De Gooijer and Kumar (1992)

(Irie and Miyake, 1988; Hormik et al., 1989; Cybenko, 1989; Funahashi, 1989; Hormik 1991, 1993).
- Are fault tolerant: In general, thanks to the distributed representation over a large number of units, together with interconnectedness among processing units, they are fault tolerant.

Challenges of Macroeconomic Forecasting[33]

Macroeconomic forecasting is a very difficult task due to the lack of an accurate, convincing model of the economy[34]. As a consequence, the reliability of economic forecasting seems increasing when fewer assumptions are made about the structure of the economy[35].
Constructing reliable time series models is challenging due to short data series, high noise levels, non-stationarities, and non-linear effects[36]. The neural network solutions for this challenging question imply balancing the bias/variance trade-off and the noise/non-stationarity trade-off. Some methods for this include hyperparameter selection (regularisation parameter and training window length), input variable selection and pruning, network architecture selection and pruning, new smoothing regularisers, and committee forecasts.
Of great interest to forecasters of the economy is predicting the ``business cycle", or the overall level of economic activity.
Business cycles are irregular in magnitude, duration, and structure. The standard measures of economic activity used by economists to track the business cycle include the Gross Domestic Product (GDP) and the Index of Industrial Production (IP).

Characteristics Micro and Macroeconomic data

The economic modelling and forecasting is challenging because the sector is characterised by:
- Model complexity
Nonquantifiable factors such as mass psychology and sociology that influence economic activity. One measure of complexity for non-linear models is P_{eff} , the effective number of parameters (Moody, 1992; Moody, 1994b).
- Short data series
This is mainly due both to the impossibility of doing controlled experiments on the economy and to the techniques employed in measuring macroeconomic series.
- Noise:
Macroeconomic time series are intrinsically very noisy and generally have poor signal to noise ratios[37].
The noise is due both to the many unobserved variables in the economy and to the survey techniques used to collect data for those variables that are measured. The

[33] Moody, (1995)
[34] Moody, (1995)
[35] In fact Moody (1995) highlights that "the most accurate models for economic forecasting, "black box"' time series models, assume little about the structure of the economy".
[36] Moody, J. (1995)
[37] Moody, J. (1995)

noise distributions are typically heavy tailed and include outliers (Geman, Bienenstock and Doursat, 1992)

- There are no "a priori" models:

A convincing and accurate scientific model of business cycle dynamics is not yet available. Thus, NNs can often outperform standard linear time series models, being data driven models. However, the lack of an "a priori" model of the economy makes input variable selection, the selection of lag structures, and network model selection critical issues.

- Non-stationarity:

Due to the evolution of the world's economies over time, macroeconomic series are intrinsically non-stationary.

The underlying dynamics of the economy change with time, and the noise distributions for the measured series vary with time also. Training on older data will induce biases in predictions.

- Non-linearity:

Traditional macroeconomic time series models are linear (Granger and Newbold, 1986; Hamilton, 1994). However, recent work by several investigators have suggested that non-linearities can improve macroeconomic forecasting models in some cases (Granger and Terasvirta, 1993; Moody et al., 1993; Natter, Haefke, Soni and Otruba, 1994; Swanson and White, 1995).

Characteristics of Financial data and Foreign Exchange Rates

Financial forecasting is an example of a signal processing problem, which is challenging due to small sample sizes, high noise, non-stationarity, and non-linearity [38].

Foreign exchange rates exhibit very high noise, and significant non-stationarity. In real problems, the evaluation of prediction algorithms is based generally on the percentage of right trend forecasts[39].

The main difference in respect to micro and macroeconomic forecasting is a larger amount of data.

Predicting Noisy, Non-Stationary Time Series

Financial time series data typically contains high noise and significant non-stationarity. The prediction of future events from noisy time series data is commonly done using various forms of statistical models[40]; also, NNs could be used. However, high noise, high non-stationarity time series prediction is fundamentally difficult for NNs[41].

Generally, also macroeconomic prediction task is difficult for the poor signal to noise ratios and erratic behaviour of target series. In fact, also in this case,

---

[38] Lawrence, Chung Tsoi , Giles, (1996)

[39] For a prediction of the direction of change in foreign exchange rates for the next business day see Weigend et al. [48]; and for the same prediction using Grammatical Inference, see: Lawrence, Chung Tsoi , Giles, (1996)

[40] C.W.J. Granger and P. Newbold. Forecasting Economic Time Series. Academic Press, San Diego, second edition, 1986

[41] Lawrence, Chung Tsoi , Giles, (1996)

significant non-stationarities and deviations from normality of the noise distributions are present.

The degree of non-linearity captured by neural network models of macroeconomic and financial data series tends to be mild (Moody et al., 1993; Levin, Leen and Moody, 1994; Rehfuss, 1994; Utans, Moody, Rehfuss and Siegelmann, 1995; Moody, Rehfuss and Saffell, 1996; Wu and Moody, 1996)[42].

Noise/non-stationarity trade-off[43]: In fact, using a short training window results in too much model variance or estimation error due to noise in limited training data while, using a long training window results in too much model bias or approximation error due to non-stationarity. So, to deal with the first question it is desirable to use as large a training set as possible, but increasing the size of the training set results in more data with statistics that are less relevant to the task at hand being used in the creation of the model.

The easy approach to deal with the non-stationarity of the signal is to build models based on a short time period only. Any inefficiency found in the market would typically not last for a long time. However, the number of training data points is generally already small. If the training set is too small, the noise makes it harder to estimate the appropriate mapping. We can choose how long the moving forward training window must be using some test.

Controlling Overfitting[44]: Similar problems are present for controlling the overfitting[45] and choosing the validation set[46]. The high noise and small data sets make the models prone to overfitting. In fact, high noise and small data sets cause random correlations between the inputs and outputs. The use of recurrent neural networks (RNNs) partially solves this problem, as they are less prone to learning random correlations, which do not occur in temporal order[47]. Then, some authors suggest converting the time series into a sequence of symbols using a self-organising map (SOM) (grammatical inference)[48]. This is a way of reducing the noise and the RNN training becomes more effective.

Neural Network Solutions

They have been investigating a variety of algorithms for neural network model selection that go beyond the vanilla neural network approach[49] for constructing models with minimal prediction risk. The main techniques that have been developed are[50]:

---

[42] Moody, J. (1995)
[43] Lawrence, Chung Tsoi , Giles, (1996)
[44] Lawrence, Chung Tsoi , Giles, (1996)
[45] Weigend, Huberman, and Rumelhart (1992)
[46] Lawrence, Chung Tsoi , Giles, (1996)
[47] Y. Bengio, (ed), (1996) noted a better account of short-term dependencies than not long-term dependencies..
[48] Zeng, Goodman, and Smyth, (1994).
[49] We define a vanilla neural network to be a fully connected, two-layer sigmoidal network with a full set of input variables and a fixed number of hidden units that is trained on a data window of fixed length with backprop and early stopping using a validation set. No variable selection, pruning, regularization, or committee techniques are used.
[50] Moody, J. (1995)

Hyperparameter selection: they are parameters that appear in the training objective function, but not in the network itself[51]. Examples include the regularisation parameter, the training window length, and robust scale parameters. Varying the regularisation parameter trades off bias and variance while varying the training window length trades off noise and non-stationarity.

Input Variable Selection and Pruning: Selecting an informative set of input variables and an appropriate representation for them is critical to the solution of any forecasting problem. There are both model-independent and model-dependent variable selection procedures: the Delta Test, a model independent procedure (Pi and Peterson, 1994); Sensitivity-based pruning (SBP) techniques are model-dependent algorithms (Mozer and Smolensky, 1990; Moody and Utans, 1994; Moody, 1994b; Utans et al., 1995).

Model Selection and Pruning: A key technique for controlling the bias / variance trade-off for noisy problems is to select the size and topology of the network: sequential network construction (SNC) (Ash, 1989; Moody and Utans, 1994; Moody, 1994b); optimal brain damage (OBD) (LeCun, Denker and Solla, 1990); optimal brain surgeon (OBS) (Hassibi and Stork, 1993) and supervised principal components pruning (PCP) method (Levin et al., 1994).

Better Regularisers: Introducing biases in a model via regularisation or pruning reduces model variance and can thus reduce prediction risk. Prediction risk can be best minimised by choosing appropriate biases. Quadratic weight decay (Plaut, Nowlan and Hinton, 1986; Hoerl and Kennard, 1970b; Hoerl and Kennard, 1970a); there are other system too: (Moody and Rögnvaldsson, 1995; Wu and Moody, 1996).

Committee Forecasts: Due to the extremely noisy nature of economic time series, the control of forecast variance is a critical issue[52]. One approach to deal with this question is to average the forecasts of a committee of models: (Granger and Newbold, 1986; Winkler and Makridakis, 1983; Clemen, 1989).

Causal and time series forecasting using NNs

For an explanatory or causal-forecasting problem, the inputs to an ANN are usually the independent or predictor variables. The functional relationship estimated by the ANN can be written as y =f(x1,x2, , xp), where x1,x2,...,xp are p independent variables and y is a dependent variable. In this sense, the neural network is functionally equivalent to a non-linear regression model.

For an extrapolative or time series forecasting problem, the inputs are typically the past observations of the data series and the output is the future value. The ANN performs the following function mapping yt+1 =f(yt,yt-1,. . . ,yt-p), where yt is the observation at time t, Thus the ANN is equivalent to the non-linear autoregressive model for time series forecasting problems. It is also easy to incorporate both predictor variables and time-lagged observations into one ANN model, which amounts to the general transfer function model. For a discussion on the relationship between NNs and general ARMA models, see Suykens et al. (1996).

---

[51] Moody, J. (1995)

[52] Moody, J. (1995)

- *The number input nodes for causal forecasting*: In this case, the number of input nodes is well defined and it is the number of independent variables associated with the problem.
- *The number of input nodes for time series forecasting*: vice versa, in this case, however, the appropriate number of input nodes is not easy to determine. Whatever the dimension, the input vector for a time series forecasting problem will be usually composed of a moving window of fixed length along the series. The total available data is usually divided into a training set (in-sample data) and a test set (out-of-sample or hold-out sample). The training set is used for estimating the arc weights while the test set is used for measuring the generalisation ability of the network.
The network training is actually an unconstrained non-linear minimisation problem.

Effectiveness of NNs for forecasting and prediction

Opinions about NNs contribution to the increase in the accuracy of forecasts and predictions in economics are mixed (Dutta et al.., *1994;* Chatfield, *1993).*
For evaluating research in this area there is a lack of clear criteria. Adya' and Collopy[53], identifyed some sufficiently general guidelines that could be used in this evaluation.
When we want to accertaint if a tool is superior to another on the base of empirical researches we must consider two main points. a) *effectiveness of validation*: evaluating how effectively the proposed technique was compared with alternatives to appropriately evaluate the predictive capabilities of the proposed network; and b) *effectiveness of implementation*: how well the technique was implemented.
The procedures must be considered in cascade. In fact, only for studies that have effectively validated the NN it is possible to consider how the proposed architecture was well implemented. In fact, there is reason to believe that it could be a contender when it will be well implemented.

Guidelines for testing effectiveness of validation and implementation:

- *Effectiveness of validation*: If a new approach is to be taken seriously, it must be evaluated in terms of alternatives that are or could be used. If such a comparison is not conducted, it is difficult to argue that the study has taught us much about the value of the tool for forecasting. Some guidelines about validation (Collopy, Adya' and Armstrong, 1994; Adya', M.; Collopy, F., 1998) are: a) comparisons of forecasts with well-accepted models; b) comparison of forecasts should be based on *ex ante* (out-of-sample) performance; c) use of a reasonable sample of forecasts.
- *Effectiveness of implementation*: Some guidelines about the effectiveness of implementation (Refenes, 1995; Adya', M.; Collopy, F., 1998) concerns the level of: a) convergence; b) generalisation; c) stability.

Following such guidelines, from the point of view of usefulness of neural nets to forecasting and prediction, studies could be classified as being of three types. a) *well implemented and well validated*: they are the most valuable because they are of interest whatever their outcome and (according with their outcome) can be used either to argue that NNs are useful in forecasting or that they are not; b) *well validated, insufficiently implemented* (in some respects): these are relevant when

[53] Adya', M.; Collopy, F. (1998)

their results are positive; c) *studies of little interest*: because their validation suffers or they produce null or negative results.
When such guidelines are applied, the superiority of NNs respect to the traditional approaches emerges. Aida' and Collopy evaluated 48 empirical studies done between 1988 and 1994 that applied NN approaches to business forecasting and prediction problems. Considering only studies that passed the tests, in 19 (86%) of them NNs outperformed alternative approaches[54].

Conclusions

Having highlighted the main characteristics of economic data and surveyed the latest works about economic forecasting using NNs, we have given some pragmatic indication about how to correctly implement a NN dedicated to economic forecasting. It seems important to follow them. In fact, the main conclusion that emerges from the evaluation by Aida' and Collopy of NN implementations for forecasting is that "when NNs are effectively implemented and validated, they show potential for forecasting and prediction". The 86% of the correctly implemented studies that they considered produced results that were favourable to NN, but many where nor effectively implemented.
NNs results in negative findings only when the minimal condition of correct implementation are not respected. Then, NNs seem to be a very powerful tool for economic forecasting that often is not effectively implemented.

References

Adya', M.; Collopy, F. (1998): How Effective are Neural Networks at Forecasting and Prediction? A Review and Evaluation, *Journal of Forecasting* 17 481 -495

Azoff E.M., (1994) Neural Network Time Series Forecasting of Financial Markets. John Wiley and Sons, Chichester.

Barron, AK, (1994). A comment on "Neural networks: A review from a statistical perspective". Statistical Science 9 (1), 33-351994;

Battiti, R., (1992): First- and second-order methods for learning: Between steepest descent and Newton's method. Neural Computation 4 (2), 141-166.

Bengio, S.; Fessant, F. and Collobert, D.. (1995): A Connectionist System for Medium-Term Horizon Time Series Prediction In *International Workshop on Applications of Neural Networks to Telecommunications*, Stockholm, Sweden,

Bengio, Y. (ed) (1996) Neural Networks for Speech and Sequence Recognition. Thomson, 1996.

Bishop C. M. and C. Legleye. Estimating conditional probability densities for periodic variables. In G. Tesauro, D. S. Touretzky, and T. K. Leen, editors, *Advances in Neural Information Processing Systems*, volume 7, pages 641648, Cambridge MA, 1995. MITPress.

Bishop, C. M. (1996): *Theoretical Foundations of Neural Networks*, Invited paper for Physics Computing 1996, Krakow, Poland. - Neural Computing Research Group - Dept. of

[54] Adya', M.; Collopy, F. (1998). Often studies do not meet all of the minimal criteria for effectiveness of validation and implementation (only 11 met the criteria in the survey by Aida' and Collopy). The most disregarded aspect seems to be that of implementation (of the remaining 38, 17 had this sort of problem).

Computer Science & Applied Mathematics - Aston University, Birmingham -Technical Report: NCRG/96/024, C.M.Bishop@aston.ac.uk; www: http://www.ncrg.aston.ac.uk/

Bishop. C. M. *Mixture density networks*. Technical Report NCRG/94/001, Neural Computing Research Group, Aston University, Birmingham, UK, 1994

Bishop. C. M. *Neural Networks for Pattern Recognition*. Oxford University Press, 1995;

Chester, D.L., (1990): Why two hidden layers are better than one? In: *Proceedings of the International Joint Conference on Neural Networks*, pp. 1265-1268

Cottrell M.;Girard, Y, Mangeas, M., Muller, C., (1995). Neural modeling for time series: a statistical stepwise method for weight elimination. *IIEE Transactions on Neural Networks 6* (6), 1355-1364et al., 1995

Cybenko, G.(1989): Approximation by superposition of sigmoidal functions. *Mathematics of Control, Signal and Systems*, 2:303--314.

de Bodt, Eric; Cottrell, Marie; Levasseur, Michel (1995): Les reseaux de neurones en finance: Principes et revue de la litterature. (Neural Networks in Finance: A Survey. With English summary.), *Finance*; 16(1), June, pages 25-91.

De Gooijer LG.. Kumar, K (1992). Some recent developments in non-linear time series modelling, testing, and forecastingInternational Journal of Forecasting 8, 135-156

El-Sharkawi, M.A., Oh, S_ Marks, R.J., Damborg, M1, Brace, C.M., 1991. Short-term electric load forecasting using an adaptively trained layered perceptron. In: *Proceedings of the Ist International Forum on Application of Neural Networks to Power Systems*, 3-6.

Falhman, S. (1989): Faster-learning variations of back-propagation: An empirical study. In: Touretzky, D., Hinton. G., Sejnowski, 1, (Eds.), *Proceedings of the 1988 Connectionist Models Summer School*, pp. 38-51.

Fessant, F. Bengio, S. and Collobert, D. (1995): On the Prediction of Solar Activity Using Different Neural Network Models. *Annales Geophysicae*

Fessant, F. Bengio, S. and Collobert, D. Use of Modular Architectures for Time Series Prediction. *Neural Processing Letters*, 1995.

Gorr, W. L. (1994): Neural Networks in Forecasting: Special Section: Editorial: Research Prospective on Neural Network Forecasting, *International Journal of Forecasting*; 10(1), June, pages 1-4.

Gorr, W. L.; Nagin, Daniel; Szczypula, Janusz (1994): Comparative Study of Artificial Neural Network and Statistical Models for Predicting Student Grade Point Averages, *International Journal of Forecasting*; 10(1), June, pages 17-34.

Granger, C.W.J.; Newbold. P. *Forecasting Economic Time Series*. Academic Press, San Diego, second edition, 1986.

Grudnitski, G.. Osburn, L., (1993): Forecasting S and P and gold futures prices: An application of neural networks. *The Journal of Futures Markets 13 (6)*, 631-643.

Hecht-Nielsen. R_ 1990. Neurocomputing. Addison-Wesley, Menlo Park, CA.

Hernandes, L. G. L.; Navaux, P. O. A.; Portugal, M. S. (1996): Previsco de series de tempo: Redes neurais artificiais e modelos estruturais., *Pesquisa e Planejamento Economico*; 26(2), August, pages 253-76.

Hill, Tim; Leorey Marquez; Marcus O'Connor, and William Remus (1994): Artificial Neural Network Models for Forecasting and Decision Making, *International Journal of Forecasting*; 10(1), June 1994, pages 5-15.

Hornik, K. Stinchcombe, M. and White. H. (1989): Multilayer feedforward networks are universal approximators. *Neural Networks*, 2: 359-366.

Hung, M.S., Denton, LW, 1993. Training neural networks with the GRG2 nonlinear optimizer. *European Journal of Operational Research* 69, 83-91.

Hwang, J. T. Gene; Ding, A. Adam (1997): Prediction Intervals for Artificial Neural Networks, *Journal of the American Statistical Association*; 92(438), June, pages 748-57.

Jacobs, 1988;

Kang, S., 1991. An Investigation of the Use of Feedforward Neural Networks for Forecasting. Ph.D. Thesis, Kent State University.

Kim, S. H.; Hak Chun, Se (1998): Graded Forecasting Using an Array of Bipolar Predictions: Application of Probabilistic Neural Networks to a Stock Market Index, *International Journal of Forecasting*; 14(3), September, pages 323-37.

Kohzadi, N., Milton S. Boyd, Iebeling Kaastra, Bahman S. Kermanshahi, and David Scuse (1995): Neural Networks for Forecasting: An Introduction, *Canadian Journal of Agricultural Economics*; 43(3), November, pages 463-74.

Lachtermacher, G., Fuller, J.D., (1995): Backpropagation in timeseries forecasting. *Journal of Forecasting* 14, 381-393.

Lasdon, L.S., Waren, A.D., (1986). *GRG2 User's Guide*, School of Business Administration, University of Texas at Austin.

Lawrence, S. Chung Tsoi , A. Giles, C. L. (1996): *Noisy Time Series Prediction using Symbolic Representation and Recurrent Neural Network Grammatical Inference*, Technical Report UMIACS-TR-96-27 and CS-TR-3625 Institute for Advanced Computer Studies, University of Maryland, College Park, MD 20742

Lee, J, Jhee, W. (1994) A two-stage neural network approach for ARMA model identification with ESACF. *Decision Support Systems 11*, 461-479.

Lippimann, R.P., (1987): An introduction to computing with neural nets, *IISEE ASSP Magazine*, April, 4-22.

McMenamin, J. Stuart; Monforte, Frank A. (1998): Short Term Energy Forecasting with Neural Networks *Energy Journal*; 19(4), pages 43-61.

Moody, J. (1995): Economic Forecasting: Challenges and Neural Network Solutions, *Keynote talk presented at the International Symposium on Artificial Neural Networks*, Hsinchu, Taiwan, Computer Science Dept., Oregon Graduate Institute, PO Box 91000, Portland, OR 97291, USA Email: moody@cse.ogi.edu FTP: neural.cse.ogi.edu; cd pub/neural/papers

Pack, D. El-Sharkawi, M.A., Marks 11, R1, (1991a): An adaptively trained neural network. *IEEE Transactions on Neural Networks* 2 (3), 334-345.

Pack. D., El-Sharkawi. M.A., Marks 11, R.J., Atlas, L.E., Damborg. M1, (1991b): Electric load forecasting using an artificial neural network. *IEEE Transactions on Power Systems* 6 (2). 442-449,

Parker, D.B., 1987. Optimal algorithm for adaptive networks: Second order back propagation, second order direct propagation, and second order Hebbian learning. In: *Proceedings of the IEEE International Conference on Neural Networks*, 2, pp. 593-6

Poli, I., R. D. Jones, (1994): A Neural Net Model for Prediction, *Journal of the American Statistical Association* - American Statistical Association, March, Vol. 89, No. 425, 117-121.

Rumelhart, D.E., Hinton, G.E.. Williams, R.L, 1986. Learning internal representation by back- propagating errors. In: Rumelhart, D.E., McCleland, LL., the PDP Research Group (Eds.), *Parallel Distributed Processing: Explorations in the Microstructure of Cognition*. MIT Press, MA.

Srinivasan, D., Liew, A.C., Chang, C.S., 1994. A neural network short-term load forecaster. *Electric Power Systems Research* 28, 227-234.

Subramanian, V, Hung, M.S., 1993. A GRG2-based system for training neural networks: Design and computational experience. *ORSA Journal on Computing* 5 (4), 386~394.

Swanson, N.; White, H. (1997): Forecasting Economic Time Series Using Flexible versus Fixed Specification and Linear versus Nonlinear Econometric Models, *International Journal of Forecasting*; 13(4), December, pages 439-61.

Takens, F. (1981): Detecting strange attractors in turbulence. In D. A. Rand and L.-S. Young, editors, *Dynamical Systems and Turbulence, Lecture Notes in Mathematics*, volume 898, pages 366--381, Warwick 1980, 1981. Springer-Verlag, Berlin.

Tang, Z., Fishwick, P.A., 1993. Feedforward neural nets as models for time series forecasting. ORSA Journal on Computing 5 (4), 374-385.

Tang, Z., Almeida, C; Fishwick, RA. (1991): Time series forecasting using neural networks vs Box-Jenkins methodology. *Simulation* 57 (5), 303-310.

Vapnik. V. N. (1982): *Estimation of Dependencies Based on Empirical Data*. Springer-Verlag, New-York, NY, USA, 1982.

Vishwakarma, K.P., 1994. A neural network to predict multiple economic time series. In: *Proceedings of the IEEE International Conference on Neural Networks*, 6, pp. 3674~3679.

Weigend, A.S.; Huberman, B.A.; Rumelhart D.E. (1992): Predicting sunspots and exchange rates with connectionist networks. In M. Casdagli and S. Eubank, editors, *Nonlinear Modeling and Forecasting*, SFI Studies in the Sciences of Complexity, Proceedings Vol. XII, pages 395-432. Addison-Wesley.

Wong, F.S., 1991. Time series forecasting using backpropagation neural networks. *Neurocomputing* 2, 147-159.

Zeng, Z.; Goodman, R.M.; P. Smyth. (1994): Discrete recurrent neural networks for grammatical inference. *IEEE Transactions on Neural Networks*, 5(2): 320-330,.

Zhang, X., 1994. Time series analysis and prediction by neural networks. *Optimization Methods and Software* 4, 151-170.

Zhang, Guoqiang; Patuwo, B. Eddy; Hu, Michael Y. (1998): Forecasting with Artificial Neural Networks: The State of the Art, *International Journal of Forecasting*; 14(1), March, pages 35-62.

A fuzzy classification for the definition of industrial district

S.Bruni-G.Facchinetti-S.Paba

Department of Economy Politic -Faculty of Economy-Via Berengario 51
University of Modena 41100 Modena Italy
Tel: 0039-059-417711 Fax: 0039-059-417937
e-mail: facchinetti@unimo.it ; paba@unimo.it

Keywords: Industrial District, Fuzzy Inference Systems

1. Introduction

Becattini in his paper of 1998 [1] says that an "industrial district", in the Marshall sense, is a social and economic entity, placed in a fixed area, characterized by an industrious presence of population and of industrial companies. Sforzi, using the data of census of population of 1981, divides the Italian territory into 955 "local work systems. Brusco and Paba in [5], using the data of census of population from 1951 to 1991, present a mathematical algorithm which takes into account the qualifications proposed by Sforzi. This algorithm defines four indexes, which take account of different details. The first regards the manufacturer quota, the second the small (less than 100 employees) manufacturer quota, the third the specialization index, the fourth the small (less than 100 employees) sector quota. All the areas in which the four indexes are simultaneously greater than one are considered "industrial districts". Unfortunately this algorithm is too restrictive, in the sense that, some "historical districts" don' satisfy this algorithm. To solve this problem, we propose in this paper a Fuzzy System.[2],[16] It is composed of five input variables and one output variable. Four of the input variables are the four indexes proposed by Brusco and Paba, the fifth is the percentage of employees engaged in the territory we analyze. The output variable is an explicit score in percentage that evaluates to what extent that territory has the peculiarity typical of a district. In this score it is possible to see three things: 1) the classical districts, those which are considered by Brusco and Paba using the crisp valuation, are present; 2) the "historical districts" that were left out of the score by a too restrictive definition are recovered; 3) the new score offers a real classification of "industrial districts" whereas the classic result is "yes or no". In this paper we have analyzed the clothing branch.

2. Crisp algorithm

Brusco and Paba [5] introduce an algorithm, that enables one to check if an area is or is not an «industrial district». The algorithm defines four indexes in this way.

If L is the total of employees in the nation, i is a sector, j is a territory, m is a manufacturing sector, they define four indexes that must be greater than one:

1 $$\frac{L_{jm}}{L_j} \Big/ \frac{L_m}{L} > 1$$ [share of manufacturing]

where L_{jm} represents the total number of employees (L) in the manufacturing sector m of a territory j. L_j is the total number of employees in the territory j. L_m is the total number of employees in the nation who work in the manufacturing sector m:

2 $$\frac{L_{jm,small}}{L_{jm}} \Big/ \frac{L_{m,small}}{L_m} > 1$$ [small firm condition]

where $L_{jm,small}$ has the same meaning as L_{jm}, but the company is «small» in the sense that it has less than 100 employees. $L_{m,small}$ has the same meaning as L_m but the company is «small».

For at least one sector i:

3 $$\frac{L_{ji}}{L_{jm}} \Big/ \frac{L_i}{L_m} > 1$$ [specialisation index]

where L_{ji} is the number of employees in a territory j for a fixed sector i. L_i is the total number of employees in the nation who work in this sector i.

For at least one sector i for which the index defined in 3 is satisfied, it is required that:

4 $$\frac{L_{ji,small}}{L_{ji}} \Big/ \frac{L_{i,small}}{L_i} > 1$$ [small firms in the specialised sector]

where $L_{ji,small}$ has the same meaning as L_{ji} but the company is «small».$L_{i,small}$ has the same meaning as L_i, but the company is «small».

If a local system satisfies conditions 1 and 2 and for at least one sector i, conditions 3 and 4 are satisfied, the local system is called «industrial district» and the sectors i , for which conditions 3 and 4 are satisfied, are called «specialization» of the district. One district may have more than one specialization. The sector, for which index 3 is highest, is called «dominant sector» and usually the district is known by that sector.

3. A new approach. Fuzzy Inference System

The crisp algorithm is debatable for two reasons. The first is that it only considers the economic characteristics of the district, for it takes account only of «small» manufacturing companies specialized in a sector. But in order for an area to be considered a district, it is important that the persons who live in that area form a community with the same values and learning. Nevertheless, it is important that companies, employees and contractors co-operate. The second is that all the inequality, expressed in 1,2,3,4, must be satisfied simultaneously. To solve these problems, we have built a Fuzzy Inference System [2],[16] with five inputs. Four of these are the indexes introduced by Brusco-Paba the fifth is a new variable «employees» which takes into account the percentage of the number of employees in every territory with respect to the national total. The reason of this choice is that the system with only the first four indexes, provides as output a debatable final

classification in which towns with a small number of employees come before towns with a great number of employees. [3]
Figure 1

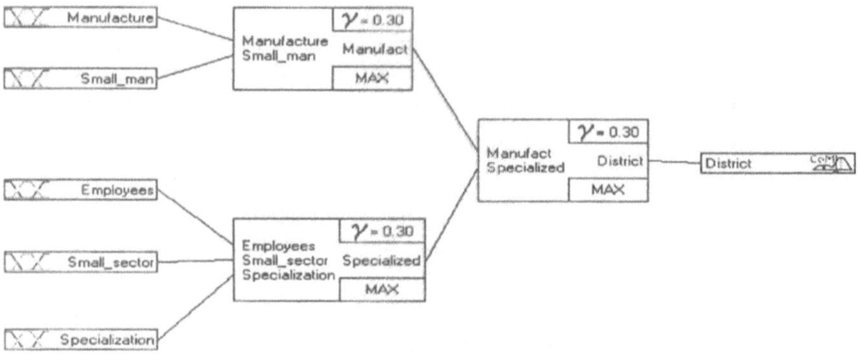

3.1. Fuzzification

The four indexes are fuzzified using three linguistic terms: low, satisfactory and high. The bounds of singular fuzzy numbers are obtained by experimental values.

Figure 2

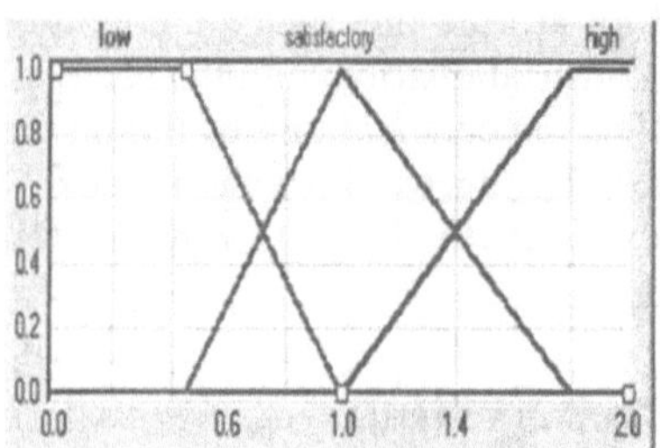

Figure 3

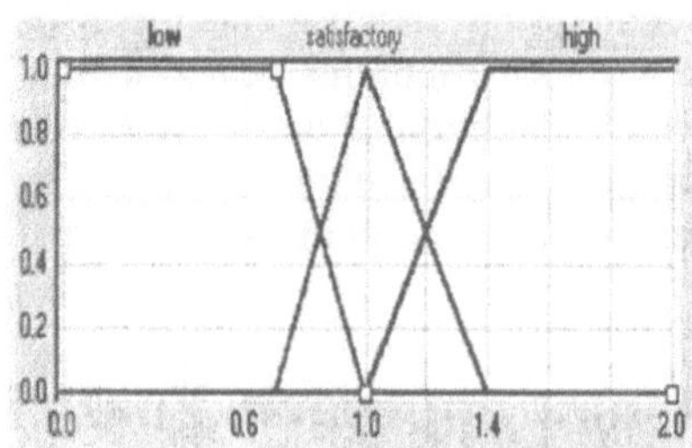

Figure 4

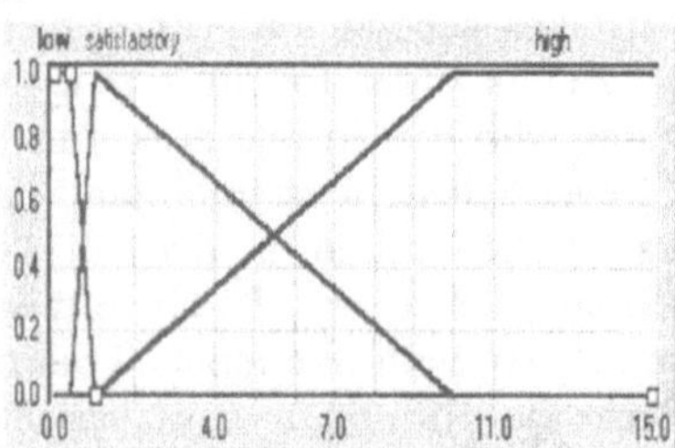

Figure 5

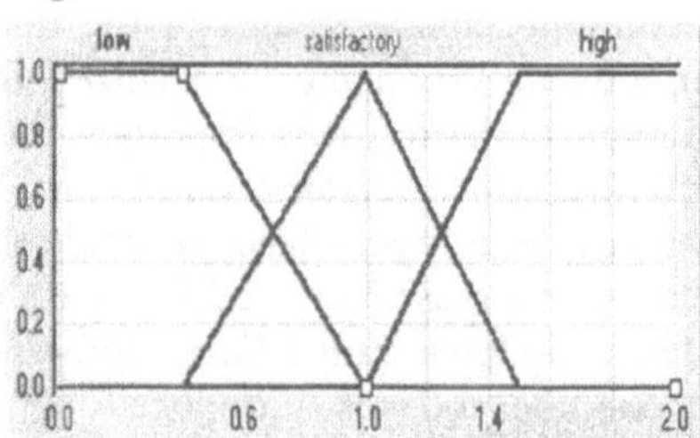

The employee's variables have two linguistic variables: few and many. Its domain is

[0,6], the experimental values suggest that the employees are few if they are less than 110 (in percentage 0,03) and many if they are over 6000 (in percentage 1,15).

Figure 6

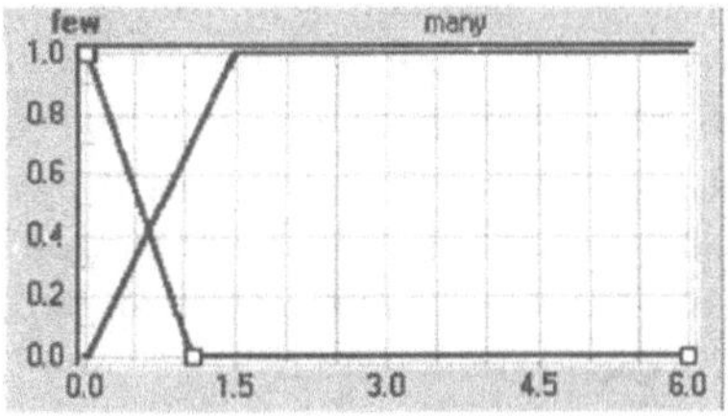

Figure 7

IF		THEN	
Manufactu	Small_man	DoS	Manufact
low	low	1.00	very_low
low		0.50	low
satisfactory	low	1.00	low
satisfactory	satisfactory	1.00	medium
satisfactory	high	1.00	high
high	low	1.00	low
high	satisfactory	1.00	high

3.2. If-Than Rules

The first Rule-Block is obtained combining the two indexes defined in 1 and 2.The output gives an evaluation on the manufacturing of every territory. (fig. 7)
The second Rule Block is obtained combining the indexes in 3 and 4 and the variable «employees» (fig. 8). The output gives an evaluation about the specialization of a local system in a particular manufacturing sector.
The two outputs so obtained turn into two new inputs that are combined, giving as output the final score of «fuzzy districts» (fig. 9).

Figure 8

IF			THEN	
Employees	Small_sect	Specializat	DoS	Specialize
	low	low	1.00	very_low
	satisfactory	low	1.00	very_low
	high	low	1.00	very_low
few	low	satisfactory	1.00	low
many	low	satisfactory	0.20	low
few	satisfactory	satisfactory	0.50	medium
many	satisfactory	satisfactory	1.00	high
few	high	satisfactory	0.50	medium
many	high	satisfactory	1.00	high
few	low	high	0.50	medium
many	low	high	0.50	medium
few	satisfactory	high	0.50	high
many	satisfactory	high	1.00	very_high
few	high	high	0.50	high
many	high	high	1.00	very_high
few			1.00	medium

Figure 9

IF		THEN	
Manufact	Specialized	DoS	District
very_low	very_low	1.00	very_negat
low	very_low	1.00	very_negat
medium	very_low	1.00	very_negat
high	very_low	1.00	very_negat
very_low	low	1.00	very_negat
low	low	1.00	very_negat
medium	low	1.00	negative
high	low	1.00	small_nega
very_low	medium	1.00	negative
low	medium	1.00	small_nega
medium	medium	1.00	zero
high	medium	1.00	zero
very_low	high	1.00	negative
low	high	1.00	zero
medium	high	1.00	small_positi
high	high	1.00	positive
very_low	very_high	1.00	small_nega
low	very_high	1.00	zero
medium	very_high	1.00	very_positi
high	very_high	1.00	very_positi

3.3. Rule of aggregation

As aggregator of IF-Then rules we have used Gamma operator [18] [19]

$$\mu = \left(\prod_{i=1}^{n} \mu_i \right)^{(1-\gamma)} \bullet \left(1 - \prod_{i=1}^{n} (1 - \mu_i) \right)^{\gamma}$$

The reason for this choice is that Min or Max operator (or weighted averaging of these) are not able to counterbalance the different membership functions. It was shown that Gamma operator represents the human decision process much more accurately than the others [21]. In this situation, we have thought that a small "defaillance" of an index may be compensated by a good "performance" of another. In most practical implementations, the value of γ lies between 0,1 and 0,4. Following the steps suggested in [16], we have found the value γ=0.3 as the best. For the aggregation of the control outputs we have used the MAX method. This geometrically means that we make the union of the fuzzy sets obtained by the aggregation of the output variables. This is equivalent to the logic OR.

3.4. Defuzzification

We have used the Center of Gravity (Center of Area) [2] [16]. This methods find the most typical value for every terms and then calculate the best «compromise» of the fuzzy inference results. All the territories that have a final valuation bigger or equal to 100% are considered a «fuzzy district».

4. Fuzzy districts for clothing sector

We have analyzed 955 territories, but for necessity of space we report only the first ones.

	Local system	Empl/ Empl. sect.	1° index	2° index	3° index	4° index	***Fuzzy District***
373	Carpi	1,56	1,6482	1,2488	3,7463	1,1038	**132,4%**
114	Gallarate	1,54	1,6558	1,1405	1,8213	1,1038	**127,9%**
323	Piove Di Sacco	1,03	1,5008	1,265	2,1429	1,0485	**127,7%**
650	Roseto Degli A.	0,97	1,2688	1,1784	4,3482	1,1038	**126,5%**
651	Sant'Egidio Alla	0,78	1,6661	1,26	5,1787	1,1038	**125,4%**
302	Castelfranco V.	0,95	1,4747	1,1709	2,6571	1,0279	**124,5%**
324	Adria	0,69	1,3345	1,3009	7,3893	1,1038	**124,0%**
426	Filottrano	0,56	1,7717	1,1463	10,417	**0,937**	**123,3%**
328	Porto Tolle	0,64	1,1454	1,3009	7,4016	1,1038	**122,8%**
474	Empoli	0,85	1,4142	1,2443	2,5955	1,1038	**122,0%**
308	Treviso	1,47	1,2008	1,0798	1,9651	**0,984**	**121,1%**
705	Putignano	0,67	1,0556	1,2559	5,4567	1,1038	**121,1%**
698	Barletta	0,8	1,1372	1,2636	2,3095	1,1038	**120,6%**
321	Padova	1,15	**0,967**	1,1602	1,1926	1,0181	**119,6%**
277	Bassano Del Grappa	0,76	1,5048	1,2267	1,7061	1,0076	**119,4%**
458	San Benedetto	0,76	1,1463	1,1544	2,7437	1,1038	**119,0%**
185	Palazzolo Sull'O.	0,73	1,6548	1,1792	1,9223	1,1038	**119,0%**

5. Conclusions

Comparing the evaluations given by the crisp algorithm and those obtained by the fuzzy system, we notice a great difference. The crisp method gives only a positive or negative reply, whereas the fuzzy method gives a real score which offers the possibility to consider all the area different from the others.

Bibliography

[1] G.Becattini, Distretti industriali e made in Italy: le basi socio-culturali del nostro sviluppo, Borighieri,(1998)

[2] G.Bojadziev, M.Bojadziev , Fuzzy Logic for Business, Finance and Management, World Scientific Publishing co, Singapore, (1997)

[3] S.Bruni, G.Facchinetti., S:Paba, A Fuzzy algorithm for a definition of industrial districts . To appear

[4] S. Brusco , Piccole imprese e distretti industriali, Rosemberg and Seller, (1989).

[5] S.Brusco, S.Paba, Per una storia dei distretti industriali italiani dal secondo dopoguerra agli anni novanta, Storia del capitalismo italiano pp.265,333 Barca F. Donzelli (1997)

[6] S.Cammarata,. Sistemi Fuzzy. RCS libri ,(1994)

[7] D.E.Cox , The Fuzzy Systems handbook. London. Academic Press (1994).

[8] G.Facchinetti , R.Ghiselli R., S.Muzzioli. Note on Ranking Fuzzy Triangular Numbers, International Journal of Intelligent Systems, (1998), vol 13 pp.613-622

[9] E.Goodmann , J.Bamford ., P.Saynor., Small firms and industrial districts in Italy, Routledge London (1989)

[10] P.Hirst , J.Zeitlin , Governing flexibility: flexible specialization, industrial districts and economic governance, Routledge London (1997).

[11] E.Iannuzzi, Impresa distrettuale: politiche di adattamento e di evoluzione. Padova, Cemar, (1995).

[12] G.Klirj., T.Folger, Fuzzy sets, uncertainty and information, New Jersey, Prentice Hall, (1998).

[13] F.Pyke , G.Beccattini , W.Sengenbergher, Industrial Districts and inter-firm co-operation in Italy, International Institute for Labour Studies, Geneva , (1990)

[14] A.Saba, Il modello italiano, Milano, Franco Angeli , (1995).

[15] R.Varaldo ,L.Ferrucci, La natura e la dinamica dell'impresa distrettuale, Ets, (1993).

[16] C.Von Altrock, Fuzzy Logic and Neurofuzzy Applications in Business and Finance. Prentice Hall, (1997).

[17] H.J.Zimmermann, Fuzzy Sets, Theory and its Applications, Third Revised Edition, Boston Kluver Academic Press, (1996)

[18] H.J.Zimmermann, Operators in Models of Decision Making, In Fuzzy Information Engineering, D:Dubois, H.Prade, R.Yager (Eds), (1997) pp.471-496 John Wiley & Sons, Inc.

[19] H.J.Zimmermann, P.Zysno, Latent Connectives in Human Decision Making. Fuzzy Set and Systems , (1980), Vol.4, pp.37-51

A Combination of tools: NLS and NN Estimation of the Expenditure in Durables. Determinants, trend and Forecasting in the Vehicles Sector.

Giovanni D'Orio
University of York (UK) - Universita' della Calabria (I)
Department of Economics - University of York YO1 5DD York (UK)
Tel: 0039(0)339 4227254; Fax: 0044 (0)1904 433759 email: gd108@york.ac.uk

Abstract

The aim of this work is to develop an equation to explain quarterly movements in personal consumption expenditure in durable goods and specifically in motor vehicles. In particular, we focus in the role of liquidity. Two sets of instruments are used. The description is done using a non-linear least squares (NLS) technique while the forecasting is done using a Neural Network. The model obtained using the NLS, even if showing a quite high goodness of fit, is quite unstable and the results are highly dependent on the starting values of some parameters. Furthermore, the high number of parameters estimated do not allow to perform a comprehensive set of econometric test. This set of reasons suggested to move to a method of forecasting able to overcome these problems. The tool used for perform the forecasting is a trained back-propagation MLP neural network.

1. The model

A stock adjustment model is commonly (see Pesaran and Wickens 1997) used to explain expenditure on durables. Spending on durable goods is viewed as the sum of two components: replacement expenditure to offset depreciation of the stock and net investment. For simplicity, in the following models, the rate of depreciation is assumed to be linear, if someone buys a car today, tomorrow he will be able to sell at the same price that he has paid. The theory of net investment rests on two assumptions:

- consumer has in mind an equilibrium stock for durables that depends on the income/wealth, interest rate and relative prices.
- the actual stock is not equal, in general, to the equilibrium stock[1].

Liquid wealth is seen as a constraint that the consumer faces when he wants to adjust his stock[2]. The change will be a constant fraction of the difference between the two stocks and it will also depend on holdings of liquid wealth.[3]

---

[1] See Caballero 1991

[2] See Deaton and Muellbauer; see Chah, Ramey, Starr; see also Norton

[3] Holding liquid assets will reduce the costs, material and psychic, of portfolio adjustment and will increase the speed of adjustment See Chah, Ramey, Starr; see Brown

$$C_t = D_t \Delta K_t e^{u_t};\; D_t = K_{t-1}^{\frac{1}{n}};\; \Delta K_t = \left(\frac{K_t^e}{K_{t-1}}\right)^f L_{t-1}^g;\; K_t^e = aY^b X^c L_{t-1}^h;\; C_t = K_{t-1}^{\frac{1}{n}} \left(\frac{Y^b X^c L_{t-1}^h}{K_{t-1}}\right)^f L_{t-1}^g e^{u_t}$$

$$\ln C_t = f \ln a + fb \ln Y_t + fc \ln X_t + (fh + g) \ln L_{t-1} + \left(\frac{1}{n} - f\right) \ln K_{t-1} + u_t \quad (1)$$

where C=expenditure in durables; D=expenditure in replacement; ΔK=net investments; K^e= equilibrium stock; L=liquid wealth; Y=personal disp. income; X=vector of relative prices.

Equation (1) can be estimated in two ways[4]: using the data on stock directly or with an autoregressive transformation since $K_t = (1 - 1/n)K_{t-1} + C_t$ and $K_{t-1} - 1/n\, K_{t-2} = C_{t-1}$. In this way we eliminate the variable "Stock" as regressor arriving to:

$$C_t = fa + fb(Y + (n-1)\Delta Y) - fc(X + (n-1)\Delta X) + (fh + g)(L_{t-1} + (n-1)\Delta L) + (1-f)C_{t-1} + \epsilon \quad (2)$$

in which the error term e is given by $e_t = nu_t - (n-1)u_{t-1}$. A moving average error process is involved[5]. So we use the first method. The problem with this method is to get some data about the variable "stock"[6]. A test for detection of correlation between the explanatory variables has shown high values. A Granger Test has shown that the null hypothesis of causation can be rejected at 5% level of significance. Here correlation does not imply "causation" in any sense of the word[7].

Variables

The data is quarterly and seasonally unadjusted. Value are expressed in logarithms

C= personal consumption expenditure in motor vehicles; prices 1990

K= depreciated capital stock for spending on vehicles; price 1990

X: real price of motor vehicles; prices 1990[8].

Y: personal disposable income pro capite; price 1990

L: cash plus current fixed and saving deposits[9].

---

[4] see Maddala,see Green.

[5] Equation (2) is sometimes used because it is difficult to find data about the stock of durable goods. The estimation is very laborious even if in practice, the moving average error process is usually ignored.

[6] To derive this variable we used the hypothesis that average life of a motor vehicle is seven years (see Smith and other). Differently to Smith's approach, the rate of depreciation is supposed to be constant each year at a rate of 14%. The stock of cars at the 31/12/69 is equal to the expenditure of the 4th quarter of 1969 plus (1-0.035) the expenditure of the 3rd quarter of '69 plus (1-0.07) the total of 2nd quarter of '69 and so on until (1-0.965) the total of the first quarter of 1963. All the car purchased before this date have lost their value and are not considered in the stock of 1970.The stock at the end of the first quarter of 1970 add to the stock at the end '69 the new cars purchased in this quarter and subtract all the quotas (3.5% of the expenditure in the quarters) of that purchased in the first quarter of each previously considered year and so on for each quarter.

[7] See Maddala and see also Green

[8] see Goldberg for some justifications about the use of this index.

2. *Methodology for the NLS Estimation*

The estimation of equation (1) has been done using a NLS estimation combined with an OLS estimation. I started using OLS estimating the equation $\ln C_t = \alpha + \beta \ln Y + \gamma \ln X + \eta \ln L_{t-1} + \theta \ln K_{t-1} + u_t$ (3) in which $\alpha = f \log a$, $\beta = bf$, $\gamma = cf$, $\eta = fh + g$, $\theta = 1/n - f$ obtaining these results:

Variable	Coefficient	Std. Error	t-value	t-prob
Constant	4.6778	1.5464	3.025	0.0031
D1	0.075626	0.046498	-1.626	0.1069
D2	-0.21937	0.045808	-4.789	0.0000
D4	-0.61898	0.045848	-13.501	0.0000
X (real price)	-0.074469	0.24595	-0-303	0.7627
Y (income)	0.91440	0.40547	2.255	0.0262
K (stock)	-0.35956	0.25457	-1.412	0.1608
L (liquidity)	0.10239	0.027767	3.688	0.0004

Table 1 Modelling equation (3) by OLS Sample is:1970 (2) to 1997 (4)[10]

The reason for starting with OLS estimation is to determine the starting values of some parameters used in the NLS. To estimate the equation and obtain at the same time convergence and a value for all the parameters, the model was rearranged in the following way:

$$\ln C_t = f \ln a + fb \ln Y_t + fc \ln X_t + Q \ln K_{t-1} + P \ln L_{t-1} + \varepsilon \qquad (4)$$

where $Q = (1/n - f)$ and $P = (fh + g)$. The best starting value for f (obtained with OLS) is 0.06 and it will conduct to the following results:

Variable	Coefficient	Std. Error	t-value	t-prob
Constant	56.807	30.535	1.860	0.0657
D1	-7.2907	3.0924	-2358	0.0203
D2	-2.5861	1.2094	-2.138	0.0349
D4	-0.89366	0.66601	-1.342	0.1826
c (real price)	-0.96448	2.9462	-0.327	0.7441
b (income)	10.742	6.5626	1.637	0.1048
Q(stock)	-4.3554	3.5470	-1.228	0.2223
P (liquidity)	1.2296	0.61800	1.990	0.0493
f	0.084833	0.035567	2.385	0.0189

Table 2 Modelling EQ(4) by NLS. Sample is: 1970 (2) to 1997 (4)[11]

The NLS has a linearized version of the autocorrelation and heteroscedasticity test since the derivatives of the function are used as regressors. Some graph tests on the residuals seem to show that there

---

[9] Notes and coins of the public plus current fixed and saving deposits with trading and saving banks.

[10] $R^2 = 0.814569$ F(7, 103) = 64.637 [0.0000] å = 0.170977 DW = 1.42; RSS = 3.011019014 8 vblss and 111 obs; ARCH 4 F(4, 95) =1.3584 [0.2542] Normality Chi(2)=1.1487 [0.5631];Xi^2; F(11, 91)=1.9818 [0.0392]; Xi*Xj F(29, 73)=1.1863 [0.2747]; RESET F(1,102)=2.192 [0.1418]

[11] $R^2 = 0.772241$ F(8, 102) = 43.23 [0.0000] å = 0.171821 DW = 1.42; RSS = 3.011277615 for 9 vbls and 111 obs.; ARCH 4 F(4, 94) = 1.3236 [0.2669] ;Normality Chi^2 (2)=1.1858 [0.5527]

is no problem of heteroscedasticity even if there is an apparent trend between the 1986 and 1990 and after that the residuals seems again random with a different trend if compared with the residuals before 1986. May be this change in the form could be interpreted as a change in the behaviour of people.
In Table 3 there are the correlation coefficients between the residuals and explanatory variables[12].

	RESIDUAL		RESIDUAL
STOCK	-0.002165	PRICE	-0.006070
MONEY	0.000619	INCOME	-0.002274

Table 3 Correlations coefficients

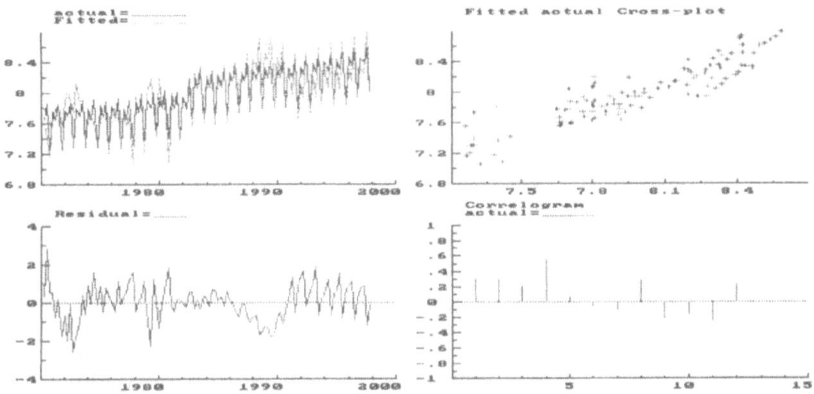

Figure 1 Results of NLS estimation

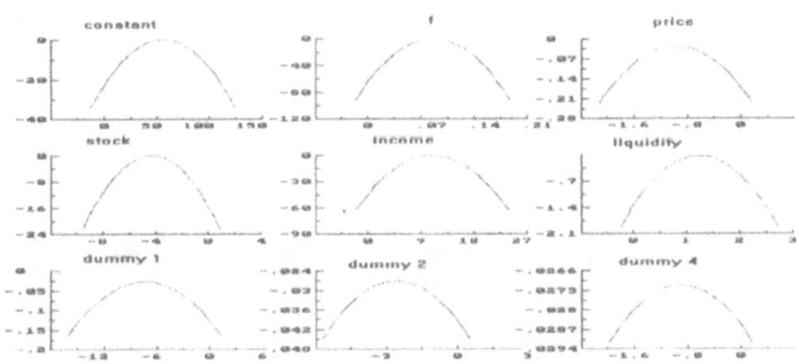

Figure 2 Grid search for eq. 4

The value of the Durbin-Watson[13] obtained, 1.42, suggests the possibility of presence of autocorrelation. A correlogram about the autocorrelation of the

[12] The values obtained in the correlation matrix are very low so we can say that, apparently, there is no correlation between the residual and explanatory variables.

[13] The program uses a linearized version of DW test. The value shows a signal for presence of autocorrelation, for the 4th value. I tried an estimation in differences and with seasonally adjusted data but there were not improvement. Since in the equation there are not lagged values of the dependent and of the

residuals shows that in the first 4 periods there is a sort of weak correlation and it seems that, even if with very low value, this behaviour is repeated with an interval of 4[14]. This particular form is probably dependent on the strong seasonality characterising the data about "C".

A graph about a grid search of the previously estimated parameters (of eq.4) shows that they fit and explain in a good way the trend of the model.

3. *Form of the Neural Network*

Given the nature of data (time series), a multilayer perceptron network(MLP) with four true input nodes, one output node, a hidden layer with two nodes and a feedback connection from the true output was used. Hidden neurons are few since the feedback connection and the presence of a fictuous input neuron as an internal network element(see Rumelhart and other). In the upper layers, each node receives a signal that is a weighted sum of the outputs of the nodes in the layer below. Generalising, to node *i* in an upper layer *l*, the total input *x* is given by

$$x_i^l = \sum_{j=0}^{N^{l-1}} w_{ij}^l y_j^{l-1}$$

N = number of nodes; w_{ij} = weight for the connection between node j and node I; y_j = output of node j.

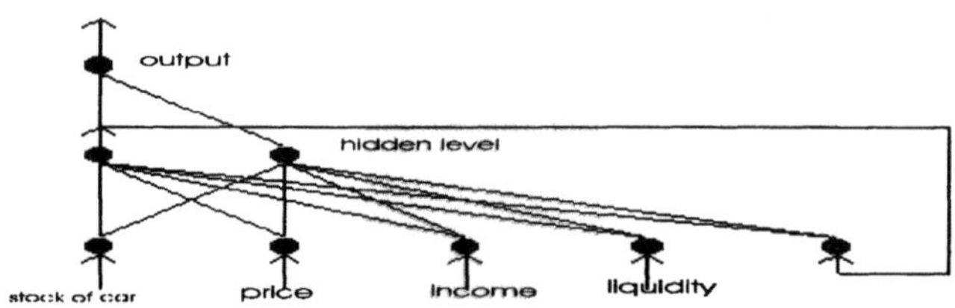

Figure. 3 The form of the network

Each node i has a bias, represented by the weight w_{i0} from a node with a constant unit activation, $y_0=1$. All the activation functions used

independent variables, I left the model in this way because the estimators are still unbiased even if this problem could cause some lose in the efficiency of the estimator. See Green, Maddala. See also Mizon, Wilson for the adjustment about autocorrelation; see White for the relevance of D-W test in Non Linear Estimation.

[14] I estimated the residual as a function of a constant and moving average terms. The clearest result was a function of four m. a. terms. The coefficent of the 2nd and the 4th component shown a significant t-ratio.

are linear with $y=x$. Data are the same of the NL estimation but in absolute (no log.) values. Prices are kept constant (1990). Since that data are homogeneous and deseasonalised, it is not useful to use a scaling process (see Haykin).
Given the time series data, a recurrent network was used (see Bulsari and Saxén 1995).

3.1. The training process

For this recurrent network the Levenberg-Marquardt method is used[15] to adjust the unknown network parameters in order to minimise the sum of squares (SSQ) of residuals, calculated as the differences between network output and target output. The problem can be analytically formulated as $\min_{\mathbf{w}} \{E = \mathbf{r}^T\mathbf{r}\}$where $\mathbf{r}$ is the residual vector and the network unknowns are collected into a vector $\mathbf{w}$. The elements of vector $\mathbf{w}$ are the weights of the two hidden neurones $w_{1,0}, w_{1,1}, w_{1,2}, w_{1,3}, w_{1,4}, w_{1,5}, w_{2,0}, w_{2,1}, w_{2,2}, w_{2,3}, w_{2,4}, w_{2,5}$, the weights to the output layer $w^2_{1,0}, w^2_{1,1}, w^2_{1,2}$ and the initial state of fictuous node.

Networks weight						
1st hid. neuron	111060,679	-47,567446	-11,27358	1,04661233	$1{,}20625211045e^{-2}$	$-5{,}32952470e^{-2}$
2nd hid. neuron	3920,77759	-213,97995	-2,244740	-16,7109109	$-4{,}4842566325e^{-2}$	-2,862109030
Output layer	-432,27302	$-7{,}993214e^{-2}$	-0,288687			
Initial state	-124325,96					
Network activation's						
Input	8.96e	2,15e4	9,39e2	3e3	-1,24e5	
1st hid. layer	-1,28e5	2,76e5				
Out	1,82e3					
des	1,82 e3					

Table 4 Network weights and initial value

At each iteration, z, the optimisation method adjust the unknown according to $\mathbf{w}^{(z+1)} = \mathbf{w}^{(z)} + \mathbf{e}^{(z)}$ where the error correction vector $\mathbf{e}$ is obtained from $\left(\mathbf{J}^{(z)}\mathbf{J}^{(z)T} + \lambda^{(z)}\mathbf{I}\right)\mathbf{e}^{(z)} = -\mathbf{J}^{(z)}\mathbf{r}$. The matrix $\mathbf{J}$ is a matrix with first derivatives of the residuals with respect to the unknowns and $\mathbf{I}$ is the identity matrix. The Marquardt parameter l is automatically adjusted during the training.

[15] The goal of the training process is to find values for the unknown, weights, bias terms and initial states for the fictuous input node, that describes the input-output relations.

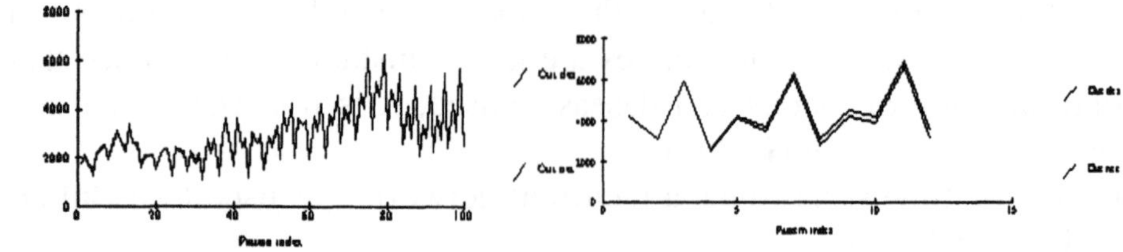

Figure 4 Training and test results results

The goodness of forecasting is less good than the training in absolute value but the explanation of the model even in Test is very high.
Initial value of the net weights are in Table nr. 4 while the results of training and forecasting in Figure 4. The training has been obtained using the first 100 observations while the Test was carried out in the last 12 periods (3 years). The goodness of fit of the training process is quite good.

4. *Comments[16] about the results:*

Expenditure in cars is characterised from a strong seasonality[17]. The price coefficient has a negative sign and since estimation is in log. it could be interpreted as a price-elasticity confirming that cars are normal goods. The *t-ratio* is not very significant and this could reflect the fact that lot of time, people prefer to pay more for a new model of the same brand that buy an old model for a lower price[18]. The income coefficient shows a positive correlation between the increase in the disposable income and the increase in the expenditure for cars even if it seems to be very high. It has to be considered that the real effect is corrected from the parameter f that it seems to be very significant[19]. The value obtained for f is very low and this shows a quite low adjustment of the actual stock to the desired stock.The stock at the beginning of the year has a negative influence on purchase during the year[20]. The liquidity stock at the beginning of the year has a positive effect on the purchase of new cars. The NLS method, using a

[16] This comments are based on the assumption previously made about the depreciation rate of stock of capital and the possible lack of efficiency due to the possible autocorrelation
[17] To buy a car in UK in the third quarter gives some facilities.See Gali, Hong, Linciano, Witt
[18] As evidence of this possible explanation see Shiratsuka and other, see also Waldmann.
[19] In a non-linear estimation of partial adjustment model, the coefficient of the variable could be seen as a "long-run" elasticity, since in the long run the value of f will tend to 1. So, if we consider together the parameter of the regressor and the parameter f, we have a sort of short-run elasticity.
[20] People that was in condition (necessity, income effects or liquidity effects) to buy a new car in the past will no buy a car for a while see Phlips, Deaton and Muellbauer

minimisation of a log-likelihood function in a recursive pattern, is quite similar to the training used from the NN to "learn" the underlying model. As consequence, a NN was used to forecast the last 3 years of consumer expenditure in cars in the model built with the NLS. The result seems quite good. This kind of approach seems to confirm the main point about econometric techniques and neural networks (see Apostolos; see also Cammarata). Using just a NN does not allow us to find a very accurate and economically significant description of consumer's behaviour. The NLS estimation instead give us a model that seems to explain quite well this point. Indeed, for the forecasting, the use of the NN shows a very good result impossible to be obtained using just a NLS estimation for the complexity of the underlying model. But the NN does not give a complete explanation of all the parameters that we need for describe this stock adjustment model. It could be really interesting to compare the two tools in the study of the complexity underlying the model. Investigating this could be the starting point for another work.

References

Apostolos P.R., (eds) ;Neural Networks in the capital market ; Wiley & son;1995

Brown,-C.; Money and consumer durable spending; Financial Sector of the American Economy series. New York and London: Garland, 1993

Brugiavini,A.; Weber,-G.; Durable and Nondurable Consumption: Evidence from Italian Household Data; Institute for Fiscal Studies (IFS) Working Paper: W92/13, 1992

Bulsari, A. H.Saxén, A Recurrent network for modeling noisy temporal sequences, Neurocomputing 7 , 1995,

Caballero,-R.-J.;Durable Goods: An Explanation for Their Slow Adjustment; National Bureau of Economic Research Working Paper: 3748, June 1991

Caballero,-R.-J.;Notes on the Theory and Evidence on Aggregate Purchases of Durable Goods;Oxford-Review-of-Economic-Policy; 10(2), Summer 1994

Cammarata S. ; Reti Neuronali : Una introduzione all'altra intelligenza artificiale ; Etaslibri 1990

Chah,-E.-Y.; Ramey,-V.; Starr,-R.; Liquidity Constraints and Intertemporal Consumer Optimization: Theory and Evidence from Durable Goods; Journal-of-Money,-Credit,-and-Banking; 27(1), February 1995

Chow G. C.; Statistical demand functions for automobiles and their use for forecasting; in Haberger, The demand for durable goods; University of Chicacgo Press;1960

Gali,-J.;Variability of Durable and Nondurable Consumption: Evidence for Six O.E.C.D. Countries; Review-of-Economics-and-Statistics; 75(3),1993

Green W.H.; Econometric analisys; Prentice Hall 1997

Hamilton,-Bruce-W.; Macauley,-Molly; Competition and Car Longevity; Johns Hopkins University, Department of Economics, Working Paper: 361, 3/96

Haykin S.; Neural Networks:a comprehensive foundation; MacMillan Publishing Company; 1994 New York

Hong,-K.; A Comment on Durable Goods Consumption;Journal-of-Monetary-Economics; 37(2), April 1996

Linciano,-N.; Household Car Usage in the U.K.; Int.-Journal-of-Transport-Economics; 24(3), October 1997

Mizon; A simple message for autocorrelation correctors: don't; Journal of Econometrics, 1995, 69

Maddala G.S.; Introduction to econometrics;Prentice Hall;1992

Norton W.E.; Broadbent J.R.; Equation for personal Consumption Expenditure; Bank of Australia; 1970

Page G.F., Gomm B., Williams D. ; (eds); Application of neural networks to modelling and control

Patterson D. W. ; Artificial Neural Networks : Theory and Applications ; Prentice Hall 1996 Singapore

Pesaran M.., Wickens M.R.,; Handbook of applied econometrics: macroeconomics; Oxford Blackwell 1997;

Peretto P. ; An introduction to the modeling of neural networks ; Cambridge University Press ;1992

Phlips L.; Applied Consumption Analysis;North Holland-American Elsevier;1982

Rogers E., Li Y. ; Parallel Processing in a control systems environment ; Prentice hall ; 1993

Rumelhart, D.E. and McClelland (eds.), Parallel Distributed Processing, MIT Press, 1986

Shiratsuka,-Shigenori; Automobile Prices and Quality Changes: A Hedonic Price Analysis of the Japanese Automobile Market; Bank-of-Japan-Monetary-and-Economic-Studies; 13(2), December 1995

Stone and Rowe; The Durability of Consumers' Durable Goods; Econometrica;1960

Takefuji Y., Wang J. ; Neural computing for optimization and combinatorics ; World Scientific 1996

Waldman,-M. ;Durable Goods Pricing When Quality Matters;Journal-of-Business; 69(4), October 1996

White,-Kenneth-J.; The Durbin-Watson Test for Autocorrelation in Nonlinear Models; University of British Columbia Department of Economics Discussion Paper: 90-04, February 1990, pages 11.

Wilson,-Arlette-C.; Weld,-Leonard-G.; Regression as an Analytical Procedure: Are Adjustments for Autocorrelation Really Necessary?; Journal-of-Accounting,-Auditing-and-Finance; 12(1), Winter 1997

Witt,-Stephen-F.; Johnson,-S.-Raymond; An Econometric Model of New-Car Demand in the UK; Managerial-and-Decision-Economics; 7(1), 1986

Estimating the Conditional Mean of a Non-linear Time Series Using Neural Networks

Francesco Giordano
(Università degli Studi di Salerno)

Cira Perna
(Università degli Studi di Salerno)

Abstract

In this paper a method is described for using neural networks to estimate the conditional mean of a non linear time series. Assuming a non constant conditional variance, the procedure allows to consider directly its estimation. The performance of the proposed approach is evaluated. A comparison with the standard technique is also presented using simulated data sets and financial time series representing daily prices of equities from the Italian stock market.

Key-words: Neural Networks, Non-linear Time Series Analysis

1. Introduction

The final objective of many time series techniques is to obtain point predictions for the observed data. So it is often of interest the estimation of the conditional first order moment of the series. (Priestley, 1981)
The aim of this paper is to evaluate the performance of neural networks in estimating the conditional mean of a non linear time series.
Assuming a non constant conditional variance, we consider a process $\{X_t\}$ generated by the model :

$$X_t = f(X_{t-1}, X_{t-2}, \ldots) + g(X_{t-1}, X_{t-2}, \ldots)\varepsilon_t \qquad (1)$$

where $\{\varepsilon_t\}$ is a sequence of independent and identically distributed (i.i.d) random variables with $E(\varepsilon_t)=0$ and $Var(\varepsilon_t)=1$ $\forall t$, $f(\cdot)$ and $g(\cdot)>0$ represent, respectively, the conditional mean and the conditional variance of the process given the past observations.
In order to estimate the conditional mean of the series (the function *f(.)*) different approaches are possible. One of the most widely used methods is a standard procedure based on a non-parametric estimator of the conditional mean, such as a neural networks estimator, without assuming any particular structure for the conditional variance. This estimation approach is simple but, considering a constant variance, it doesn't take into account the heteroskedasticity of the series. This can be important especially for some

financial time series where there is a clear evidence of changing variance. In this contest, we illustrate a second method that allows to estimate, at first, the second moment of the series.
The paper is organised as follows. In the next section, the architecture of the neural networks used in the analysis is presented. In section 3, the proposed method is illustrated and, in section 4, it is compared with the classical approach using simulated data sets and two financial time series.

2. Neural Networks

Neural networks have been used in various field to approximate complex non linear structures; recently several techniques (such as multilayer perceptron, radial basis functions) have also been applied for time series modelling (Hardle *et al.*, 1997). The common factor in neural networks techniques is that the models are constructed from simple processing units that perform non linear input-output mapping. In this study we use the *single hidden layer feedforward network* for time series modelling.
The number of input units is p and the number of hidden units is s. The notation NN(p, s) will be used to refer to a specific network structure. The weights in the connections between the input and hidden layer are denoted by α_{ij} and the weights between the hidden and output layer are denoted by β_j. In addition, the hidden and output neurons have the bias terms α_{0j} and β_0, respectively. These parameters are supposed to be 0. The activation function is chosen to be the standard normal cumulative distribution in the hidden units. The output neuron activation function is set to be the identity function. The mathematical formula for the network can be written as

$$X_t^* = \sum_{j=1}^{s} \beta_j \phi\left(\sum_{i=1}^{p} \alpha_{ij} X_{t-i} \right) \tag{2}$$

Hornik *et al.* (1989) have shown that neural networks (2) can approximate any continuos function f uniformly on compact sets, by increasing the size of the hidden layer. In addition, Barron (1993) has shown that (for sufficiently smooth f) the L_2 approximation is $O(s^{-1})$.
The parameters of (2) are determined recursively from output to input by using a chain rule and, namely, the procedure known as *backpropagation* (Haykin, 1994).

3. The Estimation of the Conditional Mean

The most widely used method for the estimation of the conditional mean of a non linear time series, using a neural networks approach, is based on the estimation of the function $f(.)$ in the model:

$$X_t = f(X_{t-1}, \ldots) + e_t \qquad (3)$$

where $\{e_t\}$ is i.i.d.

The disadvantage of this approach is that it assumes a constant variance (that is $g(.)=c$) and therefore does not take into account the possible heteroskedasticity present in the series

If the error term is $e_t = g(\cdot)\varepsilon_t$, then by estimating $f(\cdot)$ on model (3), the estimator function $\hat{f}(\cdot)$ depends from both $f(\cdot)$ and $g(\cdot)$. In order to overcome this problem, in a previous paper (Amendola, Giordano, Perna., 1998), we have proposed a new procedure for the estimation of $f(\cdot)$ and $g(\cdot)$ in the model (1) that allows to consider simultaneously the conditional mean and variance.

The step procedure is structured as follow.

I Step An estimate, $\hat{g}(\cdot)$, of the conditional variance $g(\cdot)$ is obtained minimising the objective function $\sum_{t=1}^{n}(u_t - g^*(u_t))^2$, where $u_t = (X_t - g^*(X_t))^2$ and $g^*(\cdot)$ is a generic statistic for $g(\cdot)$.

Let $Z_t = (X_{t-1}, X_{t-2}, \ldots, X_{t-p})$. Using the neural networks estimator (2), an estimator of $g(Z_t)$ is obtained. In this case β_j and α_{ij} can be determined minimising

$\sum_{t=1}^{n}(u_t - X_t^*(u_t))^2$, where $u_t = (X_t - X_t^*(Z_t))^2$.

II Step. From the model (1) it is:

$$\frac{X_t}{\hat{g}(\cdot)} = \frac{f(\cdot)}{\hat{g}(\cdot)} + \varepsilon_t + O(n^{-\alpha}) \qquad \text{where } 0<\alpha<1.$$

Let $\hat{X}_t = X_t / \hat{g}(\cdot)$ and $\upsilon(\cdot) = f(\cdot)/\hat{g}(\cdot)$, then $\hat{X}_t = \upsilon(\cdot) + \varepsilon_t$. An estimate $\hat{\upsilon}(\cdot)$ of the conditional mean $\upsilon(\cdot)$ can be obtained by neural networks.

III Step. An a estimate of $f(\cdot)$ can be now calculated by multiplying $\hat{\nu}(\cdot)$ by the previous estimate $\hat{g}(\cdot)$, obtaining $\hat{f}(\cdot) = \hat{\upsilon}(\cdot)\hat{g}(\cdot)$.

4. Empirical Study

In this section the performance of the proposed method is evaluated using both simulated and real time series. The sample size of all the series is T=505.

The first data set is a completely random and homoskedastic time series consisting of observations generated by a Normal distribution with zero mean and mean square error equal to 0.1.

The second time series is deterministic; it consists of T observations simulated by a logistic function:

$$X_t = \theta X_{t-1}(1 - X_{t-1})$$

with θ=4 and X_0 =0.4.

The third and the fourth series are real data from the Italian stock market; they are the first differences of daily prices, expressed in Euro, of two equities: Fiat and Generali. The sample covers the period from January 1997 to December 1998 involving 505 observations for each series. The ARCH test, carried out on the residuals of a linear AR(1) model, confirms the heteroskedasticity of both the series.

To estimate the conditional mean of the four series, the classical approach and the proposed one are considered; in both cases a NN(1,8) is used

Fig. 1 reports the time plot of the four time series together with the estimates of the conditional mean obtained using the standard procedure. It is evident that the estimated conditional mean of the first two data sets produces an extremely smooth curve respect to the other two series. This is due to the presence, in the first case, of a constant variance component.

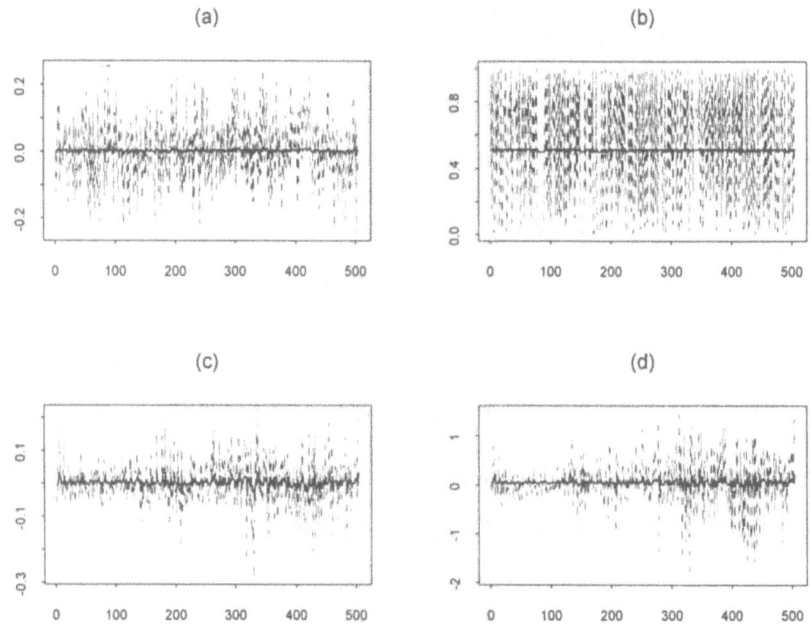

Fig.1: Observed time series (dashed line) and estimated conditional mean (solid line) with the standard procedure: (a) random data set; (b) logistic data set; (c) Fiat series and (d) Generali series

Fig. 2 shows the conditional mean estimation of the four series with the proposed procedure.
For all the four time series, the estimated conditional mean gives a smoother curve than the previous one. This is particularly true for the two real time series which are characterised by a very strong heteroskedasticity.

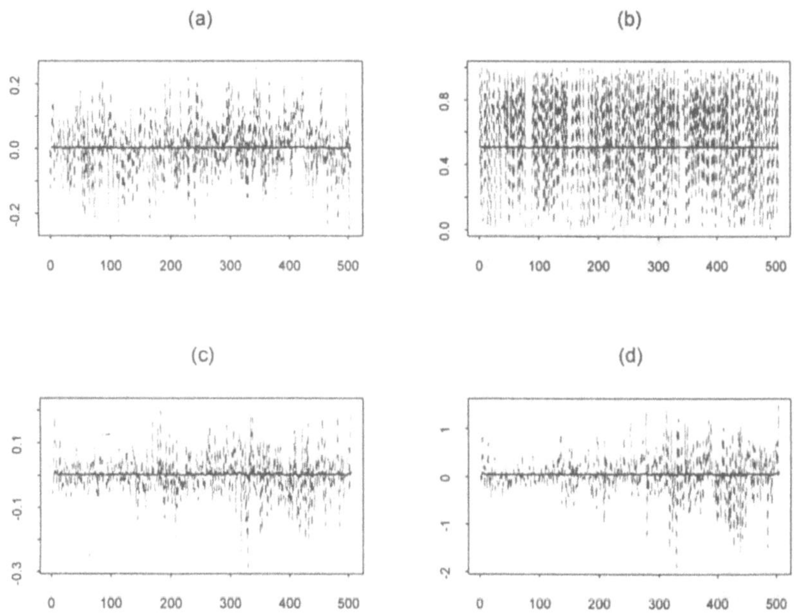

Fig.2: Observed time series (dashed line) and estimated conditional mean (solid line) with the proposed procedure: (a) random data set; (b) logistic data set; (c) Fiat series and (d) Generali series

Finally, Fig. 3 illustrates the estimated conditional variance of the four time series. The four plots are scaled for positive values of the previous ones. Relatively to the scale, a greater variability can be observed in panels (c) and (d) which confirms the presence of heteroskedacity i.e. a relevant non-constant variance component.

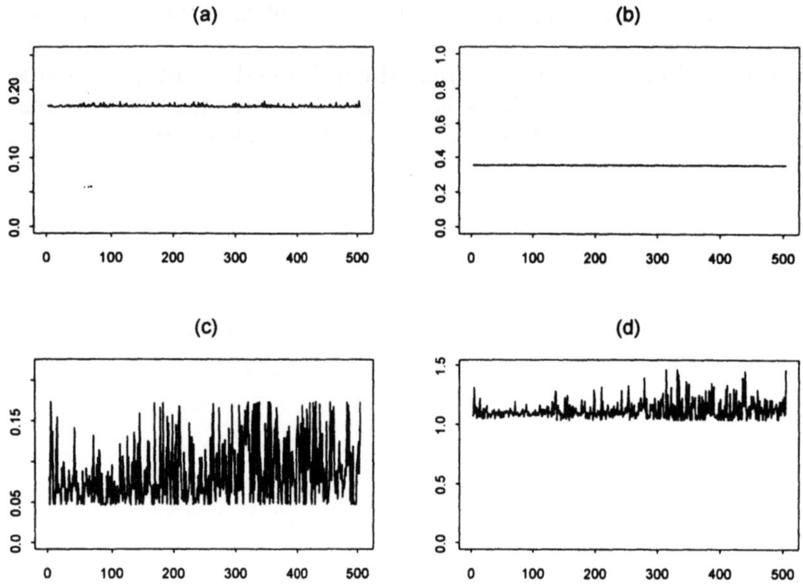

Fig.3: Estimated conditional variance of the four time series: (a) random data set; (b) logistic data set; (c) Fiat series and (d) Generali series

References

Amendola, A.; Giordano, F.; Perna, C. (1998) Parametric and Non-parametric Methods In Non-linear Time Series Analysis: A Critical Evaluation, *Pre-Proceedings of the International Seminar on "New Techniques & Technologies for Statistics"*, **2**, 25-30

Barron, A.R. (1993) Universal approximation bound for superpositions of a sigmoid function, IEEE Trans. Inform. Theory, **39**, 930-945.

Hardle, W., Lütkepohl, H. & Chen, R. (1997), A Review of Non-parametric Time Series Analysis, *International Statistical Review*, **65**(1), 49-72.

Haykin, S. (1994), *Neural Networks: a comprehensive foundation*, Macmillan, New-York.

Hornik, K., Stinchcombe, M., White, H. (1989) Multilayer feedforward networks are universal approximators, *Neural Networks*, **2**, 359-366

Priestley, M.B. (1981) *Spectral Analysis and Time Series*, Academic Press, London

Author Index

MIX
Papier aus verantwortungsvollen Quellen
Paper from responsible sources
FSC® C105338
FSC
www.fsc.org

If you have any concerns about our products,
you can contact us on
ProductSafety@springernature.com

In case Publisher is established outside the EU,
the EU authorized representative is:
Springer Nature Customer Service Center GmbH
Europaplatz 3, 69115 Heidelberg, Germany

Printed by Libri Plureos GmbH
in Hamburg, Germany